Lecture Notes in Computer Science 8607

Commenced Publication in 1973
Founding and Former Series Editors:
Gerhard Goos, Juris Hartmanis, and Jan van Leeuwen

Editorial Board

David Hutchison, UK
Josef Kittler, UK
Alfred Kobsa, USA
John C. Mitchell, USA
Oscar Nierstrasz, Switzerland
Bernhard Steffen, Germany
Demetri Terzopoulos, USA
Gerhard Weikum, Germany

Takeo Kanade, USA
Jon M. Kleinberg, USA
Friedemann Mattern, Switzerland
Moni Naor, Israel
C. Pandu Rangan, India
Doug Tygar, USA

FoLLI Publications on Logic, Language and Information
Subline of Lectures Notes in Computer Science

Subline Editors-in-Chief

Valentin Goranko, *Technical University, Lynbgy, Denmark*
Michael Moortgat, *Utrecht University, The Netherlands*

Subline Area Editors

Nick Bezhanishvili, *University of Amsterdam, The Netherlands*
Anuj Dawar, *University of Cambridge, UK*
Philippe de Groote, *Inria-Lorraine, Nancy, France*
Gerhard Jäger, *University of Tübingen, Germany*
Fenrong Liu, *Tsinghua University, Beijing, China*
Eric Pacuit, *University of Maryland, USA*
Ruy de Queiroz, *Universidade Federal de Pernambuco, Brazil*
Ram Ramanujam, *Institute of Mathematical Sciences, Chennai, India*

Margot Colinet   Sophia Katrenko
Rasmus K. Rendsvig (Eds.)

Pristine Perspectives on Logic, Language, and Computation

ESSLLI 2012 and ESSLLI 2013
Student Sessions
Selected Papers

 Springer

Volume Editors

Margot Colinet
Université Paris Diderot-Paris 7
Laboratoire de Linguistique Formelle
Paris, France
E-mail: margot.colinet@gmail.com

Sophia Katrenko
Amsterdam, The Netherlands
E-mail: sophia@katrenko.com

Rasmus K. Rendsvig
Lund Unviersity
Department of Philosophy
Lund, Sweden
E-mail: rendsvig@gmail.com

ISSN 0302-9743   e-ISSN 1611-3349
ISBN 978-3-662-44115-2   e-ISBN 978-3-662-44116-9
DOI 10.1007/978-3-662-44116-9
Springer Heidelberg New York Dordrecht London

Library of Congress Control Number: 2014942938

LNCS Sublibrary: SL 1 – Theoretical Computer Science and General Issues

© Springer-Verlag Berlin Heidelberg 2014
This work is subject to copyright. All rights are reserved by the Publisher, whether the whole or part of
the material is concerned, specifically the rights of translation, reprinting, reuse of illustrations, recitation,
broadcasting, reproduction on microfilms or in any other physical way, and transmission or information
storage and retrieval, electronic adaptation, computer software, or by similar or dissimilar methodology
now known or hereafter developed. Exempted from this legal reservation are brief excerpts in connection
with reviews or scholarly analysis or material supplied specifically for the purpose of being entered and
executed on a computer system, for exclusive use by the purchaser of the work. Duplication of this publication
or parts thereof is permitted only under the provisions of the Copyright Law of the Publisher's location,
in ist current version, and permission for use must always be obtained from Springer. Permissions for use
may be obtained through RightsLink at the Copyright Clearance Center. Violations are liable to prosecution
under the respective Copyright Law.
The use of general descriptive names, registered names, trademarks, service marks, etc. in this publication
does not imply, even in the absence of a specific statement, that such names are exempt from the relevant
protective laws and regulations and therefore free for general use.
While the advice and information in this book are believed to be true and accurate at the date of publication,
neither the authors nor the editors nor the publisher can accept any legal responsibility for any errors or
omissions that may be made. The publisher makes no warranty, express or implied, with respect to the
material contained herein.

Typesetting: Camera-ready by author, data conversion by Scientific Publishing Services, Chennai, India

Printed on acid-free paper

Springer is part of Springer Science+Business Media (www.springer.com)

Preface

It is our great pleasure to present the third volume of selected papers from the proceedings of the Student Session of the European Summer School in Logic, Language and Information (ESSLLI). The publication of the ESSLLI Student Sessions' proceedings in now a biannual tradition.

The 16 papers presented in this volume were selected among 44 papers presented by talks or posters at the Student Sessions of the 24th and 25th editions of ESSLLI, held in 2012 in Opole, Poland, and 2013 in Düsseldorf, Germany. The papers are extended versions of the versions presented, and have all been subjected to a second round of blind peer review. The papers cover vastly different topics, but each fall in the intersection of two of the three topics of ESSLLI – Logic, Language and Computation – which the volume's three part division reflects.

The chairs hold fond memories of both ESSLLI 2012 and ESSLLI 2013, of Opole and of Düsseldorf, and of the Student Sessions that co-chairs, Local Organizing Committees, FoLLI's ESSLLI Standing Committee, and not least the reviewers helped us organize. Both summer schools were meticulously organized, and accommodated the Student Session admirably. The cities each offered charms and hospitality, and gave rich opportunities for post-class socializing. Opole, with its population of 125.000, offered no rest from fellow ESSLLI participants – every cozy bar and restaurant was filled with familiar faces. Quite opposite, Düsseldorf with its close to 100 times the population, offered plenty of opportunities to get lost – from one another and in the vast city. No doubt, however, returning to the hostel where all students lodged, the 24 hour open bar and terrace showed as many ESSLLI faces as any place in Opole. Given the Standing Committee's continual efforts to make ESSLLI a yearly success by selecting overly competent Program Committees and hosts in prime locations, these fond memories do not come as a surprise. We owe a great thanks to them all, a thanks that also applies to both years' co-chairs and reviewers, without whom there would have been no Student Sessions.

Apart from continuing the tradition of providing a forum where young researchers may present their work in a friendly and supportive environment, we were also very happy to continue the tradition of concluding the Student Sessions with both entertaining academic features and an award ceremony for Best Oral and Best Poster Presentations. In 2012, we were so lucky that we could welcome Nina Gierasimczuk and Jakub Szymanik, who taught the audience about formal learning theory by playing the card game 'Eleusis' with 50+ participants – to the great satisfaction of both the participants and the local Opole TV station that came by to visit. In 2013, both laughs and games were again present when Valentin Goranko masterly illustrated the most assuming role preplay negotiations in non-cooperative games may have – at least when real people, money and golden balls marked 'Split' and 'Steal' are involved. We would like to thank Nina, Jakub and Valentin for their participation, and finally extend our gratitude to Springer for once again providing awards for the Best Oral and Best Poster Presentations.

May 2014

Margot Colinet
Sophia Katrenko
Rasmus K. Rendsvig

Organization

Program Committee 2012

Chair

Rasmus K. Rendsvig   Roskilde University, Denmark

Logic and Computation (LoCo) Co-chairs

Maxim Haddad   University of Osnabrück, Germany
Dominik Klein   Tilburg University, The Netherlands

Logic and Language (LoLa) Co-chairs

Matthijs Westera   ILLC, Universiteit van Amsterdam,
     The Netherlands
Margot Colinet   Université de Paris 7, France

Language and Computation (LaCo) Co-chairs

Anders Johannsen   University of Copenhagen, Denmark
Niels Beuck   Hamburg University, Germany

Program Committee 2013

Chair

Margot Colinet   Université Paris Diderot-Paris 7, France

Logic and Computation (LoCo) Co-chairs

Ronald de Haan   Technische Universität Wien, Austria
Michał Zawidzki   Uniwersytet Łódzki, Poland

Logic and Language (LoLa) Co-chairs

Agata Renans   Universität Potsdam, Germany
Barbara Tomaszewicz   University of Southern California, USA

Language and Computation (LaCo) Co-chairs

Pierre Bourreau   Heinrich-Heine-Universität Düsseldorf,
     Germany
Julia Zinova   Heinrich-Heine-Universität Düsseldorf,
     Germany

Table of Contents

Language and Computation

Logic and Computation

Logic and Language

Characterizing Speech Genres through the Relation between Prosody and Macrosyntax

Julie Belião

MoDyCo-UMR7114, Universit Paris Ouest, Nanterre, France
`julie@beliao.fr`

Abstract. The role of prosody and syntax in identifying basic discourse units is a recurring issue in studies of spoken language. This paper focuses on the highest-level units of macrosyntactic and prosodic structures, namely illocutionary units (IUs) and intonational periods (IPes). The study first presents macrosyntactic illocutionary units and intonational periods, and then investigates how they interact, in particular the synchronization and the relative number of their boundaries, in a corpus of spoken French. The analysis shows that it is possible to identify different types of synchronizations (total, partial, or absent) and their relative proportions, and that the combinations of these units vary according to the subgenres of the studied corpus. The results are interpreted from a functional point of view as an interaction between intonosyntax and discourse genres. It is argued that the simple features proposed here may be interesting and easily handleable and reproducible for the study of other spoken language corpora, whatever the language.

Keywords: Prosody, Macrosyntax, Illocutionary unit, Intonational period, Speech processing, Genre classification.

1   Introduction

Discourse or textual genre has been widely studied in rhetoric and literature. Many studies in "traditional" linguistics have highlighted the fact that particular situations and social contexts correspond to specific modes of production, which are associated to specific formal markers of the discourse genre at all levels, e.g. semantic, syntactic or phonological. In this context, the objectives of textual typology are threefold. First it aims at describing the diversity of discourses, e.g. literary, legal, political, religious, etc. Second, it aims at understanding their articulation into genres [1], and third at estimating their formal markers, in particular the co-occurrences of specific cues that can be considered as being typical of a genre. While it is common to consider the concept of discourse genre for written language, some recent studies have extended this concept to the oral domain, in particular to the interface of writing and speech [2, 3]. An important challenge in this context is to provide a typology of genres that is both robust and general.

Although theoretical studies concerning discourse genres have already made great progress, it remains very difficult to go beyond conventional generic types,

M. Colinet et al. (Eds.): ESSLLI 2012/2013, LNCS 8607, pp. 1–18, 2014.
© Springer-Verlag Berlin Heidelberg 2014

e.g. private, professional or public speech, and subdivisions, e.g. face to face or phone conversations, public debates, radio and TV broadcasts, spontaneous vs. planned speech, etc. [3]. In phonetics, the question of the modeling and recognition of phonostyles [4–7] is challenging for automatic speech processing. In particular, some kinds of public discourse, such as political, religious, journalistic and sport, are considered as cultural stereotypes and are related to specific expressive strategies that act as markers of a given phonostyle [8]. Identifying them is conceptually similar to discriminating between public or private discourses and is hence related to the identification of some generic types of discourse.

In all cases, the proper identification of discourse units within the discourse flow is essential for understanding and modeling how interpretation occurs, where and when inferences are made, and how each discourse component is related to the others in a (more or less) coherent way. Scholars have recently focused on defining basic discourse units": by taking into account the interaction between microsyntactic dependency and prosody, Degand and Simon proposed a typology of discourse units [9]. Whereas most studies focus on only syntactic or prosodic [6, 7] units to this purpose, very few authors have addressed the use of both kinds of units to identify genre.

Once these units have been properly defined on theoretical grounds and annotated on a corpus, it is an interesting challenge to assess whether they can be used to identify discourse genres as expected. The main contribution of the present study in this respect is first the use of recently proposed units, both prosodic with intonational periods (IPe, [10]) and syntactic with illocutionary units (IU, [11]). Second, the analysis makes use of a large corpus of spontaneous French speech, in which both IPes and IUs have been annotated. Third, results show that data as simple as the ratio between the number of IPes over the number of IUs are indeed characteristic of some genres. As the corpus is sufficiently large, these findings can be assessed on a statistical basis.

This study uses the Rhapsodie corpus which is a continuation of previous work conducted in the framework of the Rhapsodie treebank. This corpus of spoken French samples different discourse genres and is annotated both in syntax and prosody to model the intonosyntactic interface. I address this issue by focusing on the interface between macrosyntax and prosody. To this purpose, a computationally structured architecture encoding these two levels is queried. A functional point of view has been adopted, showing that it is possible to exhibit a correlation between the synchronization and number of intonosyntactic units and the types of discourse in the corpus.

This paper is structured as follows: Section 2 presents the corpus and the syntactic and prosodic units considered for the study. Since they have been proposed only recently within the Rhapsodie consortium, I briefly review related research. In section 3, the modus operandi through which this corpus was analyzed is described. Section 4 provides several quantitative results as well as a brief discussion as to how prosody and syntax may be used jointly to predict discourse genre. Section 5 concludes the paper.

2   The Treebank Raphsodie, Prosodic and Syntactic Annotations

The Rhapsodie project [12] provides a reference transcription system, based on syntactic and prosodic annotations, for the segmentation of spoken French into prosodic and syntactic units of different levels. The Rhapsodie corpus is a tree-bank of spoken French composed of 34 361 words and 87 speakers. Its purpose is the study of the interface between prosody, syntax and discourse. The three mechanisms of cohesion annotated in the corpus, i.e. syntactic cohesion, illocutionary cohesion, and prosodic cohesion, appear to operate simultaneously and independently from one another in spoken discourse. That is why these three levels were annotated separately. The first one, macrosyntax, is based on distributional constraints and syntactic tests to identify the units that compose it, regardless of prosodic information (except for a naive listening); the second one, prosody, is based only on acoustic and perceptual criteria, and the third (not taken into account in the present study), microsyntax, is based on a dependency-based approach resting on government relations. This strictly modular approach does not prejudge the reality of cognitive processes; on the contrary, it avoids circularity so as to better answer the question of the interrelationship between the two components in a situation of verbal interaction.

2.1   Corpus Design

The Rhapsodie tree-bank was created with the primary objective of proposing, implementing, and testing — on a wide coverage of different constructions — new methods of annotation and analysis to model the syntax-prosody interface in spoken French. It consists of samples collected from existing datasets, including the corpus presented in [13–17], and samples specifically recorded for the project, with a wide typological coverage (see table 1).

Table 1. Situational variables in Rhapsodie: 57 samples, 52 men, 35 women, duration = 3 h 18, 34 361 words. Monologues are coded M, dialogues D, private speech (0 or 1), and public speaking (2). Thus, the sample D2006 is a public dialogue, and '006' indicates the sequence number.

structure	monologue (M), dialogue (D)
social situation	private (0,1), professional (2)
planning	spontaneous, semi-spontaneous, planned
interactivity	interactive, semi-interactive, non-interactive
sub-genres	argumentative, descriptive, procedural, oratory

2.2 Syntax

Macrosyntactic Annotation. Combining the syntactic model proposed by the Aix School [18] and the pragmatic model developed within the Lablita experiment [19], two levels of syntactic cohesion were annotated within Rhapsodie. On the one hand, microsyntax describes the government relations which are usually encoded through dependency trees, phrase structure trees or C-Units [20]. On the other hand, macro-syntax can be regarded as an intermediate level between syntax and discourse and describes the whole set of relations holding between all the sequences — such as Bibers T-Units [20] — that make up one and only one illocutionary act. These two levels of syntactic cohesion are acknowledged by other authors (i.e. C-Units vs T-Units in Biber for instance) and the distinction between micro- and macro-syntax - even though the terminology has not been adopted, no alternative has been proposed - has been made by Blanche-Benveniste [18] and Berrendonner [21].

Some studies have addressed macrosyntax, maybe due to its recent emergence in the field of research. Most of them do not focus directly on macrosyntax but rather on the analysis of spoken language: for instance, the work on spoken French by Blanche-Benveniste [18], which proposes a grammar of speech, or the book on the intonation of French by Martin [22] which, as its name suggests, is an analysis of French prosody, presenting macrosyntax in terms of intonation. A number of articles also deal with macrosyntax. Berrendonner [21] offers a first glimpse of what a "macrosyntax" could be, while Apotheloz and Zay [23] address the challenges of the transition between "micro" and "macro" syntax. Avanzi [24] focuses on the work of the three teams specialized in this field, and shows that there is not only one but several macrosyntaxes. Deulofeu in [25] criticizes the notion of "detachment" — that is increasingly widespread in grammar — and shows that the phenomena included in this notion are inherent in "macrosyntax" and not in traditional syntax. He thereby demonstrates that it is necessary to revisit the current syntactic parsing system. Lastly, several studies in pragmatics have been devoted to the illocutionary value of IUs, but very few concern the span of IUs, and very few corpora have proposed a segmentation into IUs (see [19] and [26]).

Macrosyntax therefore seeks to describe "relationships that cannot be described by government between grammatical categories alone" [18], to characterize the organization of "certain non-governed appositions, dislocations, etc." [21], and to analyze unusual syntactic constructions that are difficult to parse by "a simple constituency grammar" [19]. If there were to remain only one motivation for the emergence of the concept of macrosyntax, it would be the inadequacy of the concept of "sentence" to cover all the phenomena of spoken discourse. Indeed, the notion of "sentence" cannot be considered as the only grammatical unit of reference for modeling spoken languages such as French for example.

However, Berrendonner and Béguelin [27] indicate that it is not so much the existence of the sentence as a unit that is questionable as the ineffectiveness of the notion of sentence for scientifically segmenting and analyzing certain discursive sequences. A sentence is defined by a set of criteria that — in the case of the analysis of spoken language — only apply to a limited number of utterances (for example (1) and (2)):

(1) vous êtes née à Cannes //[1] [Rhap-D2004 , Corpus Lacheret]
    you were born in Cannes //
(2) j'étais communiste à ce moment-là // [Rhap-D2010, Corpus Rhapsodie]
    I was a Communist at the time //

From a purely typographical point of view, the sentence begins with a capital letter and ends with a period. The presence of a period in writing is supposed to be marked by intonation in spoken language. From a semantic point of view, the sentence is expected to refer to only one meaning. Lastly, from a syntactic point of view, the sentence is a unit in which dependency relations hold between the various elements. However, if the above-mentioned criteria are applied to spoken productions such as (3) or (4), they are unable to univocally define these sequences.

(3) moi ma mère le salon c'est de la moquette // [example extracted from "Phantom Sentences", 2008 Henri-José Delofeu]
    I my mother the living-room it is carpeting //
(4) j'ai un chapeau d'homme //+[2] un feutre //+ un feutre //+ rose //+ couleur bois de rose au large ruban noir //[Rhap-D2010, Corpus Rhapsodie]
    I have a man's hat //+ a trilby //+ a trilby //+ pink //+ rosewood color with large black ribbon //

To conclude, the macrosyntactic model recognizes therefore the examples above as fully-fledged macrosyntactic units (marked-up by //) whose members are not necessarily linked by syntactic dependency relations. Just as prosody was not annotated using syntactic cues, macrosyntax was annotated in Rhapsodie without considering any theoretical prosodic information (sometimes the naive listening of a sequence was necessary to disambiguate). Therefore, macrosyntactic units were not annotated as proposed by Berrendonner [21], nor as proposed by the Florence school which defines them primarily based on prosody considerations. However, building on Cresti's proposition [19], maximal units of macrosyntax were chosen so as to coincide with the maximal extension of an illocutionary act. But, departing from Cresti, the maximal extension of an illocutionary act was not defined as a set of prosodic units, but rather as a set of units that build up to form one illocutionary act. The maximal extension of an illocutionary act was called Illocutionary Unit (IU).

2.3 Prosody

Prosodic Annotation for Spoken French. For prosody, Rhapsodie annotations are built on the theoretical hypothesis formulated by the Dutch-IPO school [28] stating that, out of the total information characterizing the acoustic

---

[1] The symbol // marks the end of an IU.

[2] The symbol + indicates a government relation that exceeds the IU frontier.

domain, only some perceptual cues selected by the listener are relevant for linguistic communication [8, 29] On this basis, only three perceptual phenomena characterizing real productions were annotated: prominences [30, 31], pauses and disfluencies [32].

Starting from this annotation, a prosodic structure was automatically generated, organized around rhythmical and melodic components. In practice, the prosodic structure, generated on the basis of the labeling of prominences, disfluencies and intonational periods, was built on the hypothesis that disfluencies, pauses and the distribution and degree of prominences define different types of prosodic cohesion. Three major levels of prosodic cohesion were identified inside the intonational period, represented as a hierarchical constituent tree, from bottom-up: metrical foot, rhythmic group and intonation package. A very noticeable feature of this annotation procedure is that these three primitives of prosodic structure were identified and annotated without any reference to syntax or pragmatics, and they are expected to be sufficiently detailed to permit the complex prosodic analysis of linguistic units.

The concept of intonation period (IPe) emerges from previous work based on the segmentation and analysis of prototypical variations in intonation and the distribution of breaks, conducted by Lacheret et al. (See [10] and [33] for the first experiment). It turns out that the intonational period is a neutral work space (the approach does not prejudge the syntactic-semantic function of the intonational forms encountered) that enables the internal organization of spontaneous speech to be analyzed on a new basis. Indeed, on the one hand the main function of prosody is to serve as a relay for syntax: since syntax cannot encode certain structural relations, prosody handles them. On the other hand, prosody is more fundamentally linked to the communicative purpose than syntax [34] and [35]. In other words, the prosodic organization alternately reflects syntactically unexpressed relations (see (8)): management information, and - when necessary - the degree of speaker involvement. Accordingly, prosody and syntax are not strictly congruent, nor necessarily redundant, although there are points of connection between the two structural levels of organization of the message.

In the Rhapsodie corpus, the intonation period (IPe) [10] is the highest intonation macro-unit in the prosodic hierarchy. Periods are calculated semi-automatically using the software Analor [36]. Prosodic segmentation of a corpus into intonation periods occurs when the following conditions exist:

(1) occurrence of a pause of at least 300 ms;
(2) detection of an $F0$ pitch movement reaching a certain amplitude, defined as the difference in height between the last $F0$ extremum and the mean $F0$ over the entire portion of the signal preceding the pause;
(3) detection of a "jump", defined as the difference in height between the last $F0$ extremum preceding the pause and the first $F0$ value following the pause.

It should be noted that the decision to recognize a periodic break is based on the principle of compensation thresholds. In other words, detection is not dependent on the exact values of the parameters above, but on their respective activation thresholds and associated weight: When a parameter is very slightly

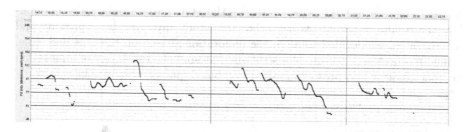

Fig. 1. Segmentation into three IPes (marked by a vertical bar, two bars give three IPes): le programme monsieur le premier ministre **(IPe)** comporte un certain nombre de projets dautant plus intrssants **(IPe)** que nous les proposons depuis longtemps — *the program Mr. Prime Minister* **(IPe)** *has a number of projects that are all the more interesting* **(IPe)** *as we have been proposing them for a long time* —[Rhap-D2006, Rhapsodie] [12]

below the selected threshold, a boundary period is detected if the other parameters have values above the threshold.

3 Methodology

Given both IPes and IUs, I aim at studying whether their relations are informative with respect to discourse genre. To this purpose and for each of the 57 samples of the corpus, I automatically extracted the total number of IUs and IPes as well as their temporal positions [37–39]. In this study, two main features were jointly considered: first, the synchronization of IUs and IPes 4.1, then their frequency within discourse 3.3.

3.1 Adopted Formalism to Request on the Corpus of Spoken Language

The study of spoken language can involve several linguistic perspectives, such as macrosyntax and prosody (among other linguistic levels). In the Rhapsodie corpus, for each one of them, the discourse was analyzed as a tree-like structure, where each node carries some specific linguistic information and features. The approach used here is named Object-Oriented Processing of Speech (OOPS), a principled computational way to aggregate and jointly study different linguistic annotations of the same data.

In the Rhapsodie project, macrosyntactic annotations are anchored in the orthographic words of the text (or tokens), while prosodic annotations are anchored in time. Previous projects developed in recent years focused on either one or the other of these supports, i.e. either only time, for which formal frameworks have been developed in phonology (e.g. Bird and Liberman [40]), or more typically, only text for the annotation of written corpora (e.g. the Prague Treebank and its three levels of annotation [41] or the Gate system (general architecture for text engineering) [42] which provides the ability to automatically manipulate the results of several parsers). For this study involving both prosody and macrosyntax, we decided to align the annotations based on the text and time, using an object architecture [43] structured around the two layers of annotation i.e. macrosyntax and prosody, making combined queries possible [39].

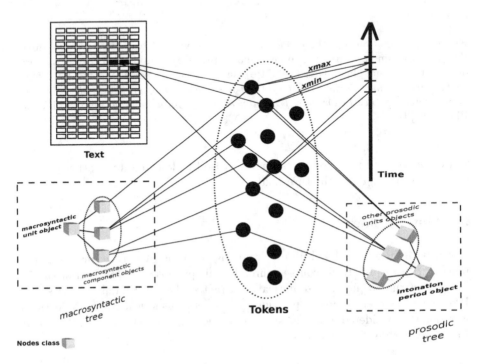

Fig. 2. Schematic articulation of the different trees considered: macrosyntactic and prosodic. Their connection to time and text is done through a galaxy of orthographic words (token objects) to which they are connected.

As seen above, a typical linguistic annotation yields different tree-like structures for the same data, one for each different perspective (or linguistic level). The main idea of my tool OOPS is to create one object for each word of the transcription, and to build a tree structure for each annotation, whose leaves are those words objects. The important point is that words are not duplicated from one perspective to another: the same objects are shared as leaves by all the trees through symbolic links. Alternatively, a perspective may operate on the speech signal as in prosodic analysis. In that case, the leaves are time intervals within the signal. Each word can be linked to the corresponding intervals in the signal by automatic alignment [44]. Therefore combined query is possible because these two levels of annotations share an anchorage on the same objects, namely tokens, which have both temporal coordinates and positions in the text.

Figure 2 shows how it is possible to jointly study both prosodic and macrosyntactic units according to time. This formalism exhibits the way in which these units are synchronized.

3.2 Synchronization of IUs and Periods

In order to describe how IUs and IPes combine to delineate basic discourse units specific to genres, we propose a study of the interactions between IPes and IUs based on the coincidence of their borders. Borders that are both IPe and IU borders will be called "synchronized borders". Borders are *out of sync* when prosodic and syntactic boundary instants do not coincide, i.e. when the boundaries of IPes are not IU borders and vice versa. It is important to note that nothing prevents an IPe whose border is aligned with an IU from containing several IUs and vice versa (see Figure 3). Finally, each IPe and IU can be fully synchronized when their right and left borders are synchronized or partially when it is only the right or left borders that are synchronized. This gives three major types of synchronization: full synchronization (IU and IPe are strictly equivalent, they share the same temporal borders (1)), right synchronization (IUs that share the same right border as IPes (2)) and left synchronization (IUs that share the same left border as IPes (3)). In addition, there are some atypical configurations, presented in (4) and (5), but these are not used in the present work. Our goal is first to report statistically different types of synchronization (partial vs total).

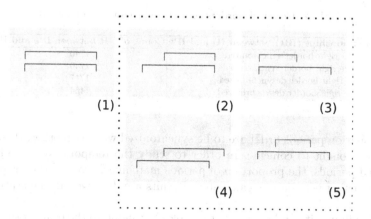

Fig. 3. Synchronization boundaries. (1) borders completely synchronized (the two units are aligned), (2) right partial synchronization, (3) left partial synchronization, (4) overlapping synchronization, (5) inclusive de-synchronization. (4) and (5) can be combined with (2) and (3).

3.3 Relative Frequency of IUs and Periods

As the proportion of IUs and IPes differs in the corpus excerpts, their relative frequency appears to be a good indicator of the discourse genre:

$$\text{Ratio (sample)} = \log \left(\frac{\text{number of IPe}}{\text{number of IU}} \right). \tag{1}$$

Another variable used for this experiment is the ratio between the frequency of IUs and the frequency of IPe. This ratio is given in log scale to make the variable symmetrical and so does not favor the rate of IUs per IPe compared to the rate of IPes per IU (because $\log(a/b) = -\log(b/a)$). This information may indicate the respective potential for inclusion of these two types of units: an IU/IPe ratio greater than 1 indicates that the sample contains more IUs than IPes, therefore IUs will probably be the unit that include IPes.

4 Results and Analysis

4.1 How IUs and IPe Are Synchronized

For each of the 57 samples of the corpus, I automatically extracted right-hand boundaries of IUs and IPes as well as their temporal positions in different configurations [37]. Synchronization and desynchronization counts as well as percentages are given in Table 2.

Table 2. Count of borders relationship between IPe and IU

Border relationships (BR) between IU and IPe	Count of BR between IPe and IU	%
IU right border synchronized	1740	50.33
IPe right border synchronized	1740	59.91
IU right border desynchronized	1717	49.67
IPe right border desynchronized	1164	40.08

If intonation periods or IU are to be synchronized with each other, their right boundaries ought to coincide. In order to study the temporal synchronization of IU and periods, the proportion of periods matching IU versus the proportion of IU matching periods was calculated. Results are displayed in the scatterplot (Figure 4).

On this figure, it can be seen that samples belonging to the oratory sub-genre (shaded in blue) differ from the others. They exhibit a high synchronization between IUs and periods (more than 80%), whereas the IPes rarely match an IU (20 to 50%), indicating that they often occur within an IU. A transcript of one of the samples of oratory speech is given in example (8) below. It is an excerpt from a speech given by former French president F. Mitterand, exhibiting many more periods than IUs. This imbalance was found to be typical of political, religious and scientific speech in the corpus. Each macrosyntactic constituent is strongly

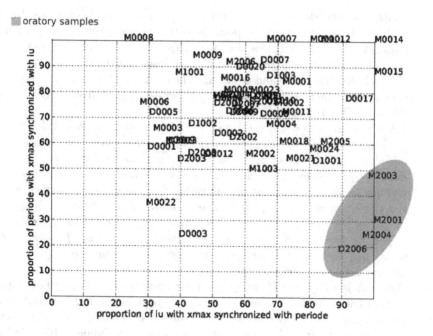

Fig. 4. Synchronization of IUs and IPs for each corpus samples

IPe marked, hence the ear perceives a very significant scansion effect. Example (9) shows the opposite phenomenon: in this case it is the intonational level which encapsulates the macrosyntactic segments. Finally example (10) gives an illustration of a perfect correspondence and synchronisation between IUs and IPes. These observations lead us to hypothesize that prosody may have a kind of relay function with respect to macrosyntax. As shown in a previous study [45], the level of prosodic cues increases with the increase in discourse markers (often meaning less syntactically elaborate phrases). between It is tempting to conclude that there is an inverse relationship between these two linguistic levels: an elaborate syntax may imply a less pronounced/marked prosody (the prosody just follows macrosyntactic units), while a less elaborate syntax generates a prosody that plays a much stronger information packaging role.

Other features of speech also stand out from Figure 4. Among them, it is noticeable that some samples located in the upper right corner of the scatterplot happen to have IUs that are closely synchronized with periods. Listening to these samples, it is noticeable that the corresponding speakers are obviously making an effort to deliver canonical speech, i.e. speech in which the prosody matches the syntax. Still, lacking objective metadata to assess this, I cannot give statistical confirmation of this point.

(8) lorsque vous semblez mettre en doute $[_{IPe}]$ notre amour des libertés $[_{IPe}]$ c'est un outrage $[_{IPe}]$ que nous n'acceptons pas $[_{IU-IPe}]$ nous sommes les héritiers de la tradition qui a instauré dans ce pays $[_{IPe}]$ la démocratie

politique et sociale [$_{IU-IPe}$] toujours [$_{IPe}$] toujours [$_{IPe}$] contre les droites coalisées [$_{IPe}$] nos combats pour la conquête du droit [$_{IPe}$] jalonnent l'histoire des deux derniers siècles [$_{IU-IPe}$] c'est à ceux de votre tradition [$_{IPe}$] que nous avons arraché le suffrage universel [$_{IPe}$] la liberté d'association [$_{IPe}$] que nous avons arraché la liberté d'association [$_{IPe}$] que nous avons arraché [$_{IPe}$] la liberté de la presse [$_{IPe}$] le droit de grève [$_{IPe}$] le droit à l'instruction [$_{IU-IPe}$]

When you seem to doubt [$_{IPe}$] our love of freedom [$_{IPe}$] this is an outrage [$_{IPe}$] which we do not accept [$_{IU-IPe}$]we are the heirs of a tradition established in this country [$_{IPe}$] the political and social democracy [$_{IU-IPe}$] Always [$_{IPe}$] Always [$_{IPe}$] Straight against the Right Wing coalition [$_{IPe}$] Our fights for the conquest of our rights [$_{IPe}$] marked the history of the last two centuries [$_{IU-IPe}$] It is against those of your tradition [$_{IPe}$] that we fought for universal suffrage [$_{IPe}$] freedom of association [$_{IPe}$] that we ripped the freedom of association [$_{IPe}$] that we ripped the freedom of the press [$_{IPe}$] the right to strike [$_{IPe}$] the right to education[$_{IU-IPe}$]

[Rhap-D2006, Broadcast corpus]

(9) alors euh je m'appelle Clara [$_{IU}$] j'ai dix-neuf ans [$_{IU}$] j'ai eu l'obtention de mon bac euh donc l'année dernière [$_{IU-IPe}$] c'est un bac euh SMS donc technologique [$_{IU-IPe}$] c'est sciences mdico-sociales [$_{IU}$] a n'a rien à voir avec euh la littrature [$_{IU-IPe}$] parce qu'en fait euh j'aime la biologie [$_{IU}$] et je suis plus euh vers la biologie et euh le social donc euh XXX sciences médico-sociales [$_{IU}$] et donc en fait euh j'ai choisi italien en deuxième choix [$_{IU}$] mon premier choix euh c'était euh psychologie [$_{IU-IPe}$]

so uh I'm Clara [$_{IU}$] I am nineteen years old [$_{IU}$] I was getting my degree last year [$_{IU-IPe}$] so uh this is a SMS baccalaureat so uh it's technic [$_{IU-IPe}$] it is a medical and social sciences degree [$_{IU}$] it has nothing to do with literature [$_{IU-IPe}$] uh because in fact uh i love biology [$_{IU}$] and I am more to biology and uh uh uh XXX social so the medico-social sciencies [$_{IU}$] and therefore actually uh I chose Italian as second choice [$_{IU}$] my first choice was uh uh psychology[$_{IU-IPe}$]

[Rhap-M1001, Broadcast corpus]

(10) alors en partant de la place Paul Vallier pour aller à la place Notre-Dame [$_{IU-IPe}$] alors j'emprunte la rue de Strasbourg [$_{IU-IPe}$] je passe par la place Vaucanson [$_{IU-IPe}$] je prends direction Maison du tourisme[$_{IU-IPe}$] euh à la Maison du tourisme je contourne enfin je prends la rue de la République en remontant la rue de la République[$_{IU-IPe}$] je tombe sur la place Sainte-Claire on va dire là où il y a la halle[$_{IU-IPe}$] [Rhap-M0014, Corpus Avanzi[14]]

then starting from the place Paul Vallier to go to the place Notre Dame [$_{IU-IPe}$] then I take the rue de Strasbourg[$_{IU-IPe}$]I pass by the place Vaucanson[$_{IU-IPe}$]I take the direction Maison du tourisme[$_{IU-IPe}$] uh at la Maison du tourisme I bypasses I finally take the rue de la République up the street of the République [$_{IU-IPe}$] I arrive on the place St. Claire instead they will say where is the hall [$_{IU-IPe}$]

[Rhap-M0014, Corpus Avanzi [14]]

4.2 How IUs and IPe Are Represented

The second step in this study was to assess whether the ratio may be a discriminating feature for characterizing some discourse genres, as was synchronization of IPes and IUs. To show the variability of the different kinds of relations between IUs and IPes, all the corpus excerpts were labeled as belonging to one of the 4 sub-genres considered: oratory, procedural, descriptive and argumentative. The studied corpus includes all 3457 IUs and 2904 IPes. On computing the ratio (1) of IPes and IUs over all the corpus samples, some detached groups of samples emerge: for example those exhibiting more IPes than IUs tend to match the oratorical genre, while argumentative ones seems to be more central, with a ratio equal to 1.

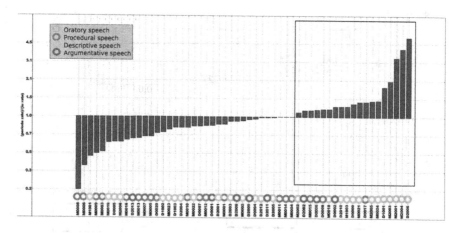

Fig. 5. Representation of the corpus samples according to their IPe/IU ratio. The samples framed on the right are those with more IPes than IUs (and vice versa for the others).

The oratory sub-genre clearly emerged as characterized by a high number of IPes per IU, highlighting the propensity of oratory to be highly rhythmic. Another fact that can be noticed on Figure 5 is that, apart from oratory and some descriptive speech, most of the Rhapsodie corpus exhibits fewer IPes than IUs. As the opposite seems characteristic of oratory, which is a typical example of planned discourse, this result is expected, since the intent in creating the Rhapsodie corpus in the first place was to build a corpus of spontaneous French speech.

To verify this observation statistically, I divided the samples into four genres according to the corpus metadata: oratory, procedural, argumentative, and descriptive, and examined whether the ratio (1) is a discriminating feature for identifying some of the groups. To this purpose, and since the data cannot be considered as having a normal distribution (thus excluding an ANOVA test), a non-parametric Kruskal-Wallis test was applied. The result of the Kruskal-Wallis test shows that groups have at least one difference: it is clear that the null hypothesis stating that the distributions of the ratio are the same in all groups can be safely rejected ($p = 5 \cdot 10^{-4}$). We can hence conclude that at least one of those distributions is significantly different from the others. To ascertain whether the difference between all the distributions is significant, a Kolmogorov-Smirnov (KS) test of comparison was carried out by dividing the alpha reference level by 6 (also called a Bonferroni correction). The test was used 6 times to compare the distributions of all four groups. The results are displayed in Figure 6.

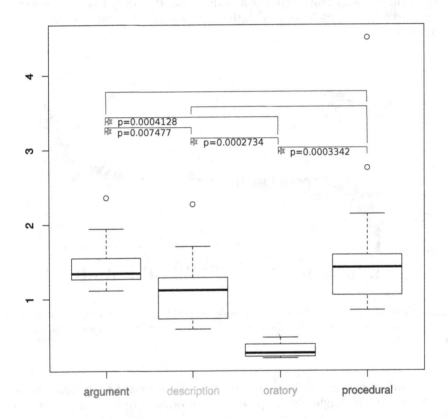

Fig. 6. Boxplot of all sub-genre groups

Oratory is significantly distinguished from the others (see Figure 6). We can also observe that argumentative speech and descriptive speech are well separated, as proved by the result of the KS test. However, procedural speech is problematic, as the KS test does not enable the null hypothesis to be rejected for the argumentative-procedural and descriptive-procedural couples, maybe due to the excessive heterogeneity of the procedural samples (some are extremely short, others much longer, etc. a data normalisation was not possible.)

5   Conclusion

This paper has examined whether prosody and syntax could be jointly used so as to characterize sub-genres of discourse such as oratory, argumentative, descriptive and procedural speech (for a similar methodology see also [46, 45]). To this purpose, I have briefly reviewed how the discourse can be split into Illocutionary Units from a syntactic perspective and into intonational periods from a prosodic perspective and have shown that these annotation processes are independent from each other.

In the domain of written language, characterizing the relations between discourse genres and textual features has been the topic of a large body of research for over half a century. In the case of spoken language, however, it is only recently that researchers have focused on the interaction between discourse genres and intonosyntax. To this end, I have exhibited syntactic, prosodic and intonosyntactic features that are characteristic not only of general discourse genres, but also of more specific types.

For this purpose, I focused on the Rhapsodie corpus, which includes more than 3 hours of spontaneous French speech, annotated both for syntax and prosody. Since the sources used to build this corpus are heterogeneous, great care has to be taken when using quantitative methods to draw conclusions about the frequency of the extracted features and the genres and types of the excerpts. Nonetheless, the study has shown that the formal features proposed are indeed relevant for the study of discourse genres. This claim has been quantified by studying the mutual information that these features share with the situational variables considered, making it possible to identify potentially redundant information and to adequately predict genres from intonosyntactic annotations. Using a corpus of spontaneous French speech, Rhapsodie, in which both intonational periods and illocutionary units are annotated, I have demonstrated that for a given sample, a much larger number of periods than of IUs is characteristic of oratory while the reverse is true of descriptive speech for example.

Furthermore, the synchronization of the prosodic and syntactic meta-units seems to be related to canonical speech, i.e. one in which syntax and prosody coincide regardless of speech genres. An interesting perspective could be to apply this framework to other languages. We hypothesize that the relative frequency of IPes over IUs is a distinguishing criterion to classify and characterize types of discourse. The difference between the observed Ipe/IU ratio and the intuitively expected ratio is illustrated by a massive production of IPes compared to the number of IUs (3457 IUs for 2904 IPes over the corpus).

To conclude, this study has shown that the interface between prosodic units (IPe) and macrosyntactic units (IU) provides simple intonosyntactic features that appear in speech as stable markers of how discourse relations are nested. All these points provide valuable feedback on the role played by the intonosyntactic interface in the identification of discourse genres in spoken French and how basic discourse units vary according to genre.

References

1. Rastier, F.: Sens et textualité. Hachette (1989)
2. Halliday, M.: Spoken and written Language. Oxford University Press (1985)
3. Biber, D.: Variation Across Speech and Writing. Cambridge University Press (1988)
4. Léon, P.: Précis de phonostylistique. Parole et expressivité. Nathan Université, Paris (1993)
5. Simon, A.-C., Auchlin, A., Avanzi, M., Goldman, J.-P.: Les phonostyles: une description prosodique des styles de parole en francais. In: Les Voix des Francais. Peter Lang (2000)
6. Obin, N., Dellwo, V., Lacheret, A., Rodet, X.: Expectations for Discourse Genre Identification: a Prosodic Study. In: Interspeech, pp. 3070–3073 (2010)
7. Obin, N.: MeLos: Analysis and Modelling of Speech Prosody and Speaking Style. Thèse de Doctorat. Ircam-UPMC, Paris (2011)
8. Lacheret-Dujour, A., Beaugendre, F.: La prosodie du francais. CNRS (1999)
9. Degand, L., Simon, A.-C.: Mapping prosody and syntax as discourse strategies: How basic discourse units vary across genres. In: Where Prosody Meets Pragmatics. Emerald Group (2009)
10. Lacheret, A., Victorri, B.: La période intonative comme unité d'analyse pour l'étude du francais parlé: modélisation prosodique et enjeux linguistiques. Verbum 1(24), 55–72 (2002)
11. Benzitoun, C., Deulofeu, J., Kahane, S., Pietrandrea, P., Bolly, C., Debaisieux, J.-M., Dister, A., Lefeuvre, F., Rossi-Gensane, N., Sabio, F., Tanguy, N.: The macrosyntactic annotation, ch. 6. Benjamins (2013)
12. Lacheret, A., Kahane, S., Pietrandrea, P.: Rhapsodie: a Prosodic and Syntactic Treebank for Spoken French. In: Studies in Corpus Linguistics. Benjamins (2015)
13. Laks, B., Durand, J., Lyche, C.: Le projet PFC (Phonologie du Français Contemporain): une source de données primaires structurées. In: Phonologie, Variation et Accents du Français Hermès, pp. 19–26. Hermès (2009)
14. Avanzi, M.: L'interface prosodie/syntaxe en francais. Dislocations, incises et asyndètes. Bruxelles, Peter Lang. PhD thesis, Université de Neuchatel (2012)
15. Branca-Rosoff, S., Fleury, S., Lefeuvre, F., Pires, M.: Discours sur la ville. Corpus de Français Parlé Parisien des années 2000 (2009)
16. Avanzi, M., Simon, A.-C., Goldman, J.-P., Auchlin, A.: Un corpus de français parlé annoté pour l'étude des proéminences, c-prom. In: Actes des 23èmes Journées d'étude sur la Parole, Mons, Belgique (2010)
17. Mertens, P.: L' intonation du français: de la description linguistique à la reconnaissance automatique. Katolieke Universiteit te Leuven (1987)
18. Blanche-Benveniste, C., Bilger, M., Rouget, C., den Eynd, K.V.: Le français parlé. études grammaticales. CNRS éditions, Paris (1990)
19. Cresti, E.: Corpus di italiano parlato. Florence, Accademia della Crusca (2000)
20. Biber, D., Johansson, S., Leech, G., Conrad, S., Finegan, E., Quirk, R.: Longman grammar of spoken and written English, vol. 2. Longman, London (1999)
21. Berrendonner, A.: Pour une macro-syntaxe. Données orales et théories linguistiques 21, 25–31 (1990)

22. Martin, P.: Intonation du français. Armand Colin, Collection U (2009)
23. Apothéloz, D., Zay, F.: Incidents de la programmation syntagmatique: reformulations micro-et macro-syntaxiques. Cahiers de Linguistique Française 21, 10–34 (1999)
24. Avanzi, M.: Regards croisés sur la notion de macro-syntaxe. Travaux Neuchâtelois de Linguistique 49, 39–58 (2007)
25. Apothéloz, D., Combettes, B., Neveu, F.: Les linguistiques du détachement: actes du colloque international de Nancy (Juin 7-9 2006). Sciences Pour la Communication, P. Lang (2009)
26. Pietrandrea, P., Kahane, S., Lacheret, A., Sabio, F.: The notion of sentence and other discourse units in spoken corpus annotation. In: Spoken Corpora and Linguistic Studies. John Benjamins Publishing Company (2014)
27. Berrendonner, A., Reichler-Béguelin, M.-J.: Décalages: les niveaux de l'analyse linguistique. Langue Française (81), 99–125 (1989)
28. Hart, J., Collier, R., Cohen, A.: A perceptual study of intonation, an experimental phonetic approach to speech melody. Cambridge University Press (2006)
29. Wightman, C.W.: ToBI Or Not ToBI? In: Speech Prosody 2002, pp. 25–29 (2002)
30. Buhmann, J., Caspers, J., van Heuven, V., Hoekstra, H., Martens, J.-P., Swerts, M.: Annotation of prominent words, prosodic boundaries and segmental lengthening by non-expert transcribers in the spoken dutch corpus. In: LREC. European Language Resources Association (2002)
31. Tamburini, F., Caini, C.: An automatic system for detecting prosodic prominence in american english continuous speech. International Journal of Speech Technology 8(1), 33–44 (2005)
32. Lacheret, A., Obin, N., Avanzi, M.: Design and evaluation of shared prosodic annotation for spontaneous french speech: from expert knowledge to non-expert annotation. In: Proceedings of the Fourth Linguistic Annotation Workshop, pp. 265–273. Association for Computational Linguistics (2010)
33. Lacheret, A., Ploux, S., Victorri, B.: Prosodie et thématisation en français parlé. Cahiers de PraxéMatique 30, 89–111 (1998)
34. Rossi, M., de phonétique d'Aix-en Provence, I.: L'Intonation de l'acoustique à la sémantique. In: Études Linguistiques, Klincksieck (1981)
35. Rossi, M.: L'intonation. le système du français: description et modélisation. In: Paris-Gap, Ophrys (1999)
36. Avanzi, M., Lacheret-Dujour, A., Victorri, B.: Analor: A tool for semi-automatic annotation of french prosodic structure. In: Speech Prosody (2008)
37. Belião, J.: Formalisation, implémentation et exploitation d'une hiérarchie objet intono-syntaxique: étude sur un treebank de francais oral spontané. Master's thesis, Université Sorbonne Nouvelle (2012)
38. Belião, J.: Création d'un multi-arbre à partir d'un texte balisé: l'exemple de l'annotation d'un corpus d'oral spontané. In: RECITAL (2012)
39. Beliao, J., Liutkus, A.: Oops: une approche orientée objet pour linterrogation et lanalyse linguistique de linterface prosodie/syntaxe/discours. In: CMLF 2014 (2014)
40. Bird, S., Liberman, M.: A formal framework for linguistic annotation. Speech Communication 33(1), 23–60 (2001)
41. Böhmová, A., Hajič, J., Hajičová, E., Hladká, B.: The prague dependency treebank. In: Treebanks, pp. 103–127. Springer (2003)
42. Cunningham, H.: Gate, a general architecture for text engineering. Computers and the Humanities 36(2), 223–254 (2002)

43. Rumbaugh, J., Blaha, M., Premerlani, W., Eddy, F., Lorensen, W.E., et al.: Object-oriented modeling and design, vol. 199. Prentice Hall, Englewood Cliffs (1991)

44. Goldman, J.-P.: Easyalign: an automatic phonetic alignment tool under praat. In: InterSpeech (2011)

45. Beliao, J., Lacheret, A.: Disfluencies and discursive markers: when prosody and syntax plan discourse. In: The 6th Workshop on Disfluency in Spontaneous Speech (2013)

46. Beliao, J.: Characterizing genres through syntax and prosody. In: The 28th ESSLLI Student Session Proceedings (2013)

Evaluating Supervised Semantic Parsing Methods on Application-Independent Data

Sebastian Beschke

Natural Language Systems Division
Department of Informatics
University of Hamburg, Germany
beschke@informatik.uni-hamburg.de

Abstract. While supervised statistical semantic parsing methods have received a good amount of attention in recent years, this research has largely been done on small and specialized data sets. This paper introduces a work-in-progress with the objective of examining the applicability of supervised statistical semantic parsing to application-independent data with linguistically motivated meaning representations. The approach discussed in this paper has three key aspects: The circumvention of data scarcity using automatic annotation, experimentation with different types of meaning representations, and the design of a suitable graded evaluation measure.

1 Introduction

We understand semantic parsing to be the task of extracting a formal meaning representation (MR) from a natural language text. Supervised statistical methods of semantic parsing are a research topic to which various approaches and formalisms have been applied over the past years. Evaluation of these methods has generally been performed on small data sets from very limited and application-specific domains. One example is Geoquery, a widely used corpus for natural language database queries on US geography [1]. A prime reason for the focus on small data sets is that the annotation of training data with full semantic MRs is laborious. These representations are even more complex than data used for many other tasks in statistical natural language processing. Therefore, fully annotated data has so far been scarce and mostly limited to application-specific data.

There is however mounting interest in application-independent semantic analysis. This task entails the creation of linguistically motivated MRs that attempt to represent certain linguistic features as completely as possible, as opposed to application-specific types of MR that only capture the amount of information that is needed for the application at hand. A prominent rule-based system performing this task is Boxer [2], while Le and Zuidema recently presented a statistical approach [3]. Both of these systems are based on Discourse Representation Theory [4].

M. Colinet et al. (Eds.): ESSLLI 2012/2013, LNCS 8607, pp. 19–25, 2014.
© Springer-Verlag Berlin Heidelberg 2014

(a) answer(count(river(loc_2(stateid('california')))))
(b) answer(A,count(B,(river(B),loc(B,C),const(C,stateid(california))),A))
Give me the number of rivers in California.

Fig. 1. The Geoquery corpus contains two styles of meaning annotations: (a) variable-free expressions, and (b) Prolog-style expressions with variables. The meaning representations correspond directly to database queries and only contain enough information to perform the task of question answering. Linguistic details that are irrelevant to this task are not represented.

some(A,some(B,some(C,and(not(some(D,and(n12thing(D),not(r1after(A,D))))),
and(r1patient(A,B),and(r1agent(A,C),and(v1demand(A),and(n1solution(B),
and(a1global(B),and(n1problem(C),a1global(C)))))))))))
After all, global problems demand global solutions.

Fig. 2. An example of the type of meaning representation created by Boxer. As Boxer's meaning representations aim to address a wider range of linguistic phenomena, they tend to be more comprehensive than typical Geoquery representations. As an example, consider the use of Neo-Davidsonian event semantics, with explicit agent and patient relations, which provides greater flexibility for semantic analysis but leads to an increase of the meaning representation size.

While the methods used for supervised semantic parsing (SSP) are in principle applicable to application-independent data, it is important to note the different characteristics of the data. While application-specific corpora such as Geoquery tend to exhibit low linguistic variability and complexity (such as consisting only of questions with short average sentence lengths), application-independent data from more open domains, such as newswire, is likely to contain longer, more varied sentences. In addition, as linguistically motivated MRs attempt to encode meaning as fully as possible, they also tend to be more complex than special-purpose MRs, which only encode information important to the application at hand. This dual increase in complexity is illustrated in Figures 1 and 2, and can also be witnessed by comparing the (application-specific) Geoquery corpus to the (application-independent) Groningen Meaning Bank [5]. It is not yet well understood how well the established SSP methods scale up to this type of data.

For this reason, we propose an experiment designed to help better understand how SSP generalizes to application-independent data. Its key aspects are the use of automatic annotation to generate open-domain test data (Section 2), experimentation on how the complexity of MRs can be adjusted to balance the expressiveness of the MR against the capabilities of the learning algorithm (Section 3), and the design of a graded measure to evaluate the performance of an SSP system (Section 4). We also present some thoughts on the possible learning framework to be used (Section 5). The paper closes with a brief discussion (Section 6).

2   Automatic Annotation

The scarcity of corpora annotated with deep semantic representations has been a significant limit for SSP research. The widely used Geoquery corpus [1] with its 880 sentences is both small in size and narrow in scope. The same applies to most other data sources used in SSP research so far.

An important recent development in this area is presented by the Groningen Meaning Bank (GMB) [5]. Its current 2.1.0 release consists of 8,000 texts with over 1 million tokens, which are annotated in Discourse Representation Theory [4]. The annotations are first created automatically by a tool pipeline and then refined by human annotators, including both experts and non-experts, wherein gamification is employed to allow the latter to contribute their linguistic knowledge [5]. The GMB is not limited to a specific domain, containing Voice of America newswire texts, country descriptions from the CIA Factbook, texts from the Open ANC [6], and Aesop's fables. As such, it is likely to become an important data source for future SSP efforts that take an open domain approach. In fact, one such effort has already been presented [3].

While the GMB thus seems to be a very suitable data source for experiments in SSP, it also has a few drawbacks. Importantly, the linguistic complexity and average sentence length of the texts is quite high, especially when compared with special-purpose corpora such as Geoquery. This might pose problems when working with algorithms whose computational performance is not yet up to par. In addition to the GMB, we therefore plan to use data annotated using the semantic parsing tool Boxer, which is also being used in the preparation of the GMB [5]. Manual inspection suggests that the MRs generated by Boxer are of sufficient quality to serve as training material for SSP systems. This allows any corpus to be used as training data, given that it can be automatically annotated. In this way, we are able to vary the training data's complexity as seems appropriate.

Automatically generated annotations are likely to be flawed. We do not suggest that training SSP models using automatic annotation will yield systems of the highest quality. Automatic annotation should rather be seen as a crutch in developing SSP methods, which will hopefully become unnecessary as more varied training data become available.

3   Experimentation on Meaning Representations

An important open question in SSP is which type of MR is most beneficial to the task. As an example, the Geoquery corpus is annotated using two distinct types of MR: variable-free functional expressions, and Prolog-style expressions using variables (see Fig. 1). While there is of course an interaction between the type of MR and the learning algorithm used in a specific system configuration, most SSP systems are designed to be somewhat independent of the MR formalism. This allows us to study this interaction experimentally.

Some of the current SSP systems can process only variable-free forms (such as [7]), while others can process both types of MR (such as [8]). As most semantic formalisms, including Discourse Representation Theory, rely crucially on variables (or, put differently, graphical structures such as those used in [3]), our preference should be on the latter type of learning framework. However, there is also recent work on the design of variable-free MRs with the same expressivity as lambda-calculus forms [9,10]. There are also underspecified semantic formalisms such as Lexical Resource Semantics [11]. Converting meaning representations into alternative formalisms would allow comparing these formalisms from the point of view of SSP performance.

Besides conversion to other formalisms, another likely way to improve SSP performance is the simplification of MRs. By this, we mean modifications that do not necessarily preserve the full content of an MR, but in some way make it easier to process. For instance, the use of nested logical connectives and quantifiers imposes a structure on MRs with which learning algorithms might struggle, so removing some or all of these phenomena may yield representations that are easier to learn (this can also be thought of as a kind of underspecification). The idea is that even if we remove some information from the MR, there may still be enough information left to fulfill some useful purpose. Therefore, we plan to also examine the effect of this progressive degradation.

4   Evaluation of SSP Performance

So far, the performance of SSP systems has generally been measured in terms of "complete matches", i.e. either the complete construction of the correct MR by the SSP system, or the construction of an MR that yields the same result when executed [1]. However, with meaning representations that are longer and more complex, complete and exact reconstruction of MRs becomes increasingly unlikely. It is therefore desirable to assign partial credit even to imperfect MRs.

Ideally, we would like to compare two meaning representations in terms of the similarity of their meaning. Since such a notion is inaccessible even from a theoretical point of view, we are left with the choice of a suitable proxy [12]. Logical equivalence is an option, but still undecidable. For lack of alternatives, we therefore decide to state a similarity measure for a pair of meaning representations in purely syntactic terms.

It seems natural to use a measure that exploits the graphical nature of MRs by searching for a node-to-node assignment between gold-standard annotation and SSP output. In fact, [13] presents such a measure, where an assignment's score is determined by matching node labels as well as the number of matching edges on nodes that are assigned to each other. The score is then defined to be the highest score achieved by any assignment. In [3], a similar measure is introduced based on a maximum common subgraph alignment.

Instead of maximum common subgraph alignment, we have opted to adopt a measure based on solving an assignment, or bipartite matching, problem. As the underlying graphical structure, we use a syntax tree of the MR. The final score is made up of two components: a node score and a variable score. Both of them are determined by the weight of an optimal assignment of certain components of the MR under evaluation to their counterparts in the gold standard MR.

In the calculation of the node score, the inner nodes – i.e., predicate names, quantifiers, and logical connectives – are assigned to each other. A weight is calculated for each pair of a single node in the test MR and a node in the gold-standard MR, based on the following factors: whether the node types match (i.e. they represent the same predicate, quantifier, or connective), whether the parents' node types match, and whether their depth in the MR syntax tree is similar.

The variable score is derived from the best assignment between the variables in the two MRs, based on the following factors: whether the variables are bound by the same type of quantifier, whether the quantifier appears in the same polarity, and how many of their occurrences match regarding name of the predicate governing the occurrence, the argument place that is filled by the occurrence, and the polarity of the occurrence.

A combined score is then derived through the multiplication of the node and variable scores. It is 1 if the MR under test equals the gold-standard, and strictly less than 1 otherwise. From manual inspection we gather that the measure seems to reflect human judgement quite well, assigning high scores to MRs that contain large sub-structures of the gold-standard.

5   The Learning Framework

Initial experimentation with the two state-of-the-art SSP systems WASP [1] and UBL [8] has revealed, not surprisingly, that the application of SSP to larger and more complex data sets requires addressing computational issues first. It will therefore be necessary to produce an implementation of an SSP system that is capable of dealing with sufficient amounts of more complex data. While this problem has prompted Le and Zuidema to invent a completely new learning framework and underlying formalism [3], we instead plan to follow the line of work represented by Kwiatkowski et al. [8]. In addition to achieving state-of-the-art performance on the Geoquery data set, it is based on combinatory categorial grammar (CCG)[14], which has a solid foundation in linguistic theory. Additionally, the existence of the rule-based Boxer system, which is also based on CCG, suggests the suitability of CCG-based models for the task.

As it is common in CCG, meaning representations are constructed using lambda-calculus. This means that any MR formalism can be used as long as it supports this construction method. Of course, this is not to say that there were no interaction between the semantic parsing model, the mode of construction, and the MR formalism used. However, as we consider CCG-based models a promising approach to SSP, we think it makes sense to evaluate the various types of MR with regards to this type of model.

The main computational problem lies in searching the space of possible splits of meaning representations over CCG items. Kwiatkowski et al. address this by limiting the size of the portion of the meaning representation that is split off. However, this strategy proves too restrictive for the large meaning representations that are generated by Boxer. We suggest that heuristics may instead be used to define the space of splits that is searched. E. g., one plausible heuristic would place split points at the boundaries of constituents generated by an external syntactic parser. This could be supplemented by a heuristic based on word-to-predicate alignment, similar to [3].

6   Discussion and Outlook

We have introduced a research project towards the extension of SSP methods to application-independent data. An important motivation is that we believe that the consideration of more complex data in SSP is crucial for its evolution to become a more general problem-solving tool. Being able to work with application-independent data means that costly annotation efforts do not need to be repeated for every potential application of semantic parsing. This will reduce the cost of exploring further potential applications.

To evaluate the applicability of a state-of-the-art semantic parsing algorithm to application-independent data, we performed a preliminary test using UBL and automatically annotated data. While annotated newswire texts proved computationally infeasible, we were able to run a test using the Geoquery dataset. The Geoquery sentences were annotated using Boxer, yielding MRs formulated in first-order logic that were considerably longer and more complex than the original Geoquery annotations. These annotations were recovered by UBL with F1-scores between 30% and 50%. Compared to the F1-score of 89% reported on the original annotations, these figures appear very low. However, we still consider this result encouraging considering that the amount of training data was very small, and that the re-annotation of the corpus increased the variance of the annotated MRs. The Geoquery corpus contains many sentences where different natural language formulations are used for expressing the same semantic content, which will however be assigned different MRs by Boxer. In addition, inspection of the parser output suggested that in some cases where MRs could not be exactly recovered, important MR components were nonetheless present.

As has already been detailed, computational issues need to be addressed when dealing with input data of higher complexity. Our current main concern is therefore the design of suitable algorithms, notably for the induction of CCGs for semantic parsing.

The results of this work will be beneficial to various endeavors related to SSP, such as improving existing SSP systems, developing new SSP methods, and applying SSP to other tasks in natural language processing. An example for such a task is the development of hybrid syntax/semantics-based machine translation systems.

References

1. Wong, Y.W., Mooney, R.: Learning synchronous grammars for semantic parsing with lambda calculus. In: Proceedings of the 45th Annual Meeting of the Association of Computational Linguistics, Prague, Czech Republic, pp. 960–967. Association for Computational Linguistics (June 2007)

2. Bos, J.: Wide-coverage semantic analysis with boxer. In: Semantics in Text Processing, STEP 2008 Conference Proceedings. Research in Computational Semantics, vol. 1, pp. 277–286. College Publications (2008)

3. Le, P., Zuidema, W.: Learning compositional semantics for open domain semantic parsing. In: Proceedings of COLING 2012, Mumbai, India, pp. 1535–1552. The COLING 2012 Organizing Committee (December 2012)

4. Kamp, H., Reyle, U.: From Discourse to Logic: Introduction to Modeltheoretic Semantics of Natural Language, Formal Logic and Discourse Representation Theory. Kluwer, Dordrecht (December 1993)

5. Basile, V., Bos, J., Evang, K., Venhuizen, N.: Developing a large semantically annotated corpus. In: Calzolari, N., Choukri, K., Declerck, T., Doğan, M.U., Maegaard, B., Mariani, J., Odijk, J., Piperidis, S. (eds.) Proceedings of the Eighth International Conference on Language Resources and Evaluation (LREC 2012), Istanbul, Turkey, pp. 3196–3200. European Language Resources Association (ELRA) (2012); ACL Anthology Identifier: L12-1299

6. Ide, N., Baker, C., Fellbaum, C., Passonneau, R.: The manually annotated subcorpus: A community resource for and by the people. In: Proceedings of the ACL 2010 Conference Short Papers, Uppsala, Sweden, pp. 68–73. Association for Computational Linguistics (July 2010)

7. Lu, W., Ng, H.T., Lee, W.S., Zettlemoyer, L.S.: A generative model for parsing natural language to meaning representations. In: Proceedings of the Conference on Empirical Methods in Natural Language Processing, pp. 783–792 (2008)

8. Kwiatkowski, T., Zettlemoyer, L., Goldwater, S., Steedman, M.: Inducing probabilistic CCG grammars from logical form with higher-order unification. In: Proceedings of the 2010 Conference on Empirical Methods in Natural Language Processing, Cambridge, MA, pp. 1223–1233. Association for Computational Linguistics (October 2010)

9. Liang, P., Jordan, M., Klein, D.: Learning dependency-based compositional semantics. In: Proceedings of the 49th Annual Meeting of the Association for Computational Linguistics: Human Language Technologies, Portland, Oregon, USA, pp. 590–599. Association for Computational Linguistics (June 2011)

10. Alshawi, H., Chang, P.C., Ringgaard, M.: Deterministic statistical mapping of sentences to underspecified semantics. In: Bos, J., Pulman, S. (eds.) Proceedings of the Ninth International Conference on Computational Semantics (IWCS 2011), Oxford, UK, pp. 15–24 (2011)

11. Richter, F., Sailer, M.: Basic concepts of lexical resource semantics. In: Collegium Logicum. ESSLLI 2003 - Course Material I. Collegium Logicum, vol. 5, pp. 87–143. Kurt Gödel Society, Wien (2004)

12. Shieber, S.M.: The problem of logical-form equivalence. Computational Linguistics 19(1), 179–190 (1993)

13. Allen, J.F., Swift, M., de Beaumont, W.: Deep semantic analysis of text. In: Bos, J., Delmonte, R. (eds.) Semantics in Text Processing, STEP 2008 Conference Proceedings. Research in Computational Semantics, vol. 1, pp. 343–354. College Publications (2008)

14. Steedman, M.: The Syntactic Process. MIT Press, Cambridge (2000)

Using Corpus Statistics to Evaluate Nonce Words

Özkan Kılıç

Department of Psychology, Lehigh University, Bethlehem PA, USA
ozkan.kilic@lehigh.edu

Abstract. Nonce words are widely used in linguistic research to evaluate areas such as the acquisition of vowel harmony and consonant voicing, naturalness judgment of loanwords, and children's acquisition of
morphemes. Researchers usually create lists of nonce words intuitively
by considering the phonotactic features of the target languages. In this
study, a corpus of Turkish orthographic representations is used to propose
a measure for the nonce word appropriateness for linearly concatenative
languages. The conditional probabilities of orthographic co-occurrences
and pairwise vowel collocations within the same word boundaries are
used to evaluate a list of nonce words in terms of whether they would
be rejected, moderately accepted or fully accepted as novel words. A
group of 50 Turkish native speakers was asked to judge the same list of
nonce words on how native-like the words sound. Both the model and
the participants displayed similar results.

Keywords: Nonce words, Orthographic representations, Conditional
probabilities.

1 Introduction

Nonce words are frequently employed in linguistic studies to evaluate areas such
as well-formedness [1], morphological productivity [2] and development [3], judgment of semantic similarity [4], and vowel harmony [5]. Nonce words are also
used to understand the process of adopting loan words. The majority of loaned
words undergo certain phonetic changes to more resemble the lexical entries of
the language into which they will be adopted [6]. For example, *television* in
Turkish becomes *televizyon* /televɪzjon/ because /jon/ is more frequent than
/ʒɪn/ in Turkish[1]. Similarly, *train* is adopted as *tren* /tren/ because, similar
to diphthongs, vowel-to-vowel co-occurrences are not usually allowed in Turkish
non-compound words. This phenomenon shows that the speakers of a language
are aware of the possible sound frequencies and collocations of their native languages, and they can make judgements on the naturalness of loan words, recently

---

[1] In the METU-Turkish Corpus, there are 181 occurrences with the segment /ʒɪn/ of
which only 30 are at the terminating word boundaries. On the other hand, there
are 5,945 occurrences with the segment /jon/ of which 3,190 are at the terminating
word boundaries, excluding the word *televizyon*.

M. Colinet et al. (Eds.): ESSLLI 2012/2013, LNCS 8607, pp. 26–35, 2014.
© Springer-Verlag Berlin Heidelberg 2014

invented words and nonce words by using their knowledge of the existing Turkish lexis. Alternatively, it can be claimed that when a loan word does not match statistical properties of a target language, the native speakers of that language either consciously change the word for a better alignment, or the speakers instead perceive the word in accordance with the sound patterns they are used to hearing in their language. It is also reported that known-word statistics is determinant in some linguistic processes [26, 27]; thus, the acceptability of nonce words can be a decision based on these statistics as in the current study.

The acceptability of nonce words can be studied by experimental investigations through phonotactic properties or factor-based analysis [7]. In the experimental investigations, it is observed that the participants accepted or rejected nonce words according to probable combinations of sounds [1, 8]. In factor-based analysis, the acceptability of nonce words is evaluated through the co-occurrences of syllables or consonant clusters locally [9] or non-locally [10–12] or through nucleus-coda combination probabilities [13].

In this study, the acceptability of nonce words was assessed using the conditional probabilities of the bigram co-occurrences of the orthographic representations locally and the pairwise collocations of the vowels within the same word boundaries. Similar models within the context of phonotactic modeling had been used for Finnish vowel harmony [14]. The model for Finnish language uses Boltzmann distribution. Yet the current study much simpler because the local bigram frequencies were used to evaluate Turkish nonce words. Two threshold values were set for the decision to reject, moderately accept and fully accept to judge how the words sound native-like. The threshold values were computed according to the length of each input string. For the evaluation of the conditional and collocation probabilities, the METU-Turkish Corpus containing about two million words was employed [15]. The list of nonce words was created intuitively by randomly combining frequent and infrequent syllables in Turkish. The same list of nonce words evaluated by the model was also given to 50 Turkish native speakers to judge the level of acceptability of each word. The 25 male and 25 female Turkish native speakers, had an average age of 31.26 ($s = 4.11$).The judgements from the native speakers and the model agreed on 82% of the words. In this paper, brief information about Turkish language and plausibility of conditional probabilities will be followed by details of the model and the results.

2 Turkish Language and Conditional Probability

Turkish has 8 vowels and 21 consonants, and it is agglutinative with a considerably complex morphology [16, 17]. While communicating, the word internal structure in Turkish is required to be segmented because Turkish morphosyntax plays a central role in semantic analysis. For example, although Turkish is considered as an SOV language, the sentences are usually in a free order. Thus, the subject and object of a verb can only be determined by the morphological markers as in (1) rather than the word order.

(1) *Köpek adam-ı ısırdı.*   *Köpeğ-i adam ısırdı.*
   Dog man-Acc bit   Dog-ACC man bit
   The dog bit the man.   The man bit the dog.

The description of Turkish word structure depends heavily on morphophono-logical constraints and morphotactics. In Turkish morphotactics, the continua-tion of a morpheme is determined by the preceding morpheme or by the stem as in (2).

(2) *ev-de-ki*   **ev-ki-de*
   house-Loc-Rel
   The one in the house

These morphotactic constraints in Turkish are captured by statistical mod-els based on conditional probabilities [18, 19]. In addition to morphotactics, the morphophonology of Turkish needs a brief explanation because nonce words have to mimic this morphophonology.

Vowel harmony is dominantly effective in Turkish morphophonology in order to preserve the roundedness and the frontness of vowels within the same word boundaries. While a morpheme with a vowel is concatenated to a string, its vowel is modified with respect to the roundedness and frontness properties of the most recent vowel in the string as in (3).

(3) *ev-ler*   *oda-lar*   *bil-di*   *duy-du*
   house - Plu   room - Plu   know - Past   hear - Past
   houses   rooms   knew   heard

Another important phenomenon in Turkish morphophonology is voicing. If some of the strings terminating with the voiceless consonant, *p, t, k, ç*, are fol-lowed by the suffixes starting with vowels, then the consonants are voiced as *b, d, ğ, c* as in (4).

(4) *sonuç*   *sonuc-um*   *kanat*   *kanad-ı*
   result   result -1S.Poss   wing   wing - Acc
                    my result   he wing

Consonant assimilation is also important in Turkish morphophonology. The initial consonants of some morphemes undergo an assimilation operation if they are attached to the strings terminating in the voiceless consonants, *p, t, k, ç, f, s, ş, h, g,*. For example, the surface forms of of the Turkish past tense *-DI* in (5) start with a *-t* because of the terminal sounds *-t* and *-ş* of the roots.

(5) *at-tı*   *konuş-tu*
   throw - Past   speak - Past
   threw   spoke

The final Turkish morphophonological phenomena that need to be briefly mentioned are deletion and epenthesis. Some of the loanwords as in (6) either lose their final vowel (deletion) or receives an additional copy of their final consonant (epenthesis).

(6) *hak*   *hakk-ım*   *isim*   *ism-im*
   right   right - 1S.Poss   name   name - 1S.Poss
          my right   my name

The Turkish morphophonological phenomena described above occur in the co-occurrences of the orthographic representations in the concatenating positions except in vowel harmony and the deletion. This results in high conditional probabilities evaluated using the frequencies of the pairs of immediately consecutive orthographic representations. Since the vowel harmony and deletion take place after or before the concatenation positions, their pairwise collocations within the same word boundaries are also required to be utilized in the statistical model.

The transition probability between A and B is simply based on the conditional probability statistics as in Formula 1.

$$P(B|A) = \text{(frequency of } AB) \,/\, \text{(frequency of } A) \tag{1}$$

Infants are reported to successfully discriminate speech segments using transitional probabilities of syllable pairs [20, 21]. Adults also make use of transitional probabilities between word classes to acquire syntactic rules [22, 28]. Similarly, transition probabilities are dominantly used in unsupervised morphological segmentation and disambiguation [18, 19], [23–25].

Statistical approaches to linguistics support the empiricist view, which states that knowledge comes only or primarily from sensory experience instead of being genetically encoded. Such approaches provide an explanatory account of some linguistic phenomena such as the one in the current study. Considering the properties of the Turkish language, using the conditional probabilities of orthographic representations and the collocations of vowels within the same word boundaries is a plausible model to decide whether nonce words or loan words will be *rejected*, *moderately accepted* or *accepted*. In the current study, it is assumed that native speakers judge nonce words mainly based on their morphotactic, morphophonological and phonotactic properties. These properties can be captured by constraints on orthographic collocations by the model explained in the next section.

3   The Model

Let s be a string such that $s = u_1 u_2 \ldots u_n$, where u_i is a letter in the Turkish alphabet. The string s is unified with the empty strings σ and ε such that $s = \sigma u_1 u_2 u_n \varepsilon$, where σ denotes the initial word boundary and ε denotes the terminal word boundary. Word boundaries are essential in the judgement process. For example, although the sound $\breve{g}$ is moderately frequent in Turkish, it never occurs as an initial sound but it is rarely the terminal letter. The overall transition probability of the string s is evaluated from the METU-Turkish Corpus using Formula 2, which is actually the product of Formula 1.

$$P_t(s) = \prod_1^{n+1} P(u_i|u_{i-1}) \qquad (2)$$

For example, using the Formula 2, $P(a|\sigma)$ gives the probability of the strings starting with the letter a, and $P(b|a)$ estimates the probability of the substring ab in the corpus. Now let v be a subset of the string s such that $v = u_{i,1}u_{j,2} \ldots u_{k,m}$ where $u_{k,m}$ is the mth vowel in the kth location of the string s. The overall vowel collocations of the string s are estimated from the substring of vowels v using Formula 3.

$$P_c(v) = \prod_2^m \frac{g(v_{i-1}v_i)}{f(v_{i-1})} \quad if \ |v| > 1$$

$$P_c(v) = \frac{f(v_i)}{CorpusSize} \quad if \ |v| = 1 \qquad (3)$$

In the Formula 3, the function $f(v_i)$ gives the frequency of the words that contain the vowel v_i as a substring in the corpus. The function $g(v_{i-1}v_i)$ gives the frequency of words in which the vowels v_{i-1} and v_i are collocating not necessarily in immediately consecutive positions but within the same word boundaries. This frequency is divided by $f(v_{i-1})$ because some Turkish words may violate Turkish vowel harmony. The division provides the model with the obedience or violation of the vowel harmony in a probabilistic manner with respect to v_{i-1}. The acceptability probability of the string s is calculated by $P_a(s) = P_t(s)P_c(v)$. The acceptability decision of the string s in the model is made by using the Formula 4.

$$Accept \quad if \qquad P_a(s) \geq 10^{-(t+v)}$$

$$Moderately \ accept \quad if \qquad 10^{-(t+v+1)} \leq P_a(s) < 10^{-(t+v)} \qquad (4)$$

$$Reject \quad if \qquad 10^{-(t+v+1)} > P_a(s)$$

where t is the number of transitions (which is *the length of the string + 1*) and v is the number of the vowel collocations (which is *the number of the vowels - 1*) in the string. If the string s has only one vowel, then $v = 1$. The threshold values are chosen to best fit the participants responses. Thus, they are changeable values depending on the size size of the corpus.

The model was applied to the list of nonce words given in the following section. The same list was also given to the 50 Turkish native speakers to evaluate the acceptability of each item. The comparison of the results from the model and the native speakers is given below.

Table 1. The results of the model and the results of the participants (Bold text indicates that the model predicted the majority of participants' responses)

Nonce Words	Results of the Model	Reject	Moderately Accept	Accept
öğtar	**Reject**	96%	4%	
söykıl	**Reject**	96%	4%	
talar	**Accept**			100%
telüti	**Reject**	64%	28%	8%
prelüs	**Reject**	84%	14%	2%
katutak	**Moderately Accept**	8%	50%	42%
par	**Accept**		14%	86%
öçgöş	**Reject**	100%		
jeklürt	**Reject**	100%		
böşems	**Reject**	88%	12%	
trüğat	**Reject**	96%	4%	
cakeyas	**Reject**	92%	8%	
çörottu	**Reject**	74%	16%	10%
döyyal	**Reject**	78%	22%	
efföl	**Reject**	92%	8%	
aznı	Reject	32%	60%	8%
fretanit	**Reject**	64%	30%	6%
erttiçe	**Moderately Accept**	36%	64%	
goytar	Reject	38%	52%	10%
hekkürük	Reject	41%	47%	12%
henatiya	**Moderately Accept**	36%	64%	
taberarul	**Reject**	84%	16%	
gövük	Reject	30%	44%	26%
sör	**Moderately Accept**		78%	22%
perolus	**Reject**	84%	16%	
kletird	**Reject**	98%	2%	
ojuçı	**Reject**	100%		
ürtanig	**Reject**	94%	6%	
lezğaji	**Reject**	100%		
lamafi	**Moderately Accept**		64%	36%
nort	Reject	38%	42%	20%
netik	**Accept**		18%	82%
meşipir	Moderately Accept		24%	76%
oblan	**Moderately Accept**		58%	42%
öftik	**Reject**	62%	34%	4%
özola	**Moderately Accept**	32%	60%	8%
ayora	Accept		72%	28%
sengri	**Moderately Accept**	32%	68%	
sakkütan	**Reject**	58%	34%	8%
şepilt	**Reject**	78%	22%	
şür	**Moderately Accept**		78%	22%
puhaptı	**Moderately Accept**	38%	44%	18%
upapık	**Reject**	54%	28%	18%
ülü	Reject	28%	52%	20%
yukta	**Moderately Accept**		74%	26%
zerafip	**Reject**	54%	34%	12%
upgur	**Reject**	70%	16%	14%
kujmat	**Reject**	90%	10%	
lertic	**Reject**	94%	6%	
düleri	Accept		64%	36%

4   Results

The nonce word *talar* is evaluated as in (7)

$$(7) \qquad P_a(talar) = P_t(\sigma talar\varepsilon)P_c(aa)$$
$$= P(t|\sigma)P(a|t)P(l|a)P(a|l)P(r|a)P(\varepsilon|r)P_c(aa)$$
$$= 7.66e - 06P_c(aa) = 7.66e - 06 * 4.75e - 01 = 3.63e - 06$$

Since $P_a(talar) \geq 10^{-(6+1)}$, in which 6 conditional probability estimations and 1 vowel collocation are evaluated, the nonce word *talar* is accepted.

The word list is evaluated by the 50 Turkish speakers. The participants are composed of 25 males and 25 females with at least undergraduate degrees. They are given the words written on a paper with a 3-level scale (A: Accept, M: Moderately accept, R: Reject), and instructed that these words need to be evaluated by native speakers because the words are going to be used as novel words to name some recently invented colors, objects and actions in Turkey. The distribution of the native speaker responses and the results of the model are given in Table 1.

For 82% of the words the Turkish native speakers' responses are in agreement with the results from the model. The model failed to simulate the responses from the participants in 18% of the results.

The nonce word *ülü* was rejected by the model but accepted by the participants. A possible reason might be that the nonce word *ülü* sounds similar to an existing Turkish word *ölü* 'death'. Similarly, the responses for the nonce word *nort* were in disagreement. This nonce word has a similar pronunciation to an English word *north* and the most of the participants also knew English as a foreign language. Therefore, the participants might also make use of their foreign language knowledge to evaluate nonce words.

5   Discussion and Conclusion

The acceptability of loan words and nonce words is mainly determined by the phonological properties of the target language and the current approaches are syllable-based [7–13]. Since there are no lexical entries for nonce words, the model in this study tries to estimate the acceptability of the words using the bigram conditional probabilities and collocations of the orthographic representations within the word boundaries, which is a simplified way of inducing Turkish morphophonology.

Although the model does not assume to utilize any property of Turkish phonology and it does not implement any phonologic filtering mechanism, it is able to mimic, in a remarkable way, a large number of the responses from the participants. Indeed, this study does not propose that acceptability is based on raw orthographic representations rather than syllables and phonemes. Instead, it underlines that simple pairwise conditional properties and vowel collocations from a corpus can give an estimation of the acceptability of a list of Turkish nonce words. This can be used by researchers that need an evaluation for the nonce words for their studies when no phonologically annotated corpus with syllables exists.

The argument in this study could be extended to grammaticality judgement in a way that the speaker does not need to store explicit rules about which rules are grammatical. Sensitivity to statistical properties of observed combinations can be enough to account for the speaker's grammaticality judgement behaviour. Yet, in this case, it is necessary to include additional steps in the model to represent how a speaker *smooths* a novel grammatical construction with zero probability by using the frequency information from known constructions in order to reject, moderately accept, or accept the novel construction.

6   Limitations and Future Work

The model needs to be tested with larger word lists to improve the results. The model is successful because there is a close correspondence between phonotactic and orthotactic in Turkish. If one wants to test the model in different languages, it requires improvements in terms of the morphophonological properties of the target languages. The model uses exact orthographic representations. Thus, it requires an additional phonological similarity measure for the representations to increase the success rate because it seems that the native speakers also make use of phonologic similarities among sounds, such as accepting the nonce word *ülü* since *ü resembles ö* in the real word *ölü* in terms of roundedness and backness.

The threshold values for the acceptability decisions depend on word lengths. They also need to be improved with respect to the target languages. The model also needs to be tested in and adapted for different languages with ablaut or umlaut phenomena such as English and German, and the templatic languages such as Arabic and Hebrew, because they are not linearly concatenative and immediate sound co-occurrences are not powerful enough to capture their morphophonological properties.

References

[1] Hammond, M.: Gradience, phonotactics, and the lexicon in English phonology. Int. J. of English Studies 4, 1–24 (2004)
[2] Anshen, F., Aronoff, M.: Producing morphologically complex words. Linguistics 26, 641–655 (1988)
[3] Dabrowska, E.: Low-level schemas or general rules? The role of diminutives in the acquisition of Polish case inflections. Language Sciences 28, 120–135 (2006)
[4] MacDonald, S., Ramscar, M.: Testing the distributional hypothesis: The influence of context on judgements of semantic similarity. In: Proc. of the 23rd Annual Conference of the Cognitive Science Society, University of Edinburgh (2001)
[5] Pycha, A., Novak, P., Shosted, R., Shin, E.: Phonological rule-learning and its implications for a theory of vowel harmony. In: Garding, G., Tsujimura, M. (eds.) Proc. of WCCFL, vol. 22, pp. 423–435 (2003)

[6] Kawahara, S.: OCP is active in loanwords and nonce words: Evidence from naturalness judgment studies. Lingua (to appear)

[7] Albright, A.: From clusters to words: Grammatical models of nonce word acceptability. Handout of talk presented at 82nd LSA, Chicago (January 3, 2008)

[8] Shademan, S.: From clusters to words: Grammatical models of nonce word acceptability. Grammar and Analogy in Phonotactic Well-formedness Judgments. Ph. D. thesis, University of California, Los Angeles (2007)

[9] Hay, J., Pierrehumbert, J., Beckman, M.: Speech perception, well-formedness and the statistics of the lexicon. In: Local, J., Ogden, R., Temple, R. (eds.) Phonetic Interpretation: Papersbin Laboratory Phonology VI. Cambridge University Press, Cambridge (2004)

[10] Frisch, S.A., Zawaydeh, B.A.: The psychological reality of OCP-Place in Arabic. Language 77, 91–106 (2001)

[11] Koo, H., Callahan, L.: Tier-adjacency is not a necessary condition for learning phonotactic dependencies. Language and Cognitive Processes 77, 1–8 (2011)

[12] Finley, S.: Testing the limits of long-distance learning: learning beyond a three-segment window. Cognitive Science 36, 740–756 (2012)

[13] Treiman, R., Kessler, B., Knewasser, S., Tincoff, R., Bowman, M.: English speakers sensitivity to phonotactic patterns. In: Broe, M.B., Pierrehumbert, J. (eds.) Papers in Laboratory Phonology V: Acquisition and the Lexicon, pp. 269–282. Cambridge University Press, Cambridge (2000)

[14] Goldsmith, J., Riggle, J.: Information theoretic approaches to phonological structure: the case of Finnish vowel harmony. Natural Language & Linguistic Theory (to appear)

[15] Say, B., Zeyrek, D., Oflazer, K., Özge, U.: Development of a corpus and a treebank for present-day written Turkish. In: Proc. of the Eleventh International Conference of Turkish Linguistics (2002)

[16] Göksel, A., Kerslake, C.: Turkish: A Comprehensive Grammar. Routledge, London (2005)

[17] Lewis, G.: Turkish Grammar, 2nd edn. University Press, Oxford (2000)

[18] Kılıç, Ö., Bozşahin, C.: Semi-supervised morpheme segmentation without morphological analysis. In: Pro. of the LREC 2012 Workshop on Language Resources and Technologies for Turkic Languages, Istanbul, Turkey (2012)

[19] Yatbaz, M.A., Yuret, D.: Unsupervised morphological disambiguation using statistical language models. In: Pro. of the NIPS 2009 Workshop on Grammar Induction, Representation of Language and Language Learning, Whistler, Canada (2009)

[20] Aslin, R.N., Saffran, J.R., Newport, E.L.: Computation of conditional probability statistics by human infants. Psychological Science 9, 321–324 (1998)

[21] Gomez, R.L.: Variability and detection of invariant structure. Psychological Science 13, 431–436 (2002)

[22] Kaschak, M.P., Saffran, J.R.: Idiomatic syntactic constructions and language learning. Cognitive Science 30, 43–63 (2006)

[23] Creutz, M., Lagus, K.: Unsupervised models for morpheme segmentation and morphology learning. ACM Tran. on Speech and Language Processing 4(1) (2007)

[24] Bernhard, D.: Unsupervised morphological segmentation based on segment predictability and word segments alignment. In: Proc. of 2nd Pascal Challenges Workshop, pp. 19–24 (2006)

[25] Demberg, V.: A language-independent unsupervised model for morphological segmentation. Ann. Meet. of Assoc. for Computational Linguistics 45(1), 920–927 (2007)

[26] Debrowska, E.: The effects of frequency and neighbourhood density on adult native spakers' productivity with Polish case inflections: An empirical test of usafe-based approaches to morphology. Memory and Language 58, 931–951 (2008)

[27] Baayen, R.H., Dijkstra, T., Schreuder, R.: Singulars and plurals in Dutch: Evidence for a parallel dual route model. Memory and Language 37, 94–117 (1997)

[28] Reeder, P.A., Newport, E.L., Aslin, R.N.: From shared contexts to syntactic categories: The role of distributional information in learning linguistic form-classes. Cognitive Psychology 66, 30–54 (2013)

XMG: A Modular MetaGrammar Compiler

Simon Petitjean

Univ. Orlans, LIFO EA 4022, F-45067 Orlans, France

Abstract. XMG (eXtensible MetaGrammar) is a metagrammar compiler which has already been used for the design of large scale Tree Adjoining Grammars and Interaction Grammars. Due to the heterogeneity in the field of grammar development (different grammar formalisms, different languages, etc), a particularly interesting aspect to explore is modularity. In this paper, we discuss the different spots where this modularity can be considered in a grammar development, and its integration to XMG.

1 Introduction

1.1 Grammar Engineering

Nowadays, a lot a applications have to deal with languages and consequently need to manipulate their descriptions. Linguists are also interested in these kinds of resources, for study or comparison. For these purposes, formal grammars production has became a necessity. Our work focuses on large scale grammars, that is to say grammars which represent a significant part of the language.

The main issue with these resources is their size (thousands of structures), which causes their production and maintenance to be really complex and time consuming tasks. Moreover, these resources have some specificities (language, grammatical framework) that make each one unique.

Since a handwriting of thousands of structures represents a huge amount of work, part of the process has to be automatized. A totally automatic solution could consist in an acquisition from treebanks, which is a widely used technique. Semi automatic approaches are alternatives that give an important role to the linguist: they consist in building automatically the whole grammar from information on its structure. The approach we chose is based on a description language, called metagrammar [1]. The idea behind metagrammars is to capture linguistic generalization, and to use abstractions to describe the grammar.

1.2 Metagrammars for Tree Adjoining Grammars

The context that initially inspired metagrammars was the one of Tree Adjoining Grammars (TAG) [2]. This formalism consists in tree rewriting, with two specific rewriting operations: adjunction and substitution. An adjunction is the replacement of an internal node by an auxiliary tree (one of its leaf nodes is labelled with ⋆ and called foot node) with root and foot node having the same syntactic category as the internal node. A substitution is the replacement of a leaf node

M. Colinet et al. (Eds.): ESSLLI 2012/2013, LNCS 8607, pp. 36–48, 2014.
© Springer-Verlag Berlin Heidelberg 2014

(marked with ↓) by a tree with a root having the same syntactic category as this leaf node.

TAG is said to have an extended domain of locality, because the adjunction operation and the depth of the trees allow to represent long distance relations between nodes: two nodes of the same elementary tree can after derivation end up at an arbitrary distance from each other. Here, we will only manipulate LTAG (lexicalized-TAG), which means each elementary tree is associated with at least one lexical element.

LTAG is traditionnaly used with respect to the Condition on Elementary Tree Minimality from [3], which means that an elementary tree only encapsulates the arguments of its anchor, recursion being factored away.

What can we do to lower the amount of work implied by the conception of the grammar ? Let us take a look at some rules:

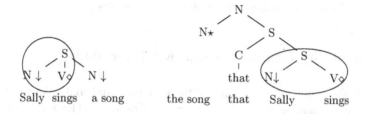

Fig. 1. Verb with canonical subject and canonical or extracted object

Those two trees share some common points: part of the structure is the same (the subject is placed before the verb in both circled parts), and the agreement constraints, given in feature structures associated to nodes (not represented here), are similar. This kind of redundancy is one of the key motivations for the use of abstractions. These abstractions are descriptions of the redundant fragments we can use everywhere they are needed.

Metagrammars are based on the manipulation of those linguistic generalizations. They consist in generating the whole grammar from an abstract description, permitting to reason about language at an abstract level.

1.3   A Need for Modularity

The metagrammatical language we will deal with here is XMG (eXtensible Meta-Grammar)[1], introduced in [4]. A new project, XMG-2[2], started in 2010 to achieve the initial goal of the compiler, extensibility, which has not been realized yet: XMG-1 only supports tree based grammars (two formalisms, Tree Adjoining

---

[1] https://sourcesup.cru.fr/xmg/
[2] https://launchpad.net/xmg

Grammars and Interaction Grammars), and includes two levels of description, the syntactic one and the semantic one.

Using this metagrammatical approach for the generation of another type of linguistic resource implies the creation of a new XMG compiler. This compiler needs to provide dedicated description languages for the needed structures. A high level of flexibility is needed so that the user can assemble by their own a new metagrammatical framework.

Our goal is to go towards two levels of modularity: we want it to be possible to assemble a grammar in a modular way, thanks to a metagrammar assembled in a modular way. The first level of modularity, provided by a compiler, allows to combine abstractions to build a linguistic resource. The second one allows to build new compilers dedicated to new grammar engineering tasks.

We will begin pointing out the modularity on the grammar side in section 2. In section 3, we will focus on a new level of modularity, a metagrammatical one. In section 4, we will give an overview of what has been done, and what remains to be done. Finally, we will conclude and give some perspectives.

2   Assembling Grammars in a Modular Way

XMG consists in defining fragments of the grammar, and controlling how these fragments can combine to produce the whole grammar. The following figure shows the intuition of the combination of fragments to produce a tree for transitive verbs. It is done by combining three tree fragments, one for the subject (in its canonical form, that we noticed redundant previously), one for the object (relative) and one for the active form.

To build a lexicon, the metagrammar is first executed in an non-deterministic way to produce descriptions. Then these descriptions are solved to produce the models which will be added to the lexicon.

2.1   The Control Language and the Dimension System

The main particularity of XMG is that it allows to see the metagrammar as a logical program, using logical operators.

The abstractions (possibly with parameters) we manipulate are called classes. They contain conjunctions and disjunctions of descriptions (tree fragments descriptions for TAG), or calls to other classes. This is formalized by the following control language:

$$Class \quad := \quad Name[p_1, \ldots, p_n] \rightarrow Content$$
$$Content \quad := \quad \langle Dim \rangle \{Desc\} \mid Name[\ldots] \mid Content \lor Content$$
$$\mid Content \land Content$$

For example, we can produce the two trees of the figure 1 by defining the tree fragments for canonical subject, verbal morphology, canonical object and relativized object, and these combinations:

$$Object \rightarrow CanObj \lor RelObj$$
$$Transitive \rightarrow CanSubj \land Active \land Object$$

This part of metagrammar says that an object can either be a canonical object or a relative object, and that the transitive mode is created by getting together a canonical subject, an active form and one of the two object realizations.

Notice that descriptions are accumulated within dimensions, which allow to separate types of data. Sharing is still possible between dimensions, by means of another dimension we call interface. In XMG's TAG compiler for example, the *syn* dimension accumulates tree descriptions while the *sem* dimension accumulates predicates representing the semantics. Each dimension comes with a description language, adapted to the type of data it will contain. For each type of description we need to accumulate, we have to use a different description languages. The first version of XMG provides a tree description langague (for TAG or Interaction Grammars) associated with the *syn* dimension and a language for semantics associated with the *sem* dimension.

A Tree Description Language. For trees in TAG, we use the following tree description language:

$$Desc := x \rightarrow y \mid x \rightarrow^+ y \mid x \rightarrow^* y \mid x \prec y \mid x \prec^+ y \mid x \prec^* y \mid x[f{:}E]$$
$$\mid x(p{:}E) \mid Desc \land Desc$$

where x and y are node variables, $\rightarrow$ and $\prec$ dominance and precedence between nodes ($^+$ and * respectively standing for transitive and reflexive transitive closures). ':' is the association between a property p or a feature f and an expression E. Properties are constraints specific to the formalism (the fact that a node is a substitution node for example), while features contain linguistic information, such as syntactic categories, number or gender.

When accumulated, the tree description in the syntactic dimension is still partial. The TAG elementary trees that compose the grammar are the models for this partial description. They are built by a tree description solver, based on constraints to ensure the well-formedness of the solutions. XMG computes minimal models, that is to say models where only the nodes of the description exist (no additional node is created).

Here is a toy metagrammar, composed of three description classes (representing canonical subject, relative object, active form) and one combination class (transitive mode):

$$CanSubj \rightarrow \langle syn \rangle \{(s_1[cat : S] \rightarrow v_1[cat : V]) \wedge (s_1 \rightarrow n_1(mark : subst)[cat : N])$$
$$\wedge (n_1 \prec v_1)\}$$
$$RelObj \rightarrow \langle syn \rangle \{(n_2[cat = N] \rightarrow n_3(mark = adj)[cat = N]) \wedge (n_2 \rightarrow s_2[cat = S])$$
$$\wedge (n_3 \prec s_2) \wedge (s_2 \rightarrow c) \wedge (s_2 \rightarrow s_1[cat = S]) \wedge (c \prec s_1)$$
$$\wedge (c \rightarrow wh[cat = wh]) \wedge (s_1 \rightarrow n_1[cat = n])\}$$
$$Active \rightarrow \langle syn \rangle \{(s_1 \rightarrow v_2[cat : V])\}$$
$$Transitive \rightarrow CanSubj \wedge RelObj \wedge Active$$

The minimal models for the classes named CanSubj, Active and Object are the trees with matching names on the previous figure. The tree Transitive is a minimal model for the description accumulated in class Transitive.

A Language for Semantics. To describe semantics, we use another description language, which is:

$$SemDesc := \ell : p(E_1, ..., E_n) \mid \neg \ell : p(E_1, ..., E_n) \mid E_i << E_j \mid E$$

where ℓ is a label for predicate p (of arity n) and $<<$ is a scope-over relation for dealing with quantifiers. To add binary relations to the semantic dimension, we can use a class of this type:

$$BinaryRel[Pred, X, Y] \rightarrow \langle sem \rangle \{Pred(X, Y)\}$$

When instantiated with $Pred=love$, $X=John$, $Y=Mary$, calling the class $BynaryRel$ accumulates the predicate $love(John, Mary)$.

2.2 Principles

Some additional sets of constraints we call principles are available. Their goal is to check some properties in the resulting models of the compilation, they are consequently dependent from the target formalism. For example, in TAG, the color principle is a way to forbid some fragments combination, by associating colors to each node.

A valid model is a model in which every node is colored either in red or black. When unifying nodes, their colors are merged: a red node must not unify, a white node has to unify with a black node, creating a black node, and a black node can only unify with white nodes. The only valid models are the ones in which every node is colored either in red or black. The following table shows the results of colors unifications.

For example, if we consider our previous example, the colored trees of the metagrammar are the following:

Fig. 2. Unification rules for colors

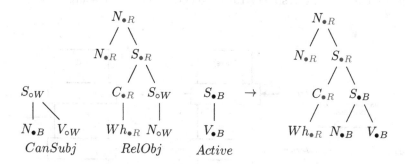

The tree description solver (ignoring the colors) will produce models where the nodes labelled S of CanSubj and Active unify with any of the two nodes labelled S in RelObj, where the nodes labelled V do not unify, etc. But when filtering with the colors principle, the only remaining model is the one of the right, which is linguistically valid, contrary to the others.

We can also cite the rank principle: we use it to add constraints on the ordering of nodes in the models of the description. In French for example, clitics are necessarily ordered, so we associate a rank property to some nodes, with values that will force the right order.

3 Assembling Metagrammars in a Modular Way

The main aim of the XMG-2 project is to make it possible for the linguist to design new metagrammatical scopes, that can accomodate a large number of linguistic theories. A modular way to realize this ambition is to provide a set of bricks the user can pick or create and combine to build the compiler he needs. Those bricks could be used to design new description languages, new principles, etc.

3.1 A Modular Architecture

XMG compiler comes with a modular processing chain. This chain is composed of two phases. The first one consists in translating the metagrammatical description into executable code.

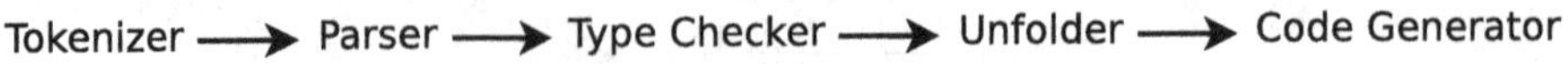

Tokenizer ⟶ Parser ⟶ Type Checker ⟶ Unfolder ⟶ Code Generator

Fig. 3. Compilation steps

First, the description is analysed and turned into an abstract syntax tree. The types into this tree are checked. The tree is then unfolded into terms of depth one, representing instructions. Instructions are finally translated into code.

The second phase corresponds to the generation of the resource.

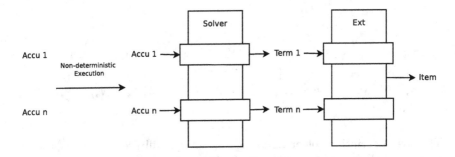

Fig. 4. Generation steps

The execution of the non-deterministic code generated by the compiler triggers accumulations in the dimensions. Each accumulation is composed of structures, and of a set of constraints over these structures. A solver extracts the models from the accumulations. The terms resulting from the solving are then translated into an output language.

The particularity of XMG is to make it possible to choose the modules that suits the best the user's metagrammar. By this mean, descriptions accumulated in different dimensions can be handled differently. For example, the end of the processing chain for TAG is a tree description solver, that builds the grammar's elementary trees from the descriptions accumulated in the syntactic dimension. The user can choose the kind of output the compiler will produce: he can interactively observe the grammar he produced, or produce an XML description of the grammar. This description can be used by a parser (for example TuLiPA [5][3] for TAG, or LeoPar[4] for IG).

The modules of the processing chains are contributed by the XMG-2 bricks. The new compiler includes bricks that recreate the two processing chains (for Tree Adjoining Grammars and Interaction Grammars) featured by XMG-1.

---

[3] https://sourcesup.cru.fr/tulipa/
[4] http://wikilligramme.loria.fr/doku.php?id=leopar:leopar

3.2   Representation Modules

As we wish to build a tool which is as universal as possible, being independent from the formalism is a priority. To achieve this goal, we need to be able to describe a large number of types of structure into XMG. We saw the dimension system was useful to separate syntax from semantics, but adding new dimensions also allows to describe and combine other levels of description. A set of dimensions, with description language, has recently been proposed.

These dimensions are packaged into XMG-2 bricks and can be used to build new compilers. Different dimensions can be built from similar sets of bricks: for example, feature structures, which can be used in a lot of formalisms, are provided by a brick. Getting the support for feature structures inside a new dimension can be done simply by plugging the feature structure brick into the new dimension brick.

Syntactic Dimensions. In [6], description languages for two syntactic formalisms, namely Lexical Functional Grammars (LFG) and Property Grammars (PG), are proposed. Here, we will focus on Property Grammars, because they differ from TAG in many aspects. PG are not based on tree rewriting but on a local constraints system: the properties. A property concerns a node and applies constraints over its children nodes. One of the interesting aspects of PG is the ability to analyse ungrammatical utterances. When parsing a utterance, its grammaticality score is lowered at every violated property. Here, we will consider these six properties:

Obligation	A: $\triangle$B	at least one B child
Uniqueness	A: B!	at most one B child
Linearity	A: B$\prec$C	B child precedes C child
Requirement	A: B$\Rightarrow$C	if a B child, then also a C child
Exclusion	A: B$\not\Leftrightarrow$C	B and C children are mutually exclusive
Constituency	A: S	children must have categories in S

A real size PG consists in an inheritance hierarchy of linguistic constructions. These constructions are composed of feature structures and a set of properties. Variables are manipulated on both sides, and can be used to share data between them. Figure 5 represents a part of the hierarchy built in [7] for French.

The V-n construction of the figure says that in verbs with negation in French, negation implies the presence of an adverb *ne* labelled with category $Adv - ng$ (*ne*) and/or an adverb labelled with category $Adv - np$ (like *pas*). We also have a uniqueness obligation over these adverbs, and an linear order must be respected (*ne* must come before *pas*). When the mode of the verb is infinitive, the verb must be placed after the adverbs.

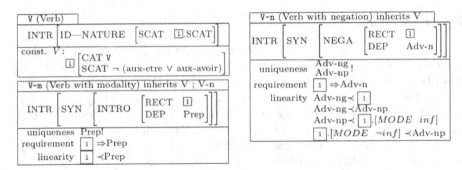

Fig. 5. Fragment of a PG for French (basic verbal constructions)

To describe a PG, we need to be able to represent encapsulations, variables, feature structures, and properties. We can notice that XMG classes can be seen as encapsulations, and that variables and feature structures were already used for TAG descriptions. Considering that, the XMG description language for PG can be formalized this way:

$$Desc_{PG} := x = y \mid x \neq y \mid [f{:}E] \mid \{P\} \mid Desc_{PG} \wedge Desc_{PG}$$
$$P := A : \triangle B \mid A : B! \mid A : B \prec C \mid A : B \Rightarrow C \mid A : B \not\Rightarrow C \mid A : B$$

where x, y correspond to unification variables, $=$ to unification, $\neq$ to unification failure, : to association between the feature f and some (possibly complex) expression E, and $\{P\}$ to a set of properties. Note that E and P may share unification variables.

The translation of the linguistic construction for V-m in XMG would be:

$$V{-}m \;\rightarrow\; (Vclass \vee V{-}n) \wedge \langle PG \rangle \{[\mathsf{INTR}{:}[\mathsf{SYN}{:}[\mathsf{INTRO}{:}[\mathsf{RECT}{:}X, \mathsf{DEP}{:}\mathsf{Prep}]]]]$$
$$\wedge (V : \mathsf{Prep}!) \wedge (V : X \Rightarrow \mathsf{Prep}) \wedge (V : X \prec \mathsf{Prep})\}$$

Here, inheritance is made possible by calls of classes. The control language even allows to do disjunctive inheritance, like it happens in class V-m. The end of the compilation process for PG will differ from TAG's one. We don't need any solver for descriptions, the accumulation into PG dimension is the grammar. To get the properties solved for a given sentence, the solution is to use a parser as a post processor for the compiler.

Morphological Dimension. For the needs of the study of verbal morphology in Ikota [8], a morphological dimension based on the notion of topological fields [9] was proposed. The description language available inside this dimension is the following:

$$Desc_{Morph} := f \leftarrow c \mid attr = val \mid Desc_{Morph} \wedge Desc_{Morph}$$

where f is a field, declared for the whole metagrammar, c is a contribution, and $\leftarrow$ corresponds to the accumulation of a contribution into a field. attr = val means that the feature composed of this pair will be part of the accumulated description.

The execution of the metagrammar starts with the ordering of fields. This solving has to be done only once in this dimension because in the chosen morphological theory, positions are fixed. For every solution of the execution of the classes, strings are accumulated into the fields, and morphosyntactic information into features.

The output of the compilation process is not a grammar strictly speaking, but a lexicon of fully inflected forms, basically obtained by concatenation of the fields contents.

Frame Semantics Dimension. A dimension handling a second formalism for semantics was proposed in [10]. The dedicated description language allows to describe frames, which are representations of mental concepts [11] and can be represented as feature structures. The unification of frames implies the unification of their types, which belong to a type hierarchy. This specific type unification is handled by the frame compiler brick. The description language for frames is the following:

$$Desc_{Frame} := f(\ t,\ [a_1 = f_1,\ \ldots a_n = f_n]\ )\ |\ Desc_{Frame} \wedge Desc_{Frame}$$

where f is an optional variable labeling the frame, $a_1 \ldots a_n$ are attributes of the frame, and $f_1 \ldots f_n$ are frame associated to these attributes. The execution of the frame dimension leads to the accumulation and combination of frame fragments.

The output for this dimension is a set of frames, that should be associated to syntactic structures. The interface between TAG trees and frames in discussed in [12].

Including a specific representation module to the XMG-1 compiler could be seen as an ad-hoc solution. This is why allowing the linguist to build their own dimension, begining with the choice of a description language, is a central feature of the new version of XMG. A XMG-2 brick corresponding to a new representation module is composed of the definition of the language used by the brick (the dedicated description language) and of the compilation modules to handle this language.

3.3   Specific Virtual Machines

During the generation of the linguistic resource, objects corerponding to the described structures are manipulated. The main operation between structures is unification, triggered explicitly (by using the equal sign) or implicitely (by importing variables from other classes). For most of these structures, standard unification is adequate, but for some of them, specific engines have to be used. For example, feature structures (like the ones used in TAG) need a dedicated unification algorithm, corresponding to set union.

A XMG-2 brick for a new description language has to include the set of specific virtual machines needed to handle the unification of its structures. For the frame semantics dimension for example, a dedicated virtual machine handling the unification of typed feature structures is contributed by the brick.

3.4 Principle Bricks

The notion of principles defined in XMG was too restrictive for our aims. Their specificity for the target formalism, for example, is incompatible with the multi-formalism ambition. An interesting way to handle principles is the one of [13], both allowing the linguist to create his own principles or to use a subset of the ones already defined. An example is the tree principle, which states that the solution models must be trees.

What we aim to provide is a meta-principles library: generic and parametriz-able principles the user can pick and configure. For example, the color principle provided for TAG could be an implementation of a generic polarity principle, parametrized with the table of figure 2. Another example of meta-principle is called unicity and was already implemented in XMG-1. It is used to check the uniqueness of a specific attribute-value pair in each solution, and thus is not specific to any linguistic theory.

Principles are also packaged into XMG-2 bricks. This means that for any new metagrammatical scope where trees have to be solved, the tree principle brick just has to be plugged into the new (or existing) dimension brick.

For the morphological dimension discussed early, a principle brick handling linear ordering constraints between fields was created.

3.5 Dynamic Definition of a Metagrammar Compiler

To build their own metagrammatical scope, one only has to create and configure the dimensions he needs and the properties he wants to check on them. Building a compiler consists in picking and combining independent modules, which we call compiler bricks. XMG-2 provides a compiler builder, that assembles the needed parts of the compiler according to a description of the connections between the bricks. The tokenizer and the parser for the metagrammatical language are automatically generated from this description, and each brick contributes its own compilation modules.

One of the main advantages of this modular approach is that the specific part of the compiler is mostly written automatically, and new features could be added just for experiments. A user can either use an existing compiler or assemble parts to build their own. Defining the principles would just consist in taking meta-principles out from the library and instantiate them.

Building a metagrammar compiler in this way allows to deal with a large range of linguistic theories, or even to quickly experiment while creating a new grammar formalism.

4 Current State of the Work

XMG project started in 2003 with a first tool, that has been used to produce large TAG grammars for French [14], German [15] and English, and a large

Interaction Grammar for French [16]. The compiler was written in Oz/Mozart, a language which is not maintained any more and not compatible with today's architectures (64 bits). It was also important to restart from scratch, in order to build a compiler more in adequation with its ambitions : modularity and extensibility.

Consequently, a new implementation started in 2010, in YAP (Yet Another Prolog) with bindings with Gecode for constraints solving. XMG-2 is currently the tool used for modeling the syntax and morphology of various African and Creole languages, and is compatible with the previous large metagrammars. It also includes the support for the dimensions discussed in this article.

5   Conclusion

In this paper, we showed how modularity, together with a metagrammatical approach, eases the development of a large scale grammar. This modularity is essential for reaching the main goal of XMG, that is to say extensibility. Getting to that means taking a big step towards multi-formalism and multi-language grammar development, and then offers new possibilities for sharing data between different types of grammar, or even for comparing them.

Two levels of modularity are given by XMG. The first one is the grammatical modularity, which makes it easier to generate and maintain large scale grammars thanks to the definition and the combination of abstractions. The second level of modularity is metagrammatical: XMG-2 provides a way to build new compilers by defining and combining elementary parts of compiler, called compiler bricks. The users have different options: they can use existing compilers (the one for TAG and 'flat' semantics for example), combine bricks to build a new type of compiler (like a compiler having two TAG dimensions, for two different languages), or create their own bricks, to combine them with existing ones (a brick for dependency grammars for example).

References

1. Candito, M.: A Principle-Based Hierarchical Representation of LTAGs. In: Proceedings of COLING 1996, Copenhagen, Denmark (1996)
2. Joshi, A.K., Schabes, Y.: Tree Adjoining Grammars. In: Rozenberg, G., Salomaa, A. (eds.) Handbook of Formal Languages. Springer, Berlin (1997)
3. Frank, R.: of Pennsylvania. Institute for Research in Cognitive Science, U.: Syntactic Locality and Tree Adjoining Grammar: Grammatical, Acquisition and Processing Perspectives. IRCS report. University of Pennsylvania, The Institute for Research in Cognitive Science (1992)
4. Duchier, D., Le Roux, J., Parmentier, Y.: The Metagrammar Compiler: An NLP Application with a Multi-paradigm Architecture. In: Van Roy, P. (ed.) MOZ 2004. LNCS, vol. 3389, pp. 175–187. Springer, Heidelberg (2005)

5. Kallmeyer, L., Lichte, T., Maier, W., Parmentier, Y., Dellert, J., Evang, K.: TuLiPA: Towards a Multi-Formalism Parsing Environment for Grammar Engineering. In: Coling 2008: Proceedings of the Workshop on Grammar Engineering Across Frameworks, Manchester, England, pp. 1–8. Coling 2008 Organizing Committee (2008)
6. Duchier, D., Parmentier, Y., Petitjean, S.: Cross-framework Grammar Engineering using Constraint-driven Metagrammars. In: CSLP 2011. Karlsruhe, Allemagne (2011)
7. Guénot, M.L.: Éléments de grammaire du français pour une théorie descriptive et formelle de la langue. PhD thesis, Université de Provence (2006)
8. Duchier, D., Magnana Ekoukou, B., Parmentier, Y., Petitjean, S., Schang, E.: Describing Morphologically-rich Languages using Metagrammars: a Look at Verbs in Ikota. In: Workshop on "Language Technology for Normalisation of Less-resourced Languages", 8th SALTMIL Workshop on Minority Languages and the 4th Workshop on African Language Technology, Istanbul, Turkey (2012)
9. Stump, G.T.: On the theoretical status of position class restrictions on inflectional affixes. In: Booij, G., van Marle, J. (eds.) Yearbook of Morphology 1991, pp. 211–241. Kluwer (1992)
10. Lichte, T., Diez, A., Petitjean, S.: Coupling Trees and Frames through XMG. In: ESSLLI 2013 Workshop on High-level Methodologies for Grammar Engineering (HMGE 2013), Duesseldorf, Germany (2013)
11. Fillmore, C.J.: Frame semantics. In: The Linguistic Society of Korea. Linguistics in the Morning Calm, pp. 111–137. Hanshin Publishing (1982)
12. Kallmeyer, L., Osswald, R.: Syntax-driven semantic frame composition in Lexicalized Tree Adjoining Grammar. Journal of Language Modelling 1, 267–330 (2013)
13. Debusmann, R.: Extensible Dependency Grammar: A Modular Grammar Formalism Based On Multigraph Description. PhD thesis, Saarland University (2006)
14. Crabbé, B.: Représentation informatique de grammaires fortement lexicalisées: Application à la grammaire d'arbres adjoints. PhD thesis, Université Nancy 2 (2005)
15. Kallmeyer, L., Lichte, T., Maier, W., Parmentier, Y., Dellert, J.: Developing a tt-mctag for german with an rcg-based parser. In: LREC. ELRA (2008)
16. Perrier, G.: A French Interaction Grammar. In: RANLP, Borovets, Bulgaria (2007)

Locative Alternation in English and Russian: A Frame Semantic Analysis

Yulia Zinova

Heinrich Heine University Düsseldorf, Germany
zinova@phil.uni-duesseldorf.de

Abstract. In this paper, an analysis of locative alternation phenomena in Russian and English within the framework of compositional frame semantics and Lexicalized Tree Adjoining Grammars (LTAG) is proposed. It features a compositional approach to locative alternation in both discussed languages and takes advantage of the possibility of separating construction meaning from the meaning of lexical elements provided in LTAG. As an additional decomposition step, metagrammar descriptions for both syntactic and semantic representations are given. Moreover, for Russian the decomposition goes further towards the morphology-semantics interface which makes it possible to account for the differences in the behavior of verbs with different prefixes (or no prefix) with respect to locative alternation.

1 Introduction

There are a number of formalisms that capture the idea that the meaning of a verb-based construction depends both on the lexical meaning of the verb and on the construction in which the verb is used (Goldberg, 1995; Van Valin and LaPolla, 1997). The key question is how exactly the components of the meaning are distributed and how they combine.

In Kallmeyer and Osswald (2012) introduces a combination of Lexicalized Tree Adjoining Grammars (Joshi and Schabes, 1997) and Frame Semantics. It is shown that the resulting framework is very flexible with respect to the factorization and combination of lexical and constructional units at the syntax and semantics level and is also suitable for computational processing. The approach is further motivated and developed in Kallmeyer and Osswald (2013), where a more detailed description and formalization for both semantic and metagrammar parts is offered.

Although a number of different approaches to semantic composition using LTAG (Joshi and Vijay-Shanker, 1999; Frank and van Genabith, 2001; Kallmeyer and Joshi, 2003) already exist as well as approaches that combine other syntactic formalisms with Frame Semantics (Frank, 2004, 2007), the novel combination of an LTAG and Frame Semantics benefits from both LTAG's extended domain of locality and the underspecification allowed by frames.

In this paper, I present an analysis of locative alternation that takes advantage of the flexibility offered by this novel framework.

M. Colinet et al. (Eds.): ESSLLI 2012/2013, LNCS 8607, pp. 49–68, 2014.
© Springer-Verlag Berlin Heidelberg 2014

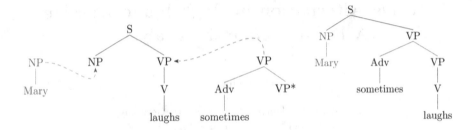

Fig. 1. Example of a TAG derivation

2   Tree Adjoining Grammar

Tree Adjoining Grammar (TAG, Joshi and Schabes (1997)) is a tree-rewriting grammar formalism. TAG consists of a finite set of *elementary trees* with labeled nodes and two operations on them: *substitution* and *adjunction*. All elementary trees are either *auxiliary* or *initial trees*. An *auxiliary tree* is a tree which has exactly one *foot node* – a leaf that is marked with an asterisk (see Fig. 1). Leaf nodes can be labeled with terminals and other nodes are labeled only with non terminals. The derivation process starts from an initial tree. In the final *derived tree* all the leaves must be labelled by terminals.

With *substitution* a non terminal leaf is replaced with a new tree and *adjunction* is used for replacing an internal node with an auxiliary tree. Adjunction is allowed if the root and foot nodes of the adjoining auxiliary tree have the same label. Figure 1 shows an example of a derivation: the initial tree for *Mary* substitutes into the subject slot of the elementary tree for *laughs*, and the *sometimes* auxiliary tree for the VP modifier adjoins to the VP node.

We will use *feature-structure based TAG*, or FTAG. It is a variant of TAG in which elementary trees are enriched with feature structures (Vijay-Shanker and Joshi, 1988). In an FTAG each node has a top feature structure and all the nodes except substitution nodes have a bottom feature structure. Feature unification happens when adjunction and substitution take place. Due to the extended domain of locality, nodes within one elementary tree can share features, making it possible to express constraints among dependent nodes easily.

For natural languages, a specific version of TAG called *lexicalized TAG*, or LTAG, is used. In an LTAG, each elementary tree must have at least one non empty lexical item, called a *lexical anchor*. Another important principle for a natural language TAG is that every elementary tree where the lexical anchor is a predicate must contain slots (leaves with non terminal labels) for all arguments of this predicate, including the subject, and for nothing else (*θ-criterion for TAG*, Frank 2002: 55).

The facts that LTAG has an extended domain of locality and that elementary trees are lexicalized and contain slots for all the predicate's arguments, makes it a good candidate for combination with frame-based compositional semantics (Kallmeyer and Osswald, 2012). In the approach proposed in

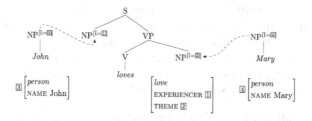

Fig. 2. Syntactic and semantic composition for *John loves Mary*

Kallmeyer and Osswald (2012), a single semantic representation (a semantic frame in this case) is linked to the entire elementary tree. When coupling an elementary tree with a semantic frame, syntactic arguments can be directly linked to their counterparts in the semantics. The described approach is similar to those used in Gardent and Kallmeyer (2003) and Kallmeyer and Romero (2008), but uses different kinds of semantic representations.

Semantic composition is modeled by unification triggered by adjunctions and substitutions. Figure 2 provides an illustration of syntactic and semantic composition. In this example, substitutions trigger unifications between ①and ③and ②and ④which leads to correct insertion of argument frames into the frame of *loves*. The introduction to the frame semantics approach and the details on the syntax-semantics interface will be provided in Section 4.2.

Elementary trees can be further factorized by a so-called metagrammar, thereby capturing important linguistic generalizations. There are two factorization steps, that are important for this paper: *unanchored elementary trees* are specified separately from lexical anchors; trees are organized into *tree families* which represent different realizations of one subcategorization frame. This allows the definition of a meaning for sets of unanchored elementary trees, i.e., a meaning of constructions.

3   The Data

3.1   Previous Approaches

(1) and (2) show basic examples of locative alternation in English and Russian. Despite the fact that in English both constructions have a PP that can be omitted without losing the meaning of the construction, let us call the first variant ((1-a), (2-a)) *prepositional phrase construction,* or *PPC,* and the second variant ((1-b), (2-b)) – *instrumental case construction,* or *ICC,* for convenience when referring to them[1].

(1)   a.   John$_x$ loaded the hay$_y$ onto the truck$_z$.   (PPC)
       b.   John$_x$ loaded the truck$_z$ with hay$_y$.   (ICC)

---

[1] There is no established way to refer to these two constructions.

(2)   a.   Ivan$_{\boxed{x}}$ zagruzil^{PF} seno$_{\boxed{y}}$ v vagon$_{\boxed{z}}$.
           Ivan   PREF.load.PST.SG.M hay.SG.ACC in wagon.SG.GEN
           'Ivan loaded the hay onto a/the wagon.'
      b.   Ivan$_{\boxed{x}}$ zagruzil^{PF} vagon$_{\boxed{z}}$ senom$_{\boxed{y}}$.
           Ivan   ZA.load.PST.SG.M wagon.SG.ACC hay.SG.INSTR
           'Ivan loaded the wagon with hay.'

PPCs are traditionally analyzed as having a change of location meaning and ICCs as having a change of state meaning (Kageyama, 1997; Levin and Rappaport Hovav, 1998; Goldberg, 1995). An analysis for (1) following Kageyama (1997) is provided in (3). It demonstrates that there is a difference between the two constructions, but it is reduced to the difference in the perspective. In the sense of information contributed by the two constructions, they seem to be indistinguishable if described by expressions in (3).

(3)   a.   X CAUSE [BECOME [hay BE ON truck]]
      b.   X CAUSE [BECOME [truck$_{\boxed{z}}$ BE [WITH [hay BE ON z]]]]

The analysis proposed in Levin and Rappaport Hovav (1998), which can be found under (4), provides more detailed information about the difference between PPCs and ICCs. (4-a) tells us that the hay changes its location as a result of the loading event, while (4-b) describes that the result is a change in the state of the wagon. One can notice that in (3) there is no explicit reference to the verb itself and the only component that is taken from the verb meaning is that the result of the loading is that the THEME is on the LOCATION in the end.

(4)   a.   [[x ACT] CAUSE [y BECOME P$_{loc}$ z] [LOAD]$_{MANNER}$]
      b.   [[x ACT] CAUSE [z BECOME []$_{STATE}$ WITH-RESPECT-TO y] [LOAD]$_{MANNER}$]

A first version of a frame-based analysis of locative alternation was proposed in Zinova and Kallmeyer (2012). The current paper presents an updated and more detailed analysis (especially with respect to Russian data) that benefits from the formal description of the frame-based semantics, developed in Kallmeyer and Osswald (2013).

3.2   Russian and English Data in More Detail

If one looks carefully at what the sentences in (1) and (2) mean, the question that arises is whether it is really the case that there is no change of state in PPC examples. In fact, any loading activity leads to both a change of location of the content and some change of state of the container (if it is specified), and the difference between the two constructions is that

- different components of the effect become more salient;
- in the case of ICC, the initial and result states of the container are specified.

In order to understand how the meaning of verbs and constructions should be represented, let us look at the range of verbs in English and Russian that allow locative alternation. Pinker (1989) provides the following classification for English.

Content-oriented classes:

(a) simultaneous forceful contact and motion of a mass against a surface (brush);
(b) vertical arrangement on a horizontal surface (heap);
(c) force is imparted to a mass, causing ballistic motion in a specified spatial distribution along a trajectory (inject);
(d) mass is caused to move in a widespread or non directed distribution (scatter).

Container-oriented classes:

(e) a mass is forced into a container against the limit of its capacity (crowd);
(f) a mass of size, shape, or type defined by the intended use of a container is put into the container, enabling it to accomplish its function (load).

From the description of verb classes that allow locative alternation in English, one can see that the result state of the container in the case of ICC is such that the action cannot be performed any longer. There is no result state common for all the cases, so it depends on the verb, i.e., on how the change of location happens. The easiest way to solve this would be to assume different construction meanings for different verb classes (e.g., one with an EFFECT where the RESULT_STATE of the THEME is 'full' and another one where the value of the same attribute is 'covered'), but let us first look at some Russian data.

In Russian many verbs allow only one of the constructions, i.e., a change of construction requires a different verbal prefix (Dudčuk, 2006). However, some of the verbs remain the same in both prepositional and instrumental constructions. Such verbs can be organized in three groups: the first one is similar to the (f) group in English (see example (2)), the second one is similar to group (a) in English, like in (5), and the third class can be thought of as a combination of the first and the second: a mass is put into a container, enabling it to accomplish its function, or on a container, covering its surface (6).

(5)   a.   On namazal^{PF} maslo   na hleb.
            He na.spread.pst.sg.m butter.sg.acc on bread.sg.acc
            He distributed butter over a piece of bread.
    b.   On namazal^{PF} hleb   maslom.
            He na.spread.pst.sg.m bread.sg.acc butter.sg.instr
            He covered a piece of bread with butter.

(6)   a.   On zasypal^{PF} sahar   v banku.
            He za.fill.pst.sg.m sugar.sg.acc in can.sg.acc
            He put sugar in a/the can.
    b.   On zasypal^{PF} banku   saharom.
            He za.fill.pst.sg.m can.sg.acc sugar.sg.instr
            He covered/filled the can with sugar.

Verbs of the third group exhibit the following interesting behaviour: while in the case of the PPC example (6-a) there is a preposition which tells us that the content goes *in* the container, in the case of the ICC example (6-b) two different readings are possible: the content can be put in the container or the content can cover the container. In both cases there is a clear result state: either the container is full or the container's surface is fully covered with content. This means that the verb *zasypat'* ('to fill/to cover') does not provide information about how the THEME is positioned with respect to the GOAL. In the case of the PPC this information comes from the preposition used (both *v* ('in') and *na* ('on') are possible) and in the ICC the ambiguity can be resolved only using world knowledge. So (6) demonstrates conclusively that there should be one construction accounting for different result states of the theme and allowing different interpretations of one verb due to underspecification of how the change of location process goes.

ICCs in English are claimed to give rise to the "holistic effect": the interpretation that the container is full or covered at the end of the event (Fillmore, 1970; Anderson, 1971; for more details see Iwata, 2008). As for Russian, the same effect is due to the perfective verb, not to the construction itself: while (2-a) means that all the hay was loaded and (2-b) that the wagon was fully loaded, examples in (7) lack those entailments. The same effect is observed if one compares (5) with (8): the former triggers a holistic interpretation, while the latter does not.

(7)   a.   Ivan gruzil^{IPF} seno   v vagon.
           Ivan load.pst.sg.m hay.sg.acc in wagon.sg.gen
           Ivan was loading/loaded the hay into a/the wagon.
      b.   Ivan gruzil^{IPF} vagon   senom.
           Ivan load.pst.sg.m wagon.sg.acc hay.sg.instr.
           Ivan was loading/loaded the wagon with hay.

(8)   a.   On mazal^{IPF} maslo   na hleb.
           He spread.pst.sg.m butter.sg.acc on bread.sg.acc
           He distributed butter over a piece of bread.
      b.   On namazal^{IPF} hleb   maslom.
           He na.spread.pst.sg.m bread.sg.acc butter.sg.instr
           He covered a piece of bread with butter.

As illustrated by (7) and (8), some Russian verbs can participate in both ICCs and PPCs independently of the presence of a prefix[2], but this is not true for all the verbs. For example, the unprefixed variant of the verb *zasypat'* 'to fill up' (ex. (6)), which is *sypat'* 'to pour,' can be used in the PPC (9-a), but not in the ICC: (9-b) is uninterpretable, as the direct object must have the semantic role of content, not a container.

(9)   a.   On sypal^{IPF} sahar   v banku.
           He fill.pst.sg.m suggar.sg.acc in can.sg.acc
           He put sugar in a/the can.

---

[2] Some prefixes restrict this ability, but describing the behavior of all the prefixes goes far beyond the scope of the current paper.

b. $^{??}$On sypal^{IPF} banku   saharom.
He fill.pst.sg.m can.sg.acc sugar.sg.instr

In such cases, the presence of the prefix "removes" a certain constraint from the verbal frame. This fact was captured in Zinova and Kallmeyer (2012) by introducing a special underlined value that rewrites the value of some attribute instead of unification of the values. This non monotonic operation is not present anymore in the current approach.

Only certain prefixes are able to change the set of constructions a verb can participate in: for example, the addition of *na-* leads to the holistic effect in (10-a), but the sentence similar to (9-b) remains ungrammatical (10-b). The source of the ungrammaticality of (10-b) is not the prefix, but the initial verb, as is clear from the comparison of (10) with (5). Like in the case of (9-b), in (10-b) the direct object must be something that can be poured, and not a container.

(10)   a.   On nasypal^{PF} sahar   v banku.
He na.fill.pst.sg.m suggar.sg.acc in can.sg.acc
He put sugar in a/the can.
b. $^{??}$On nasypal^{PF} banku   saharom.
He na.fill.pst.sg.m can.sg.acc sugar.sg.instr

In sum, the analysis must be able to capture the following facts: in English, using the ICC leads to the holistic effect; in Russian, the presence of the holistic effect is dependent on the verbal aspect and a class of verbs can participate in the ICC only when prefixed with *za-*.

4   Locative Alternation: The Analysis

4.1   Syntactic Representation

In examples (1), (2) and (5)–(10) from the previous sections both container and content are specified. However, both ICCs and PPCs may be uttered when only the direct object is realized; in this case, they will have the same difference in semantics, being syntactically the same. Using LTAG and metagrammar decomposition, one can obtain the tree family in Fig. 3 for the PPC and tree family in Fig. 4 for the ICC (the second NP_{INSTR} stands for both NP in instrumental case in Russian and PP with the preposition *with* in English).

4.2   Proposed Frame Semantics

The first goal of this section is to provide richer and more explicit representations of (1) and (2) than those of (4). The next step is taking apart the construction meaning from the meaning of lexical elements.

Frames, as they were introduced in Fillmore (1982), are cognitive structures that represent situations or states. A basic example can be found in Fig. 2, where the central frame for the verb represents a situation of the type *love* and

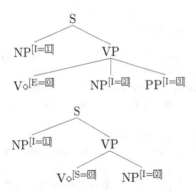

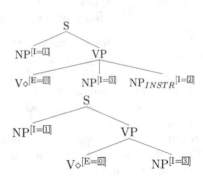

Fig. 3. Unanchored trees for the PPC   **Fig. 4.** Unanchored trees for the ICC

the semantic roles of two participants: EXPERIENCER and THEME. Such kinds of frames are used in the Berkeley FrameNet project (Fillmore *et al.*, 2003).

Kallmeyer and Osswald (2013) formalize frame semantics and propose a meta-grammar architecture that allows to combine pairs of elementary morphosyntactic trees and elementary meaning structures. Without going into the details of the logic used, I will informally introduce those notions that are relevant for the understanding of the current paper (formal definitions can be found in Kallmeyer and Osswald, 2013).

Frames are typed feature structures with multiple base nodes: nodes, labeled with base labels ($\boxed{0}$, $\boxed{1}$, $\boxed{2}$,...), from which all the other nodes are accessible via attributes. To shorten the notation, boxed letters are used to express path-identities instead of referring to a node via attributes of a base node. Relations between nodes are allowed ($\leq$, *part-of*, etc.). All the frames must satisfy the general constraints, e.g., type constraints. Any two frames can be unified. The result is either undetermined (if the constraints are not satisfied) or unique. Lexicalized elementary trees are obtained by insertion of the lexical items into the anchor nodes and combining the frame description with the general constraints. The semantics of the lexical element is the minimal model computed after this operations.

Let me first provide the frames that we want to obtain as the resulting frames for the sentences in (1) and (2). The frame for (1-a) (Figure 5 on the left) consists of a *causation* event that has the attributes ACTOR, THEME, GOAL and EFFECT. The value of the EFFECT attribute is of the type *scalar_change_of_location*, which means that it has PATIENT (identified with the THEME of the *causation* event), GOAL (identified with the GOAL of the *causation* event) and MEASURE attributes. The value of the MEASURE attribute provides information about the type of the scale along which the change happens[3].

---

[3] Note that the principal distinction between content- and container-oriented constructions is viewed in this analysis as a distinction between the measure dimensions of the direct objects: content is viewed as something that can be characterized in terms of *amount*, while the container must have a either *volume* (for containers that can be filled) or *area* (for something that can be covered) dimension.

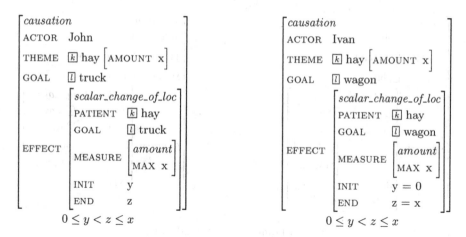

Fig. 5. Frames for (1-a) and (2-a)

In this specification, I follow the ideas from Osswald and Van Valin, Jr. (2012) where one can find a discussion of the representation of attributes, events, and results while implementing frame semantics. I introduce attributes of initial and result states and a scale which is determined by its type, start and end points. The change of state is either a decrease or an increase of the value on an ordered scale (a discussion of the analysis of scalar change can be found in Kennedy and Levin, 2008). The direction is given either by the additional constraint (see the constraint under the frame description) or by the values of attributes MIN and MAX if both of them are specified. This replaces the LESSER attribute of ordering proposed in Osswald and Van Valin, Jr. (2012).

In our example, the type of the scale is *volume*. The maximum value on the scale is equal to the total amount of hay (x), the minimum value is not specified. INIT and END are points on this scale that determine the beginning and the end of the event: such a frame means that the *scalar_change* started when the amount of PATIENT (hay) in the truck was y and ended when it was z. The value of x is the total amount of hay, but the exact amount cannot be specified because it is not explicitly stated in the sentence. y and z may have any value, as long as the constraint specified under the frame is satisfied. The GOAL of the *scalar_change_of_loc* subframe is the THEME of the main *causation* frame and the PATIENT of the subframe is the MEAN of the main event. The MEAN role is a unified role for CONTENT and COVERAGE. It is needed to capture the difference between the ICC and PPC constructions. Note, that MEAN cannot be replaced with an INSTRUMENT role: an instrument can be additionally specified.

The frame for the corresponding Russian sentence (2-a) (Figure 5 on the right) differs from the one described above by the values of INIT and END attributes: they are now not variables, their values are specified as 0 (inserted by default because no MIN value is specified in the MEASURE) and x, that is the amount of hay and the MAX of the scale. This means that the *causation* event begins when

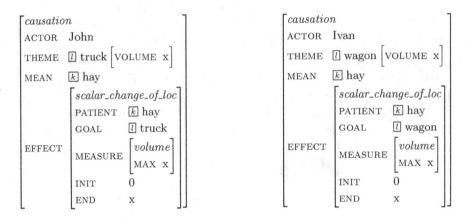

Fig. 6. Frames for (1-b) and (2-b)

no hay is in the truck and ends when all the hay is loaded. Now that y is equal to 0 and z is equal to x, the constraint is trivially satisfied.

The frames for (1-b) and (2-b) differ only in the values of the ACTOR and the THEME (that is also the GOAL of the *scalar_change*) attributes and are represented in Figure 6. In these frames, the INIT and END attributes also correspond to 0 and the MAX value of the MEASURE attribute, but now the type of MEASURE and the value of MAX is identified with the parameter of the "goal-type" theme, so it is of type *volume*. Thus, the *causation* event begins when the occupied volume of the truck is equal to 0 (it is empty) and ends when it is equal to x (the truck is full).

The next step is decomposing the frames in Figure 5 and Figure 6 into frames for the PPC and ICC and frames of the lexical items. As the holistic effect depends on the prefix in Russian in the same way as on the construction in English, the right frame in Figure 7 represents the semantics of both the English verb *load* and the unprefixed Russian verb *gruzit'* 'to load' (as lexical anchors that yet have to be inserted in the elementary trees).

The frame paired with the upper tree in Figure 3 (where both arguments are realized) is the one on the right of Figure 7. It contains slots for all the attributes that are linked to the syntactic tree and a more explicit mechanism of copying the relevant scale type and MAX value from the THEME into the MEASURE attributes. Two achieve this, two new conventions that are not present in Kallmeyer and Osswald (2013), are introduced: type identities and attribute identity under the condition of existence.

Type identity is used to access and copy the type of the value of some attribute without performing the value identity. It is denoted by the squared variable inside the square brackets which is $\boxed{k}$ in the right frame in Figure 7 (as opposed to the squared variable in front of the square brackets that indicates value identity).

The second mechanism is labelled by squared Greek letters and is a subkind of identity: if both attributes exist, their values get identified. If one of them does not exist (equals $\perp$), the values are not identified. This is used to copy only those

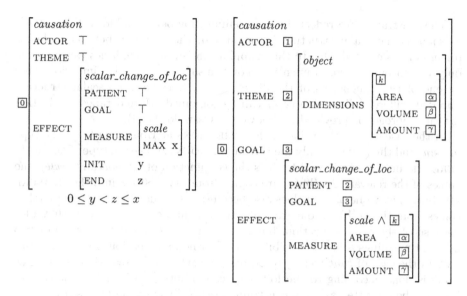

Fig. 7. Frames for *load/gruzit'* and the PPC

attributes that are allowed by the type[4]. A type with another set of constraints
can appear after performing type identity operations. E.g., in the PPC case,
when the direct object is introduced, only those attributes of the DIMENSIONS
get filled that are present in the object. All the other attributes receive the value
⊥ due to the type constraints. The partial type hierarchy (only those types that
we need for the current analysis are represented) is represented in Fig. 8.

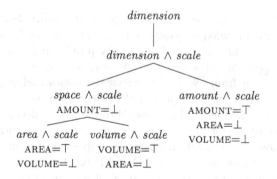

Fig. 8. Dimension types

---

[4] Note that both type identity and attribute identity under existence do not increase
the expressive power of the logic since the number of types and the number attributes
are finite.

The constraints in 8 reflect the "one delimitation per event" idea (Filip, 2000): as soon as the scalar structure is introduced, the mapping between different dimensions is created. Due to this mapping, the event boundaries can be given only within one selected dimension. For transitive verbs, this dimension must be one of the dimensions of the direct object. If there is no scalar structure imposed, different dimension types can be conjoined. More detailed motivation for this approach is provided in Zinova and Osswald (2014).

In the case of *hay*, the value of the DIMENSIONS attribute will be of the type *amount* and thus contain only the attribute AMOUNT. Once the type of the MEASURE attribute is specified (here it is the conjunction of *amount* and *scale*), the values of the relevant attributes are copied from DIMENSIONS into the MEASURE attribute. Later one of the values also becomes the value of MAX attribute (that comes from the verb). In our example, there is just one attribute of MEASURE because of the type restriction, but for sentences like (9-b) the source of two different meanings is in the possibility of either noun dimension (AREA and VOLUME in the case of *banka* 'can') to become the MAX value. Insertion of the MAX value is done according to the following constraints (specified informally here but may be rewritten as Horn constraints, see Kallmeyer and Osswald 2013):

- the value of the attribute MAX must be $\top$;
- the value of the attribute MAX may be equal to the value of the attributes VOLUME or AREA or AMOUNT.

Frames for the ICC differ from those for the PPC in the argument structure: the direct object remains the THEME of the *causation*, but the second argument is MEAN instead of GOAL. For English, the frame (Fig. 9 on the left) also specifies INIT and RESULT attributes that lead to the holistic effect in sentences like (1-b).

5   Metagrammar and Morphological Decomposition

The beauty of the formalism used in the paper is its ability to capture various linguistic and world knowledge generalizations. The frames for the PPCs and ICCs proposed in the previous section already provide a mechanism for measuring the event and obtaining the relevant interpretations by unification of a universal construction frame with the verbal semantics and subsequently the verbal arguments. However, it is clear that the common part of the frames for the two constructions is huge and it is useful to do further decomposition of the elements of the constructions to provide a more compact semantic description.

To illustrate how the metagrammatical decomposition is done, I provide the description of the classes used in the ICC and PPC classes in Russian, as there are more phenomena to model. However, most of the classes can be also used for English, sometimes with slight modifications (fixing the linear order, changing a case-marked NP into a PP and so on).

For the metagrammar description, the syntax of the tree descriptions from XMG (Crabbé *et al.*, 2013) and the feature structure descriptions in the attribute-value language of Kallmeyer and Osswald (2013) are used. Syntactic

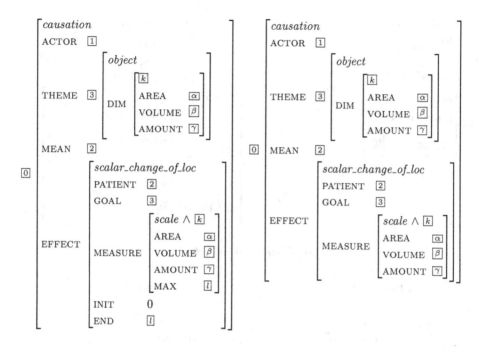

Fig. 9. Frames for ICCs in English and Russian

nodes are connected to the semantic frames via interface features I ("individual") and E ("event"). The identities between the I features and the thematic roles in the semantic frame provide the correct argument linking. Equating different E values on the V nodes is used to unify the corresponding event frames.

The first class, VSpine (Figure 10 on the left), is responsible for passing verbal restrictions on the direct object types. As it is not the set of the types allowed by the verb, but the set of types allowed in the given sentence that has to be restricted, a type-identical attribute DIM_RESTR (dimension restriction) is created from the dimensions attribute of the verb (that will be inserted when a verb fills the verbal anchor place)[5].

In addition to this, let us introduce a special class for Russian verbs prefixed with *za-*: ZaVerb. This class does two things: it introduces the holistic effect (for English, this is done by the construction) and creates the attribute for dimension restriction (DIM_RESTR) that in this case is not dependent on the verb, but may be filled with the restriction provided by the prefix if the dimension hierarchy gets expanded during the exploration of other phenomena. With the dimension hierarchy as presented in Figure 8, the DIM_RESTR attribute is redundant and can be omitted. Note that this class does not use the *VSpine* class, while, for example, the class for the verbs prefixed with *na-* would differ in this respect,

---

[5] All the classes used in the current decomposition export the event variable E, identified with ⓪.

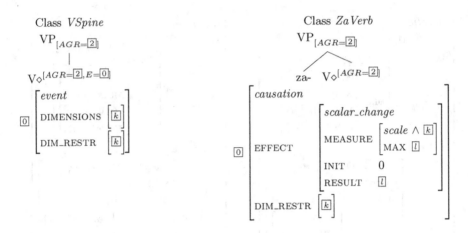

Fig. 10. *VSpine* and *ZaVerb* classes

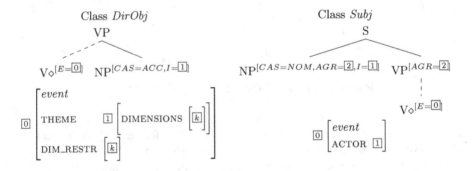

Fig. 11. *DirObj* and *Subj* classes

because the constraints on the dimensions that are introduced by the verb must be preserved.

The class for the direct object (Figure 11 on the left) tells us that the NP in the direct object position becomes the THEME of the event and the type of the value of its DIMENSIONS attribute is copied into the DIM_RESTR attribute of the event. The subject class (Figure 11 on the right) assigns the ACTOR role in the event frame to the NP in the nominative case.

The next class, *n0V*, uses 2 classes: one of the *VSpine* and *ZaVerb* classes plus the *Subj* class. There is no additional tree or frame associated with the class, it is only a conjunction of another classes. Due to the disjunction of the *VSpine* and *ZaVerb* classes, two minimal models presented in Figure 12 are computed for the *n0V* class[6].

---

[6] The syntactic agreement features are omitted in the minimal models representations.

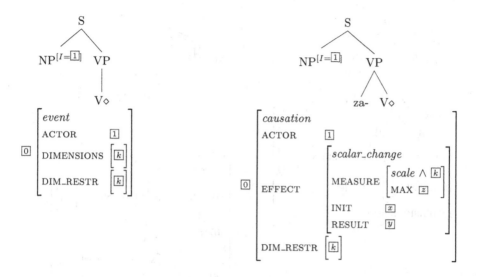

Fig. 12. Minimal models for the *n0V* class

The class *n0Vn1* is the conjunction of the *n0V* and *DirObj* classes also without any additional syntactic or semantic components. Similar to the previous class, there are two minimal models of the *n0Vn1* class, represented in Figure 13.

The classes in Figure 14 contain the final tree fragments needed to construct the full PPC and ICC classes. They determine the role of the second verbal argument. Note that these classes can be used in any other place in the grammar, they are not construction-specific: the identification of the GOAL attribute of the main event with the GOAL attribute of the *scalar_change_of_location* or the MEAN attribute with the PATIENT attribute will be done by the additional frame description that is associated with the relevant construction and the general constraints.

As the holistic effect in Russian arises not due to the construction meaning, but due to the verbal aspect (which we modeled with *ZaVerb* class), different roles of the send arguments are introduced by separate *PP* and *NPinstr* classes and the relevant dimension constraints are already assembled in one attribute, there is no need to introduce different classes for the two constructions: the *PPC/ICC* class uses an *n0Vn1* class conjoint with one of the *PP* and *NPinstr* classes. The additional frame description for the *PPC/ICC* class is on Figure 15 on the left. The following constraints[7] ensure that the correct attributes are copied from the main event frame into the *scalar_change_of_loc* subframe:

---

[7] Similar to the constraints on types described above, I provide informal descriptions that can be reformulated in terms of Horn constraints. These constraints can be replaced by overt specification of distinct frames with different semantic roles of the direct object. I believe that having just one frame for the verb plus the constraints is more plausible from the cognitive point of view.

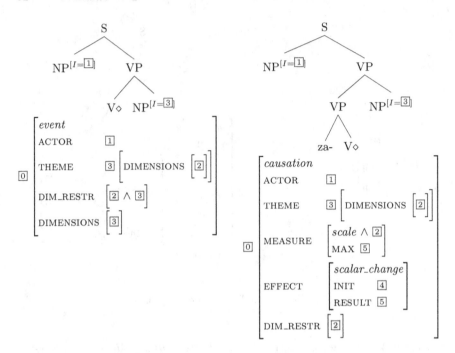

Fig. 13. Minimal models for the *n0Vn1* class

1. the value of the attribute THEME is equal to the value of either attribute GOAL or attribute PATIENT of the EFFECT;
2. if there is no AMOUNT among the attributes of the DIMENSIONS of the THEME, then the THEME of the *causation* event cannot be the PATIENT of the *scalar_change_of_loc* event;
3. if there is no VOLUME nor AREA among the attributes of the DIMENSIONS of the THEME, then the THEME of the *causation* event cannot be the GOAL of the *scalar_change_of_loc* event;
4. if the attribute MEAN is present, its value is also the value of the attribute PATIENT of the EFFECT and the THEME of the *causation* is the GOAL of the EFFECT.
5. if the attribute GOAL is present, its value is at the same time the value of the attribute GOAL of the EFFECT and the THEME of the *causation* is the PATIENT of the EFFECT.

These constraints are proposed to ensure that the THEME of the *causation* frame is identified with the correct attribute of the subevent frame. In cases in which both arguments are realized, this is easy to do, because the role of the second argument allows us to distinguish between the constructions (constrains 4 and 5). The difficulty occurs if only the first argument is overtly realized. In this case, the first argument can be either the *patient* or the *goal* of the subevent (constraint 1). One of the two possibilities can be blocked in case there is no appropriate dimension. If the THEME lacks AMOUNT dimension, it cannot be the

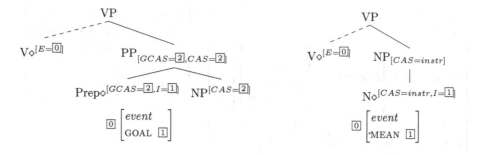

Fig. 14. *PP* and *NPinstr* classes

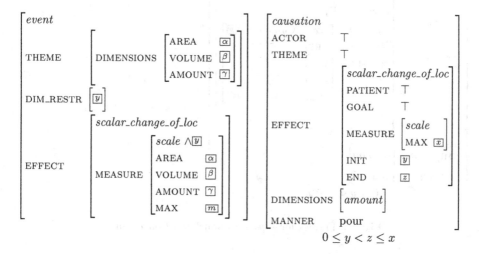

Fig. 15. Frames for the PPC/ICC class and the verb *sypat'* 'to pour' in Russian

patient of the *scalar_change_of_location* (constraint 2) and if it does not have neither volume not area, it cannot have the GOAL role (constraint 3).

Let me show that the proposed frame decomposition leads to the desired result for (9-b) and (6-b). Similar to the frame for the verb *load* in Figure 7, the frame for the unprefixed imperfective Russian verb *sypat'* 'to pour' is constructed (Figure 15 on the right). The important difference between the two verbs is that the latter has a specified type of the value of the DIMENSIONS attribute.

After the verb gets inserted as the lexical anchor in the ICC, the frame in the right-hand side of Figure 16 is obtained. The type of the value of the MEASURE attribute is the conjunction of types *scale*, *amount* and the type that will come from the object. As we see in Figure 16 (on the left), this type is *area* ∧ *volume*. Due to the constraints on the scale hierarchy, the resulting type leads to the ⊥ value of the whole sentence.

Conversely, when the verb *sypat'* 'to pour' gets prefixed with *za-*, the usage of the metagrammar *ZaVerb* class leads to a frame without an *amount* type

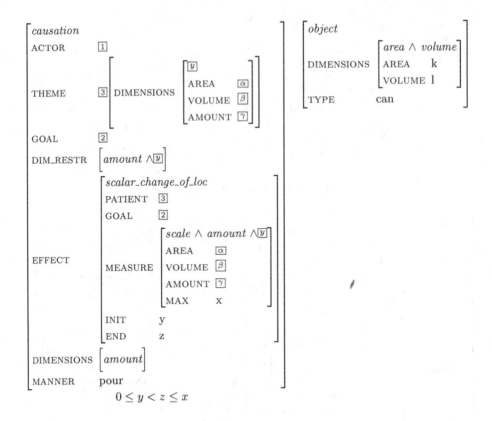

Fig. 16. Frames for *sypat'* 'to pour' and *banka* 'can'

conjunct in the DIM_RESTR and, consequently, the MEASURE attribute of the *scalar_change_of_loc*. The resulting type in this case allows the values of either AREA or VOLUME to be used as the value of MAX attribute, leading to the two desired interpretations of the sentence.

6   Conclusion

In this paper I propose an analysis for the locative alternation phenomenon in English and Russian. The important property of the approach is the high factorization level: the meaning of the lexical entries is separated from this of the constructions and the latter is decomposed on the metagrammar level. Furthermore, a step towards providing a morphology-semantics interface is made: the metagrammatical description for Russian includes a class that contains frame semantics for a prefix usage that is assumed to behave noncompositionally and was analysed in a nonmonotonic way in Zinova and Kallmeyer (2012).

The framework used in the current paper is novel: it was introduced in Kallmeyer and Osswald (2012) and formally described in Kallmeyer and Osswald

(2013). For the proposed analysis of locative alternation phenomena, two additional operations on the frame semantics side are introduced. These operations do not increase the expressive power of the logic, but provide a convenient way of capturing a range of phenomena and produce underspecified descriptions.

Acknowledgements. I would like to thank the audiences and organizers of ESSLLI 2012 Student Session and the TAG+11 conference. Special thanks to six anonymous reviewers for their comments and suggestions. This paper could not have been written without the support of Laura Kallmeyer, who devoted a lot of her time to discussing numerous problems that arose on the way to the analysis presented here. I am also grateful to Timm Lichte and Rainer Osswald for comments and discussions related to the topic of this paper. The research reported here has been financially supported by the German Science Foundation (DFG) as part of the SFB 991.

References

Anderson, S.R.: On the role of deep structure in semantic interpretation. Foundations of Language 7(3), 387–396 (1971)

Crabbé, B., Duchier, D., Gardent, C., Le Roux, J., Parmentier, Y.: XMG: eXtensible MetaGrammar. Computational Linguistics 39(3), 591–629 (2013)

Dudčuk, P.: Locativnaja alternacija i struktura glagol'noj gruppy (Locative alternation and the structure of the VP). Diploma thesis, Moscow State University (2006)

Filip, H.: The quantization puzzle. In: Events as Grammatical Objects, pp. 3–60. CSLI Press, Stanford (2000)

Fillmore, C.: The grammar of hitting and breaking. In: Jacobs, R., Rosenbaum, P. (eds.) Readings in English Transformational Grammar, pp. 120–133. Ginn, Waltham (1970)

Fillmore, C.J.: Frame Semantics. In: Linguistics in the Morning Calm, pp. 111–137. Hanshin Publishing Co., Seoul (1982)

Fillmore, C.J., Johnson, C.R., Pertuck, M.R.L.: Background to FrameNet. International Journal of Lexicography 3(16), 235–250 (2003)

Frank, A.: Generalizations over corpus-induced frame assignment rules. In: Proceedings of the LREC 2004 Workshop (2004)

Frank, A.: Question answering from structured knowledge sources. Journal of Applied Logic 5(1), 20–48 (2007)

Frank, A., van Genabith, J.: GlueTag. Linear logic based semantics for LTAG – and what it teaches us about LFG and LTAG. In: Butt, M., King, T.H. (eds.) Proceedings of the LFG 2001 Conference, Hong Kong (2001)

Frank, R.: Phrase structure composition and syntactic dependencies. MIT Press, Cambridge (2002)

Gardent, C., Kallmeyer, L.: Semantic Construction in FTAG. In: Proceedings of EACL 2003, Budapest, pp. 123–130 (2003)

Goldberg, A.E.: Constructions. A Construction Grammar Approach to Argument Structure. In: Cognitive Theory of Language and Culture. The University of Chicago Press, Chicago (1995)

Iwata, S.: Locative alternation: A lexical-constructional approach, vol. 6. John Benjamins Publishing (2008)

Joshi, A.K., Schabes, Y.: Tree-Adjoining Grammars. In: Rozenberg, G., Salomaa, A. (eds.) Handbook of Formal Languages, pp. 69–123. Springer (1997)

Joshi, A.K., Vijay-Shanker, K.: Compositional semantics with lexicalized Tree-Adjoining Grammar (LTAG): How much underspecification is necessary? In: Blunt, H.C., Thijsse, E.G.C. (eds.) Proceedings of the Third International Workshop on Computational Semantics (IWCS-3), Tilburg, pp. 131–145 (1999)

Kageyama, T.: Denominal verbs and relative salience in lexical conceptual structure. In: Kageyama, T. (ed.) Verb Semantics and Syntactic Structure, pp. 45–96. Kurioso Publishers, Tokyo (1997)

Kallmeyer, L., Joshi, A.K.: Factoring Predicate Argument and Scope Semantics: Underspecified Semantics with LTAG. Research on Language and Computation 1(1-2), 3–58 (2003)

Kallmeyer, L., Osswald, R.: A frame-based semantics of the dative alternation in lexicalized tree adjoining grammars. Submitted to Empirical Issues in Syntax and Semantics 9 (2012)

Kallmeyer, L., Osswald, R.: Syntax-driven semantic frame composition in lexicalized tree adjoining grammars. Journal of Language Modelling 1(2), 267–330 (2013)

Kallmeyer, L., Romero, M.: Scope and situation binding in LTAG using semantic unification. Research on Language and Computation 6(1), 3–52 (2008)

Kennedy, C., Levin, B.: Measure of change: The adjectival core of degree achievements. In: McNally, L., Kennedy, C. (eds.) Adjectives and Adverbs. Syntax, Semantics, and Discourse. Oxford University Press, Oxford (2008)

Levin, B., Rappaport Hovav, M.: Morphology and lexical semantics. In: Spencer, A., Zwicky, A.M. (eds.) Handbook of Morphology, pp. 248–271. Blackwell Publishers, Oxford (1998)

Osswald, R., Van Valin Jr., R.: Framenet, frame structure, and the syntax-semantics interface. Manuskript Heinrich-Heine Universität (2012)

Pinker, S.: Learnability and cognition: The aquisition of argument structure. MIT Press, Cambridge (1989)

Van Valin, R.D., LaPolla, R.J.: Syntax. Cambridge University Press, Cambridge (1997)

Vijay-Shanker, K., Joshi, A.K.: Feature structures based tree adjoining grammar. In: Proceedings of COLING, Budapest, pp. 714–719 (1988)

Zinova, Y., Kallmeyer, L.: A Frame-Based Semantics of Locative Alternation in LTAG. In: Proceedings of the 11th International Workshop on TAG and Related Formalisms (2012)

Zinova, Y., Osswald, R.: A frame-semantic analysis of Russian verbs of motion. Presentation, Szklarska Poreba Workshop (2014)

On a Formal Connection between Truth, Argumentation and Belief

Sjur Dyrkolbotn

Durham Law School, Durham University, UK
s.k.dyrkolbotn@durham.ac.uk

Abstract. Building on recent connections established between formal models used to study truth and argumentation, we define logics for reasoning about them that we then go on to axiomatize, relying on a link with three-valued Łukasiewicz logic. The first set of logics we introduce are based on formalizing so called skeptical reasoning, and our result shows that a range of semantics that are distinct for particular models coincide at the level of validities. Then, responding to the challenge that our logics do not capture credulous reasoning, we explore modal extensions, leading us to introduce models of three-valued belief induced by argument. We go on to take a preliminary look at some formal properties of this framework, offer a conjecture, then conclude by presenting some challenges for future work.

1 Introduction

There are close formal connections between argumentation, truth and kernels in directed graphs. This was first observed in [11] (truth and kernels) and [13] (kernels and argumentation), and all three were considered together in [18], where a formal link to Łukasiewicz three-valued logic was noted, see also [42]. Here we build on this work, first by characterizing the propositional validities of skeptical argumentation, and then by proposing a modal extension that allows us to capture credulous reasoning. This leads us to introduce three-valued models of belief induced by argumentation frameworks, and we offer a preliminary investigation of their properties.

The structure of the paper is as follows. In Sect. 2 we survey the connections between kernel theory and models used to study truth and argumentation, and we introduce basic concepts and notation. Then in Sect. 3 we introduce various argumentation-based logics for reasoning about these models, using Ł3 to provide axiomatizations of their validities. In Sect. 4 we develop modal extensions of these logics. We argue that three-valued $KD45$ logic is a good candidate for reasoning about truth and argumentation, and we conjecture that the classes of serial models induced from argumentation frameworks under preferred and semi-stable semantics respectively are in fact canonical for this logic. In Sect. 5 we offer a conclusion.

2 Truth, Kernels and Argumentation

To set the stage for the novel work we present in later sections, we now give a summary of the connections that have been established previously. This serves both as necessary background and further motivation for the questions we address later.

M. Colinet et al. (Eds.): ESSLLI 2012/2013, LNCS 8607, pp. 69–90, 2014.
© Springer-Verlag Berlin Heidelberg 2014

2.1 Truth

The search for truth is often carried out on the basis of an implicit and imprecise understanding of what exactly constitutes it. In most academic fields, there are methodolog-
ical principles and procedural safe-guards in place to ensure that accepted results are
indeed truthful, but these are dependent on context and differ from field to field. In philosophy, on the other hand, the search for a unified theory of truth has a long tradition
and it is often carried out analytically, by assuming as few primitive notions as possible
and abstracting away from contextual factors to the greatest possible extent. Indeed, a
large body of work is devoted to the *formal* study of truth, much of which is based on
logically examining this Aristotelian principle[1], an approach due to [39]:

T: *A statement is true if, and only if, what it says is the case*

As a theory of truth this might seem like an uninformative truism, but hard philosophical problems arise already at this level. The paradigmatic example is the *liar statement*:
"this statement is false". If it is true, then it must be false according to T, since this is
what it says. On the other hand, if it is false, then we must conclude, again by T, that
it is true. This is problematic, particularly to those who think truth and falsehood are
mutually exclusive, as one would expect from how these notions are ordinarily used.

For formal theories of truth, semantic paradoxes such as the liar occupy pride of
place. Indeed, they are the first obstacle that arises, even for the most rudimentary theoretical accounts.[2] This presents us with a surprising problem: either something is wrong
with the rules of classical logic, or else something is wrong with the intuitively obvious
principle T.

To address this problem using logic, it is common to formalize T in some system
of predicate logic, with truth as a predicate. Much work has been carried out in this
tradition, often focusing on the question of how to modify T to arrive at a theory which
does not lead to paradoxes such as the liar. However, according to some philosophers,
most notably Kripke, the paradoxes do not serve to demonstrate fault with principle T,
they merely show that truth is *partially defined* [25].

It has long been accepted wisdom that referential patterns play a crucial role in the
emergence of paradox. The liar, for instance, is viciously circular, explicitly negating
itself. It is tempting to depict referential patterns using graph-structures. For instance,
the liar can be pictured as a directed graph with a single vertex pointing to itself:

Much formal work on truth makes use of graphs as pictures, an idea that was first
developed formally in [3]. Here, non-wellfounded sets are used to define the semantics
for self-referential statements, and graphs are used to depict such sets, following the
work of [1]. More recently, it has been observed that graph-structures can also be used

---

[1] See Aristotle's *Metaphysics*, 1011b, 26.

[2] In this paper we think of a paradox as a contradiction which we arrive at from premises that
we think are uncontroversial. This only captures what Quine calls the falsidical paradoxes
[33], but one might argue that the other kind he proposes - the veridical ones - are not really
paradoxes at all, but merely surprising *facts*.

to represent the semantic content of statements directly, as an alternative to a more traditional formulation in predicate logic. This idea is due to [11], who noted that one might as well interpret edges in a directed graph (digraph) as negations and branching as conjunction.

Towards formalization, assume we have a collection Π of atoms, thought of as statement names, including a constant $\mathbf{1}$ denoting some arbitrary true statement. Then for any index set I, a *truth-theory* of cardinality $|I|$ is a collection $\bigcup_{i \in I} \{x_i \leftrightarrow \bigwedge\{\neg x \mid x \in X_i\}\}$ where $x_i \in \Pi, X_i \subseteq \Pi$ for all $i \in I$. A truth-theory is *finitary* if X_i is finite for all $i \in I$. Truth-theories encode instantiations of the principle T, applied to concrete sets of statements referring to each other.[3] The reader might worry that the form assumed for formulas appearing on the right of an equivalence is overly restrictive, but in [5] it was shown that truth-theories provide a normal form for propositional theories, so the format is in fact fully general.[4]

We can now *define* paradox, saying that a truth-theory is paradoxical just in case it is classically inconsistent [18]. This captures the liar: the truth-theory $\{p \leftrightarrow \neg p\}$ is obviously an inconsistent theory. Truth-theories might seem trivial and uninteresting, but the connection we can set up with digraphs makes them very useful. In the next subsection, we will argue for this in some depth, showing how the combinatorial perspective provides a great template for further exploration of when principle T becomes problematic.

2.2 Kernels

A directed graph over Π is a set $N \subseteq \Pi \times \Pi$ of directed edges. When $(x, y) \in N$, we write $y \in N(x)$ and $x \in N^-(y)$ (so N^- is the converse of N) and we extend this notation to sets, e.g., such for $X \subseteq \Pi$ we have $N(X) = \bigcup_{x \in X} N(x)$. We say that a digraph is finite if N is finite and *finitary* if $N(x)$ is finite for all $x \in \Pi$. The set $\Pi(N)$ is used to denote $\{x \mid N(x) \cup N^-(x) \neq \emptyset\}$, the set of atoms that stand in a relation to some other atom in N. Moreover, a digraph N' is said to be a *subdigraph* of N if $N' = \{(x, y) \in N \mid x, y \in \Pi(N')\}$.

The connection between truth-theories and digraphs can now be expressed in two simple equations. First, for all digraphs N we let $sinks(N)$ denote the set of atoms without outgoing edges. Then we form the corresponding truth-theory defined as follows:

$$\mathsf{T}(N) = \bigcup_{x \in \Pi(N) \backslash sinks(N)} \{x \leftrightarrow \bigwedge_{y \in N(x)} \neg y\} \cup \{\bigcup_{x \in sinks(N)} x \leftrightarrow \mathbf{1}\} \quad (2.1)$$

Conversely, if T is a truth-theory indexed by I we define the digraph N_T:

$$N_T = \bigcup_{i \in I} \{(x_i, x) \mid x \in X_i \setminus \{\mathbf{1}\}\} \quad (2.2)$$

---

[3] We remark that their concreteness means that we might as well omit explicit representation of truth as a predicate, e.g., not bother to write $T(p) \leftrightarrow \neg T(q) \wedge \neg T(r)$ (interpreting p, q, r as constants).

[4] This means that work on this formalism also has potential importance to the study of boolean satisfiability, as explored in [41].

When does N correspond to a non-paradoxical truth-theory? It is straightforward to verify that an assignment $f : \Pi(N) \to \{1,0\}$ satisfies $\mathsf{T}(N)$ under classical logic if, and only if, we have the following for all $x \in \Pi(N)$:

$$f(x) = 1 \Leftrightarrow f(y) = 0 \text{ for all } y \in N(x) \tag{2.3}$$

Translating this into the language of directed graphs it follows that N is non-paradoxical if, and only if, it admits some set $K \subseteq \Pi(N)$ such that:

$$N^-(K) = \Pi(E) \setminus K \tag{2.4}$$

As observed in [11], sets satisfying the above equation are known as *kernels* in graph theory. They were introduced by Von Neumann and Morgenstern to provide an abstract solution concept in cooperative game theory [40], and have attracted quite some theoretical interest, see [7] for an overview of the field. The connection with kernels means that the problem of paradox can be addressed graph-theoretically. In particular, let us write $Kr(N)$ for the set of kernels in a digraph N. Then the problem of paradox can be rephrased as follows: for what N do we have $Kr(N) \neq \emptyset$?

Many results on this have been obtained in kernel theory, most of which provide sufficient conditions for the existence of kernels [14,15,21]. Sufficient conditions actually tend to ensure something stronger than existence of kernels, namely *kernel perfectness*: the existence of kernels in all induced subdigraphs.

The first non-trivial result that was established states that a finitary digraph with no odd-length cycle is kernel perfect, due to [38].[5] The original proof is rather complicated, but was greatly simplified by the introduction of the notion of a *semikernel* [30]. A semikernel in a digraph N is a set $S \subseteq \Pi$ such that:

$$N(S) \subseteq N^-(S) \subseteq \Pi \setminus S \tag{2.5}$$

In other words, S is a semikernel if everything it points to is outside it and points back into it.[6] In particular, if S is a semikernel in N then it is a kernel in the subdigraph induced by $N(S) \cup S$. In other words, S witnesses to the fact that by restricting attention to this set and the statements it refers to, paradox can be avoided.

Given a digraph N, we use $Lk(N)$ to denote the set of all semikernels in N. Notice that $\emptyset \in Lk(N)$ for any N, and that the loop does not have any non-empty semikernel. A digraph can have a non-empty semikernel without having a kernel, however, as illustrated by the following digraph N:

---

[5] This does not hold for infinitary digraphs, the standard example being Yablo's paradox $\bigcup_{i \in \mathbb{N}} \{(i,j) \mid j > i\}$ [43]. The study of conditions applying to infinitary digraphs is harder and less progress has been made (but see [28,20,34,5]).

[6] In truth-theory terms, all statements negated by S are outside and in turn negate at least one member of S. Notice that if we assume such a collection to consists only of true statements, their truth can be verified by a constructive form of circular reasoning: all statements they negate are indeed false since they in turn negate a statement assumed to be true. In particular, the assuming their truth is perfectly consistent, not a paradox.

$$\circlearrowleft x \longrightarrow y \underset{\displaystyle\longleftarrow}{\overset{\displaystyle\longleftarrow}{}} z \tag{2.6}$$

Since they are self-negating, neither x nor y can be in a kernel. But then as x only negates y it follows that x could not possibly negate a member of any kernel, so $Kr(N) = \emptyset$. But we have $Lk(N) = \{\{z\}\}$, since z both negates and is negated by y. The technical importance of semikernels stems from the following result.

Theorem 2.7 *[30] A digraph N is kernel perfect if, and only if, every non-empty induced subdigraph of N admits a non-empty semikernel.*

In light of this result, it is possible to establish conditions that ensure kernel perfectness by showing that they ensure existence of non-empty semikernels for every nonempty induced subdigraph. Since semikernels are formulated locally on the digraph, this can be very helpful, and it is the approach followed in most work in kernel theory. The following theorem summarizes the most significant results. Recall that a *chord* on a cycle is an edge connecting two non-consecutive vertices.

Theorem 2.8 *For all digraphs N, we have that $Kr(N) \neq \emptyset$ if every odd cycle in N has one of the following*

1. *at least two symmetric edges [14],*
2. *at least two crossing consecutive chords [15] or*
3. *at least two chords with consecutive targets [21].*

As an example, consider the digraph N depicted on the left below. It has a kernel, and this is ensured by all points of Theorem 2.8.

$$\tag{2.9}$$

In fact, N has two kernels: $\{x\}$ and $\{y\}$. We notice that one of these, $\{x\}$, is also a kernel in N'. But this does not follow from any of the results from Theorem 2.8. This is interesting because it suggests that some simple cases are not covered, motivating further work. However, a natural conjecture stating that a digraph has a kernel if every odd cycle has *one* reversible edge is not true, as witnessed by the following digraph:[7]

$$\tag{2.10}$$

---

[7] It holds for the special case of a single odd cycle, however: take the target of some symmetric edge, skip two vertices, and from then on take every other vertex as you move along the cycle. You end up with a kernel, the only kernel admitted by this digraph.

The problem is that the odd cycles in this digraph interact in ways that make it impossible to solve them all simultaneously. This problem of *compatibility* is the essence of what makes the search for sufficient conditions both interesting and difficult. Semikernels and inductive arguments to establish kernel perfectness is the standard way to address it, but we mention that a new approach was recently introduced in [17]. This paper introduced the following notion, which is useful for proving sufficient conditions for the existence of kernels in digraphs that are not kernel-perfect.

Definition 2.11 *A* solver *for a digraph N is a sequence of induced subdigraphs and semikernels $\langle N_i, S_i \rangle_{1 \le i \le n}$ such that:*

1. $N_1 = N$
2. S_i *is a semikernel in N_i for all $1 \le i \le n - 1$*
3. $N_{i+1} = N_i \setminus (S_i \cup N^-(S_i))$ *for all $1 \le i \le n - 1$*
4. S_n *is a kernel of N_n.*

Solvers are useful because of the following result.

Theorem 2.12 ([17]) *A digraph has a kernel iff it has a solver.*

Using solvers, it is possible to show that a range of various conditions is sufficient to ensure the existence of kernels in digraphs that are not kernel perfect. The conditions are rather complicated to state, so we omit them here. However, we think this work deserves to be mentioned because it identifies new heuristics we can follow when attempting to map the logical consequences of truth-theories.

As an example, consider the following two digraphs below:

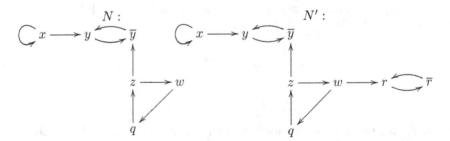

In both N and N', there are two odd cycles: (x, x) and (z, w, q, z). It is tempting to look at $y, \overline{y}$ for a possible resolution. There is a problem, however, namely that they can only solve one of the sequences in question. In both N and N', we have semikernels $\{y\}$ and $\{\overline{y}, w\}$, corresponding to whether we use them to solve (x, x), or use them to solve (z, w, q, z). In N, this is where the story ends — it is not possible to resolve both, and we conclude that $Kr(N) = \emptyset$. In N', on the other hand, it *is* possible to solve both, but only if you solve (x, x) first by choosing y. This, in particular, no longer precludes solving (z, w, q, z), since it is possible to choose r and obtain the kernel $\{y, r, z\}$, as predicted also by Theorem 2.6 from [17].

Examples such as these show how sufficient conditions for existence of kernels can be interpreted as describing circumstances under which the truth of some statements can lead to the resolution of problematic referential patterns. It is tempting, in particular, to think that y and r must be regarded as true *because* they resolve odd cycles. It seems necessary to accept their truth not because they cannot be refuted, but because accepting them is needed in order to resolve problems with (implicit) self-negation affecting other parts of the network. A basic intuition in much work on truth has been that semantic judgments should conform to classical logic to the greatest possible extent.

This involves accepting that some statements are true not because of what they say about the world, but because of what other statements say about them. Such statements are different from both truth-tellers and liars. They are not paradoxes and they are not undetermined. Rather, they must be assigned a unique value to resolve referential patterns that would otherwise become problematic. For instance, consider the following sentence A: "this sentence and the truth-teller B are both false". If B is a standard truth teller, stating "this sentence is true", it seems that B *must* be regarded as true in this referential network, since otherwise A becomes paradoxical. If we are committed to the idea that truth satisfies the property that paradox is avoided whenever possible, it seems to follow that our conclusion that B *must* be true is sufficient, in such a case, to conclude also that it *is* true.

2.3   Argumentation

The desire to arrive at some general notions of what counts as a logically correct argument seems to arise naturally in all human societies. If there is interaction there is argument, and some preliminary agreement on what is required for an argument to count as *successful* is of great importance, if nothing else then for pragmatic reasons.[8]

Following Frege and the formal turn in logic, the study of argumentation was largely seen as distinct from the formal study of correct reasoning. At best, it belonged to the informal branch. The search for logical perfection would famously flounder over results on incompleteness and undecidability, however, and since then the trend has been turning. Following the increasing popularity of non-classical logics, in particular defeasible logics [36,31], argumentation and logic have moved closer to each other.

This development took a particularly interesting turn with the seminal work of [16], who established a nice formal connection between argumentation on the one hand and non-monotonic reasoning and logic programming on the other.[9] Since then, abstract argumentation has attracted much attention, particularly in the AI-community [35]. The theoretical part of this work centers around the following question: Given some collection of arguments and some model of their content, how do we judge which arguments we should accept?

The novel move made in [16] was to rely on directed graphs as models, often referred to as argumentation frameworks (AFs) in this context. In argumentation theory, the

---

[8] In recent work from cognitive science it is even suggested that human reasoning may have evolved primarily because it proved useful in the context of argumentation [27].

[9] We also mention [12], a less cited work that did not involve the concept of argumentation, but which nevertheless has close connections to Dung's work. In particular, this work was the first, of which we are aware, to observe the close connection between kernel theory, logic programming and default logic.

atoms Π are thought of as arguments, and edges are thought of as attacks between them, such that e.g., $(p, q) \in N$ expresses that p is attacking q. By using digraphs to model the content of arguments, it becomes possible to give a range of argumentation semantics using intuitive graph constructions.

Given an AF N the task of such a semantics is to identify sets of arguments that can be held successfully together, typically called *extensions* in the literature. Most semantics are based on the intuition that a set of arguments should be internally consistent and able to defend itself against attack from other arguments. Different semantics differ about the details, but they all share the same overall aim: they give an answer, for any $p \in \Pi$, whether p should be accepted in the argumentative scenario represented by N. In particular, they all have the same signature, they are defined as an operator ϵ which takes an N and returns a set of sets $\epsilon(N) \subseteq 2^{\Pi}$.

To the best of our knowledge, all the semantics that have been studied share the property that arguments in an acceptable set should be free of internal conflict. Formally, for all semantics ϵ, all AFs N and all $A \in \epsilon(N)$, we have $N^-(A) \subseteq \Pi \setminus A$: no two arguments in A attack each other.

At first sight it seems we are working with a binary notion of acceptance: for a given argument, it is accepted or it is not. However, a moment's thought will show that this perspective fails to do justice to the nature of the structure (N, ϵ) in two important ways. First, there is the question of whether it is correct to say that p is accepted on N under ϵ when there *exists* some $A \in \epsilon(N)$ such that $p \in A$, or whether we should require $p \in A$ for *all* such A. Both notions of acceptance have been studied, and the former is typically dubbed *credulous* acceptance while the latter is referred to as *skeptical*.[10]

The second sense in which acceptance is not a binary notion has to do with the structure of N. In particular, given any $A \in \epsilon(N)$ the status of p with respect to A can be any of the following:

$$1 : p \in A \qquad\qquad 2 : p \in N(A)$$
$$3 : p \in \Pi \setminus (A \cup N(A)) \tag{2.13}$$

Notice that by conflict-freeness of A, it follows that if $p \in N(A)$ then $p \notin A$. Hence when the focus is on the status of individual arguments, we might as well view $\epsilon(E)$ as a set of partitions of Π into three disjoint sets or, equivalently, as a collection of so called *(Caminada) labellings*, functions $c : \Pi \to \{1, 0, \frac{1}{2}\}$ such that for all $x \in \Pi$:

$$c(x) = 0 \Leftrightarrow \exists y \in N^-(x) : c(y) = 1 \tag{2.14}$$

For any AF E we let $cf(N)$ be the set of all labellings for E, and we define $c^1 = \{x \in \Pi \mid c(x) = 1\}$, $c^0 = \{x \mid c(x) = 0\}$ and $c^{\frac{1}{2}} = \{x \in \Pi \mid c(x) = \frac{1}{2}\}$. This defines a semantics for argumentation such that for all N, we regard $A \subseteq \Pi$ as acceptable if there is some $c \in cf(N)$ such that $c^1 = A$.[11] In applications of argumentation theory,

---

[10] See [35, p. 32]. The terminology goes back to Dung [16], who in turn borrowed it from non-monotonic logic, where it is used to describe two notions of entailment, corresponding to existential and universal quantification over the possible extensions of a theory, see e.g., [22, p. 398].

[11] Hence it is not hard to see that values assigned by labellings correspond to the three points of (2.13) whenever we restrict attention to conflict-free sets of accepted arguments. Notice, in particular, that $p \in c^0 \Leftrightarrow p \in N^+(c^1)$ and $p \in c^{\frac{1}{2}} \Leftrightarrow p \in \Pi \setminus (c^1 \cup c^0)$.

this is usually considered too permissive, and a range of various restrictions has been considered, each giving rise to a new semantics, the most well-known of which are defined in Fig. 1.

Admissible:	$a(N) = \{c \in cf(E) \mid N^-(c^1) \subseteq c^0\}$
Complete:	$c(N) = \{c \in cf(N) \mid$
	$c^1 = \{x \in \Pi \mid N^-(x) \subseteq c^0\}\}$
Grounded:	$g(N) = \{\bigcap c(N)\}$
Preferred:	$p(N) = \{c_1 \in a(N) \mid \forall c_2 \in a(N) : c_1^1 \not\subset c_2^1\}$
Semi-stable:	$ss(N) = \{c_1 \in a(N) \mid \forall c_2 \in a(N) : c_1^{\frac{1}{2}} \not\supset c_2^{\frac{1}{2}}\}$
Stable:	$s(N) = \{c \in a(N) \mid c^{\frac{1}{2}} = \emptyset\}$

Fig. 1. Various semantics, defined for any $N \subseteq \Pi \times \Pi$

First, the *admissible* semantics [16] is obtained by restricting attention to conflict-free labellings c for which all those arguments that attack c^1 are in turn attacked by c^1. Hence the semantics captures the intuition that a set of acceptable arguments should be able to defend itself against attacks. The *complete* semantics [16] adds a further restriction, which captures the intuition that all arguments that are not disputed should be accepted. Hence, in addition to conflict-freeness it is also required that c^1 is equal to the set of those arguments that it defends. The *grounded* semantics [16] encodes a skeptical attitude, since it prescribes a unique labeling, namely the smallest complete labeling. This labeling always exists and is computable in linear time, starting from the labeling where all arguments are assigned $\frac{1}{2}$ and then iteratively labeling arguments by the boolean values, starting with those that are not attacked by any argument. The least fixed point of such a process will be the set of acceptable arguments under the grounded semantics, as explained in [16] and [8] for the labeling formulation.

The *preferred*, *semi-stable* and *stable* semantics all capture variants of the intuition that labellings should not only be admissible, but also allow us to reach a definite conclusion about the status of as many arguments as possible. According to the preferred semantics, which was first defined in terms of extensions rather than labellings [16], this amounts to maximizing the number of accepted arguments. According to the semi-stable semantics [10], it amounts to maximizing the number of boolean-valued arguments, while according to the stable semantics it amounts to requiring that no argument whatsoever is assigned the value $\frac{1}{2}$. This, however, is sometimes impossible, making the stable semantics the only one that sometimes fails to produce a labeling. This happens, for instance, on the AF $\{(x, x)\}$, corresponding to the liar statement, where all the other semantics admit $\{(x, \frac{1}{2})\}$ as the only permissible labeling.

The semantics are all defined as labellings, but (2.14) establishes an obvious one-to-one correspondence between a set of labellings and a set of extensions (sets of arguments assigned 1). Hence in the following we will allow ourselves to switch freely between these two representations, without introducing redundant notation to distinguish them.

In Fig. 2 we give two AFs, F and F′, that serve as examples. In F, every argument is attacked by some argument, and from this it follows that we have $g(F) = \emptyset$, i.e., the

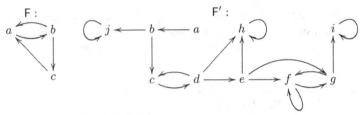

Fig. 2. Two argumentation frameworks

grounded extension is the empty set. The non-empty conflict-free sets are the singletons $\{a\}, \{b\}$ and $\{c\}$, but we observe that a does not defend itself against the attack it receives from c (since there is no attack (a, c)), and that c does not defend itself against the attack it receives from b. So the only possible non-empty admissible set is $\{b\}$. It is indeed admissible; b is attacked only by a and it defends itself, attacking a in return. In fact, since b also attacks c, the set $\{b\}$ is the unique stable set of this framework. It follows that $s(\mathsf{F}) = p(\mathsf{F}) = ss(\mathsf{F}) = \{\{b\}\}$ and $a(\mathsf{F}) = c(\mathsf{F}) = \{\emptyset, \{b\}\}$.

For a more subtle example, consider F'. The first thing to notice here is that we have an unattacked argument a, so the grounded extension is non-empty. In fact, the framework is such that all semantics from Figure 1 behave differently. It might look a bit unruly, but there are many self-attacking arguments that can be ruled out immediately (since they are not in any conflict-free sets), and it is easy to verify that the extensions of F' under the different semantics are the following:

$$g(\mathsf{F}') = \{a\}, s(\mathsf{F}') = \emptyset, ss(\mathsf{F}') = \{\{a, d, g\}\}$$
$$a(\mathsf{F}') = \{\emptyset, \{a\}, \{a, c\}, \{a, c, e\}, \{d\}, \{a, d\}, \{a, d, g\}, \{d, g\}\}$$
$$p(\mathsf{F}') = \{\{a, d, g\}, \{a, c, e\}\}, c(\mathsf{F}') = \{\{a\}, \{a, d, g\}, \{a, c, e\}, \{a, d\}\}$$

It is easy to see that the semantics for argumentation are closely connected to kernels and semikernels of digraphs. In particular, let $\overline{N} = \{(x, y) \mid (y, x) \in N\}$, so that $\overline{N}$ is the digraph obtained from N by reversing the direction of all edges. Then it is trivial to verify the following for all AFs $N, A \subseteq \Pi$.

$$A \in a(N) \Leftrightarrow A \in Lk(\overline{N}) \ \& \ A \in s(N) \Leftrightarrow A \in Kr(\overline{N}) \tag{2.15}$$

This connection was first observed in [13] but does not appear to have received much attention in the literature on argumentation. However, it follows from it that much work done in kernel theory, highly theoretical in nature, can be applied in argumentation theory. All the results mentioned in Sect. 3 detail circumstances when AFs admit non-empty stable sets, and the proofs are also mostly constructive, and identify scenarios where such sets can be computed quickly.[12] In particular, the connection to kernel theory gives us a taxonomy of different case types and different forms of inconsistency. We

---

[12] The decision problems in argumentation tend not to be tractable, and except for the grounded semantics, even computing the set of extensions is hard [35, Part I, Chap. 5]. Hence it is worth noting that many proofs from kernel theory provide computational information about *how* to argue in order to make sure that a given argument turns out to be accepted. For instance, the notion of a minimal semikernel, used in [17], can be understood in this way.

think combinatorial techniques developed in graph theory can be very helpful in future work that aims to shed light on the patterns underlying successful argumentation.

In the other direction, we note that while kernel theory can be understood as focusing on the question of classical consistency, corresponding to the existence of stable sets, argumentation theory has developed semantics which aim to facilitate reasoning about scenarios where classical consistency cannot be achieved. To assess these semantics from a philosophical perspective on truth, and a technical perspective on digraphs, seems like a very fruitful avenue for future research.

One crucial question concerns the logical foundations of these various semantics, and there has recently been quite some work devoted to this, most of which focuses on finding neat ways to *define* argumentation semantics, see [24,23] which relies on modal logic, and [2] which uses quantified boolean formulas.[13] While we think this work is interesting, we note that there has so far been a shortage of logics designed to permit reasoning *about* AFs, and to study meta-logical properties.

We can certainly attempt to use logics that are expressive enough to define various semantics in the object-language, but such an approach easily runs the risk of complicating matters to the extent that interesting results become hard or impossible to obtain. In particular, it will typically require us to use (fragments of) very powerful logics that may not admit any straightforward axiomatization, if they are decidable at all. In the next section we propose another route, focusing on extending the connection between propositional logic and semantics formulated on digraphs. In particular, we show that Łukasiewicz logic can be used to reason about AFs, and that a strong correspondence can be established for skeptical reasoning, whereby the validities of argumentation coincide for all non-stable semantics defined in Fig. 1. In particular, we show that they are all axiomatized by Wajsberg's rules for Łukasiewicz's three-valued logic.

3 Logics for Reasoning about Argumentation and Truth

In this section we will talk about digraphs using the following language $\mathcal{L}$:

$$\phi := p \mid \neg\phi \mid \phi \to \phi$$

where $p \in \Pi$. Since argumentation semantics are formulated in terms of three-valued labellings, we already have in place a corresponding interpretation of atomic formulas from $\mathcal{L}$. The semantic value of p, in particular, is one among $\{1, 0, \frac{1}{2}\}$. This is not a novel proposal, merely a logical reformulation of what is already commonplace in the literature on argumentation, see e.g., [35, Chapter 2]. However, we will now extend labellings inductively to provide a three-valued interpretation of the whole language $\mathcal{L}$. This involves a new construction, but it is easy enough to motivate once we consider the intended reading of formulas in $\mathcal{L}$.

We will think of $\mathcal{L}$ as containing meta-arguments addressing the semantic status that arguments *should* obtain in an AF. The connectives are read intuitively as follows: the

---

[13] For completeness, we also mention [19,9] which develops similar ideas by exploiting (other) ways to define argumentation semantics in modal logic, and [42], which relates argumentation to three-valued labellings for logic programing.

formula $\neg\phi$ is the argument that ϕ should be rejected, while the argument $\phi \to \psi$ is the argument that it should be at least as easy to accept ψ as it is to accept ϕ. On such a reading it seems clear that for all AFs N and all semantics ϵ, the following inductive definition of   : $\epsilon(N) \times \mathcal{L} \to \{1, 0, \frac{1}{2}\}$ appropriately extends any $c \in \epsilon(N)$ to any $\phi \in \mathcal{L}$:

$$\overline{c}(\phi) = \begin{cases} c(\phi) \text{ if } \phi = p \in \Pi \\ 1 - \overline{c}(\psi) \text{ if } \phi = \neg\psi \\ min\{1, 1 - (\overline{c}(\psi_1) - \overline{c}(\psi_2))\} \text{ if } \phi = \psi_1 \to \psi_2 \end{cases} \tag{3.1}$$

To illustrate the definition, assume we have a labeling $c = \{p \mapsto 0, q \mapsto \frac{1}{2}\}$. In this case, it is intuitively clear that the argument that it should be at least as easy to accept q as it is to accept p is itself acceptable. This is also the outcome prescribed by (3.1), since $\overline{c}(p \to q) = min\{1, 1 - (c(p) - c(q))\} = min\{1, 1.5\} = 1$. For a different example, suggesting also that formulas of $\mathcal{L}$ should not be read as stating that ϕ *is* accepted, consider the same meta-argument when the labeling is $c = \{p \mapsto 1, q \mapsto \frac{1}{2}\}$. In this case, it seems clear that we cannot accept the argument that q is at least as easy to accept as p. However, since the status of q is undetermined, we cannot reject the argument that this *should* be the case. Hence it seems that the meta-argument itself should be regarded as undetermined, which, indeed, is what (3.1) ensures. To further illustrate that this analysis is appropriate, consider an AF N which admits *two* labellings, $c_1 = c$ and $c_2 = \{p \mapsto 1, q \mapsto 1\}$. In this case, it is still not correct to say that $p \to q$ is acceptable on N, but it is also wrong to say that it has been rejected, since it only fails to be acceptable when q is undetermined, and is acceptable in all other cases. In fact, it seems that $p \to q$ not only should be accepted, but *must* be accepted, since neither of the two possible assignments entitle us to reject it.

This distinction introduces a modal flavor to $\mathcal{L}$, and in the list below we give some useful expressions along with three definable non-trivial modalities.[14]

- $\top := p \to p$ where $p \in \Pi$ is arbitrary.
- $\bot := \neg\top$.
- $\phi \vee \psi := (\phi \to \psi) \to \psi$.
- $\phi \wedge \psi := \neg(\neg\phi \vee \neg\psi)$.
- $\phi \leftrightarrow \psi := (\phi \to \psi) \wedge (\psi \to \phi)$.
- $\Box\phi := \neg(\phi \to \neg\phi)$ (meaning ϕ is accepted).
- $\Diamond\phi := \neg\phi \to \phi$ (meaning ϕ is not rejected).

---

[14] The observation that they are definable in terms of $\{\neg, \to\}$ was made by Tarski in 1921, who was Łukasiewicz's student at the time, see [26, p. 167]. Our preferred reading of these modalities is slightly non-standard. In particular, we will often think of truth normatively as providing permissions and/or obligations to accept claims as being true. According to the T-scheme, a permission to accept ϕ arises only when ϕ is the case, i.e., when it has the value 1. On the other hand, the T-scheme also implies that a permission to reject ϕ arises only when it is not the case, i.e., when ϕ has value 0. Moreover, we think of truth as prescribing the *norm* that a statement should be either accepted as true or rejected as false. The paradoxes show that this is sometimes impossible, and hence we think of the value $\frac{1}{2}$ as signifying that one has an obligation to accept (or reject), yet no permission to do so. Hence, our modal reading lets us think of semantic paradoxes as a form of normative conflict. In future work, we would like to explore this point of view further.

– $\triangle\phi := \phi \leftrightarrow \neg\phi$ (meaning ϕ is neither accepted nor rejected).

To better understand the behavior of the modal operators, consider the unpacking of the inductive definition of $\bar{c}$ for these formulas, shown below and easily established against (3.1), for any labeling c.

$$\bar{c}(\phi \vee \psi) = max\{\bar{c}(\phi), \bar{c}(\psi)\} \qquad \bar{c}(\phi \wedge \psi) = min\{\bar{c}(\phi), \bar{c}(\psi)\}$$

$$\bar{c}(\Box\phi) = \begin{cases} 1 \text{ if } \bar{c}(\phi) = 1 \\ 0 \text{ otherwise} \end{cases} \qquad \bar{c}(\Diamond\phi) = \begin{cases} 0 \text{ if } \bar{c}(\phi) = 0 \\ 1 \text{ otherwise} \end{cases}$$

$$\bar{c}(\triangle\phi) = \begin{cases} 1 \text{ if } \bar{c}(\phi) = \frac{1}{2} \\ 0 \text{ otherwise} \end{cases}$$

Notice that all the modal expressions have the property that they evaluate to boolean values. Intuitively, this is reasonable: if someone says of some argument "that argument has been accepted" he is not reiterating it, but claiming that it has as a matter of fact been accepted. This, unlike the acceptability of the argument itself, seems natural to interpret in boolean terms.

With an extension of labellings to formulas in place, it is straightforward to associate a formal logic with every argumentation semantics. In particular, let $\mathcal{AF}$ denote the set of all AFs over Π. Then we define a class of argumentation logics as follows.

Definition 3.2 *For all argumentation semantics ϵ, we define $\models_\epsilon \subseteq \mathcal{AF} \times 2^{\mathcal{L}}$ such that for all $N \in \mathcal{AF}, \phi \in \mathcal{L}$:*

$$N \models_\epsilon \phi \text{ if, and only if } \forall c \in \epsilon(N) : \bar{c}(\phi) = 1$$

We write $\models_\epsilon \phi$ just in case $N \models_\epsilon \phi$ for all $N \in \mathcal{AF}$, in which case we say that ϕ is valid in the logic ϵ.

Intuitively, we think of $N \models_\epsilon \phi$ as encoding that it is true that the meta-argument ϕ should be skeptically accepted on N, according to ϵ. To illustrate the behavior of some argumentation logics, consider the AF from Fig. 3.

Fig. 3. An AF E such that $\Pi(E) = \{p, q, q', p'\}$

Below we list some truths about this N, under various logics corresponding to semantics from Figure 1.

$$N \models_s \bot \text{ since } s(N) = \emptyset$$
$$N \models_g \triangle v \text{ for all } v \in \{p, q, q', p'\}$$
$$N \models_x \Box q \vee \Box\neg q \text{ for } x \in \{p, ss, s\}$$
$$N \not\models_x \Box q \vee \Box\neg q \text{ for } x \in \{a, c, s\}$$
$$N \models_x \triangle p \to \neg q \text{ for } x \in \{p, ss, s\}$$
$$N \models_x \triangle p \to \Diamond\neg q \text{ for } x \in \{a, c, g, p, ss, s\}$$

The first point illustrates that the stable semantics is in a special position since it requires boolean labellings. In particular, for all AFs that do not admit such a labeling, skeptical reasoning gives rise to deductive explosion – *all* arguments are skeptically acceptable. This behavior is captured and extended by the corresponding logic, which judges every formula to be true on such AFs. Next, let us turn the last two points in the list. They express variants of the intuition that in the scenario described by N, it is acceptable to argue that it is as easy to reject q as it is to leave p undetermined. For the preferred and semi-stable semantics this is true since the only labeling which leaves p undetermined involves rejecting q. For the remaining non-stable semantics, it could be that *both* p and q are undetermined, meaning that rejecting q is harder than leaving p undetermined. However, the weaker form expressed in the last formula is true for all semantics.

- Having formally defined logics based on argumentation semantics, we are ready to formally investigate the question of characterizing the validities of ϵ, the set of formulas ϕ such that $\models_\epsilon \phi$.

3.1   The Validities of Propositional Argumentation

Out of all the formalisms we consider in this paper, three-valued Łukasiewicz logic, Ł$_3$, has the longest history. It was introduced by the Polish logician Jan Łukasiewicz in the 1920s and is still studied both theoretically and from the point of view of applications. It is standardly defined for the language $\mathcal{L}$ and the semantics can be provided using three-valued functions $\rho : \Pi \to \{1, 0, \frac{1}{2}\}$, see e.g., [29]. These functions are extended to provide an interpretation for any $\phi \in \mathcal{L}$ in exactly the same way as detailed in (3.1), and the difference between Ł$_3$ and the argumentation logics arising from Definition 3.2 is that in Ł$_3$, a model is a single three-valued function ρ, not an AF which defines a *set* of such functions. Moreover, *any* three-valued function counts as a model, regardless of whether or not it is possible to induce it by an argumentation framework. Let Ł $= \{1, 0, \frac{1}{2}\}^\Pi$ denote all functions from Π to $\{1, 0, \frac{1}{2}\}$. Then we can give the following formal definition.

Definition 3.3 *The logic* Ł$_3$ *is defined as* $\models \subseteq 2^{\text{Ł}} \times \mathcal{L}$ *such that for all* $\Phi \in 2^{\text{Ł}}, \psi \in \mathcal{L}$

$$\Phi \models \psi \Leftrightarrow \forall \rho \in \text{Ł} : \big((\forall \phi \in \Phi : \overline{\rho}(\phi) = 1) \Rightarrow \overline{\rho}(\psi) = 1\big)$$

When $\Phi = \emptyset$ *we write* $\models \psi$ *and say that* ψ *is valid.*

The following deduction system is sound and complete for Ł$_3$, see e.g., [29]:

Axioms

Inference rule

1. $\phi \to (\psi \to \phi)$
2. $(\phi \to \psi) \to ((\phi \to \gamma) \to (\psi \to \gamma))$ – Modus ponens:
3. $(\neg\psi \to \neg\phi) \to (\phi \to \psi)$
4. $((\phi \to \neg\phi) \to \phi) \to \phi$

$$\frac{\phi \to \psi \quad \phi}{\psi} \; \text{(MP)}$$

Given some set Φ, we let $\Phi \vdash \phi$ denote that ϕ can be derived in this reasoning system from the premises in $\Phi \subseteq \mathcal{L}$. In case Φ is empty we write simply $\vdash \phi$ and say that ϕ

is a theorem of Ł. Soundness and completeness of the system can then be expressed as follows (see [29] for a proof of general completeness for Ł3).

$$\Phi \models \phi \Leftrightarrow \Phi \vdash \phi \tag{3.4}$$

Notice that the standard deduction theorem, $\phi \vdash \psi \Leftrightarrow \vdash \phi \rightarrow \psi$, fails for Ł3. However, the following restricted version is easy to verify.

$$\phi \vdash \psi \Leftrightarrow \vdash \phi \rightarrow (\phi \rightarrow \psi) \tag{3.5}$$

We now show that all non-stable semantics from Fig. 1 give rise to the same validities as Ł3. The most straightforward route to such a result would be to show, for each semantics, that every $\rho : \Pi \rightarrow \{1, 0, \frac{1}{2}\}$ is included in the set of labellings for *some* AF. This, however, does not hold. Consider, in particular, the assignment $\rho : \Pi \rightarrow \{1, 0, \frac{1}{2}\}$ defined by $\rho(p) = 0$ for all $p \in \Pi$. It is easy to see that it never obtains, for any of the semantics in Fig. 1. In particular, there can be no AF in which all arguments are rejected, since no argument would then be left to successfully attack them.

But for all non-stable semantics, it is not hard to show that there is an argumentation framework that induces it under this semantics. Then since all formulas from $\mathcal{L}$ contain only finitely many atoms, our result follows. For an arbitrary function $f : X \rightarrow Y$, let $f|_A = \{(x, y) \in f \mid x \in A\}$ denote its restriction to $A \subseteq X$. Then the sketch above can be formalized as follows.

Theorem 3.6 *For all semantics $\epsilon \in \{a, c, g, p, ss\}$ and all formulas $\phi \in \mathcal{L}$ we have*

$$\models \phi \Leftrightarrow \models_\epsilon \phi$$

Proof. ($\Rightarrow$) Follows trivially from Definition 3.2 since all labellings are three-valued assignments.

($\Leftarrow$) Let $\epsilon \in \{a, c, g, p, ss\}$ be arbitrary and assume $\models_\epsilon \phi$. By Definition 3.2 this means that for all AFs N and all $c \in \epsilon(N)$ we have $\bar{c}(\phi) = 1$. Let $\rho : \Pi \rightarrow \{1, 0, \frac{1}{2}\}$ be arbitrary. Then all we need to conclude the proof is to show that $\bar{\rho}(\phi) = 1$. Clearly, the value of $\bar{\rho}(\phi)$ only depends on $\rho|_{\Pi(\phi)}$ – the values assigned to arguments that appear in ϕ. Hence we are done if we can show that there is an AF N with some $c \in \epsilon(N)$ for which $c|_{\Pi(\phi)} = \rho|_{\Pi(\phi)}$, since then $\bar{\rho}(\phi) = \bar{c}(\phi) = 1$ will follow from $\models_\epsilon \phi$. To construct such an AF, we let $r \in \Pi \setminus \Pi(\phi)$ be some argument not appearing in ϕ. Then the following AF will prove the claim, for any $\epsilon \in \{a, c, g, p, ss\}$:

$$N = \{(r, x) \mid x \in \Pi(\phi) \text{ and } \rho(x) = 0\} \cup$$
$$\{(x, x) \mid x \in \Pi(\phi) \text{ and } \rho(x) = \frac{1}{2}\}$$

It is easy to verify that the only non-empty labeling in $\epsilon(N)$ is c, defined for all $x \in \Pi$ as follows:

$$c(x) = \begin{cases} \rho(x) \text{ if } x \in \Pi(\phi) \\ 1 \text{ otherwise} \end{cases}$$

Hence we obtain $c|_{\Pi(\phi)} = \rho|_{\Pi(\phi)}$ as desired and this concludes the proof.

We obtain the following as a simple corollary.

Corollary 3.7 *For all semantics $\epsilon \in \{a, c, g, p, ss\}$ and all formulas $\phi \in \mathcal{L}$, we have*

$$\models_\epsilon \phi \Leftrightarrow \vdash \phi$$

For the stable semantics, it follows already from the correspondence between kernels and truth-theories (and the fact that the latter provide a normal form for propositional theory) that the stable validities are exactly those of propositional logic. Hence if we use $\models_b$ to denote logical consequence in classical propositional logic, we can complete the picture as follows.

Theorem 3.8 *For the stable semantics and all formulas $\phi \in \mathcal{L}$, we have*

$$\models_\epsilon \phi \Leftrightarrow \models_b \phi$$

We think that the axiomatizations provided here are important observations regarding the theoretical foundations of argumentation, and we also believe they can be useful in practical applications and further developments of argumentation theory. If we allow users of this theory to make use of Ł3 in order to reason about AFs, it will permit them to make more subtle claims about their properties, allowing also the precise formal study of the acceptability of such meta-arguments. Indeed, we have identified a reasoning system for establishing validity of such arguments, allowing us to identify patterns of reasoning about AFs that can *always* be relied on. We remark that other reasoning systems have also been developed for Łukasiewicz logic, and these may be more efficient in practice than using Wajsberg's calculus, see [6].

Before we conclude, we consider the question of what happens when we interpret truth-theories using Łukasiewicz logic. Is the correspondence to AFs preserved? In [18] is was shown that complete labellings for AFs are three-valued models of the corresponding truth-theory and vice versa. This means that for the complete semantics we can use truth-theories to simulate the behavior of an AF, in place of the explicit encoding of the labeling as provided in the proof of Theorem 3.6. The advantage of doing this is that truth-theories corresponding to an AF can be computed quickly, in linear time by naive application of (2.1). Hence for the complete semantic it holds that the search for extensions in AFs is reducible in linear time to the problem of determining satisfiability of theories in Ł3.

If we switch to classical logic, this gives us a linear time *equivalence*, since we can decide satisfiability of arbitrary propositional theories by studying the kernel problem in associated digraphs. This no longer holds for Ł3, for any of the argumentations semantics from Fig. 1. To see this, note that *no* truth-theory is inconsistent in Ł3. In particular, the grounded labeling (which is also complete), witnesses to this.[15] Hence the behavior or truth-theories under Ł3 is fundamentally different from the behavior of such

---

[15] Also, the reader can easily verify that this assignment takes linear time to compute, by inductively inducing values from unattacked arguments and assigning $\frac{1}{2}$ to all remaining ones, as described, e.g., in [18].

theories under classical logic: truth in $Ł_3$ is a consistent notion, while in classical logic it is not.[16]

In the next section we consider modal reasoning about truth and argumentation, leading to the study of what propositions can rationally be believed on the basis of semantic information that it is possible to encode in a digraph.

4   Rational Belief on the Basis of Argument – A Modal Extension

The significance of our results so far is limited by the fact that we only cater to skeptical reasoning about AFs. A meta-argument is true if it holds for *all* acceptable labellings, and we lack the resources to express that a given argument can be credulously accepted (that there *exists* some acceptable labeling for which it is true).[17]

This is a shortcoming that we can address by modalizing our approach to skeptical reasoning, so that credulous reasoning arises as its dual. Notice that taking the truth-functional dual of Łukasiewicz logic, by letting $\frac{1}{2}$ count as a designated value, will not suffice.[18] Credulous acceptance of ϕ involves quantifying over all labellings under a given semantics, asking if ϕ evaluates to 1 in *one of them*. Hence, no truth-functional approach will give us what we want. However, if we think of labellings as possible worlds, we can capture credulous reasoning using a Kripkean approach.[19] In particular, we can associate to any AF a corresponding three-valued Kripke model, as follows:

Definition 4.1 *Given a semantics ϵ, an AF N and a set of states Q,*

- *An* evaluation frame *over Q is a function $V : Q \to \{1, 0, \frac{1}{2}\}^{\Pi}$, mapping states to labellings.*
- *For any evaluation frame V, the associated* Kripke model *is a tuple $\mathcal{M}(\epsilon, N, V) = (Q, V, R)$ where $R \subseteq Q \times Q$ such that for all $q, q' \in Q$:*

$$(q, q') \in R \Leftrightarrow V(q') \in \epsilon(N)$$

---

[16] We omit lengthy discussion of "revenge" issues, the worry that "stronger" paradoxes always tend to undermine attempts at regaining consistency in this way (for a collection on papers on revenge, see [4]). However, we mention that one strategy for countering revenge objections in the present context is to follow [25] who argued that the gap, the value $\frac{1}{2}$, should not to be seen as a semantic value at all, but merely as an expression of truth's partiality (so that, for instance, saying of a sentence in a gap that it is "not true" is akin to a category mistake, all the while truth as a concept does not apply to that sentence, i.e., it is like saying "the cheese is not true").

[17] In terms of truth, we are only able to address the truths that are *necessary* given the truth-theory; the mere *possible* truths, those that are contingent on the world beyond principle T, can not be talked about.

[18] Such a logic would bring us into paraconsistent territory, resulting in a system that stands to Łukasiewicz logic as Priest's LP stands to Kleene's three-valued logic [32].

[19] Importantly, we do not here ask for modal logics that encode the AF as such. This has been done already [24,9], resulting in logics where one talks directly about the structure of the digraph, rather than its meaning under a given semantics. What we want, rather, is to form three-valued meta-arguments that mix the credulous and skeptical modes of reasoning about AFs.

The definition builds models where all the states pointed to are required to correspond to acceptable labellings under an argumentation semantics. Intuitively, they are doxastic models such that the plausible states are taken to be those that cannot rationally be excluded on the basis of the argumentation semantics applied to the underlying AF. Indeed, notice that all relations R will automatically come out as both transitive and euclidian ($K45$ relations). Moreover, for non-stable ϵ we can use the axiom $\blacklozenge(p \rightarrow p)$ to restrict attention to serial relations, obtaining three-valued $KD45$ Kripke models.

In the present context, $\blacklozenge(p \rightarrow p)$ intuitively amounts to restricting attention to models where at least one state can be rationally entertained on the basis of the underlying AF. For the stable semantics this will lead us to discard some AFs, since some of them give rise to no rational beliefs under this semantics (only paradox). For non-stable ϵ, on the other hand, the fact that $\epsilon(N) \neq \emptyset$ ensures that $KD45$ models always exist.

To reason about three-valued Kripke models, we use a modal language with implication $\mathcal{L}_\blacklozenge$:

$$\phi ::= p \mid \neg\phi \mid \phi \rightarrow \phi \mid \blacklozenge\phi$$

where $p \in \Pi$. Truth can now be defined standardly, by first defining an appropriate three-valued evaluation of formulas at states. In particular, for all $M = \mathcal{M}(\epsilon, N, V)$ we define an associated three-valued labeling $\overline{M} : Q \times \mathcal{L}_\blacklozenge \rightarrow \{1, 0, \frac{1}{2}\}$ as follows:

$$\overline{M}(q, \phi) = \begin{cases} \epsilon(q)(\phi) \text{ if } \phi = p \in \Pi \\ 1 - \overline{M}(q, \psi) \text{ if } \phi = \neg\psi \\ min\{1, 1 - (\overline{M}(q, \psi_1) - \overline{M}(q, \psi_2))\} \text{ if } \phi = \psi_1 \rightarrow \psi_2 \\ \overline{M}(q, \phi) = max_{q' \in R(q)}\{\overline{M}(q', \psi)\} \text{ if } \phi = \blacklozenge\psi \end{cases} \qquad (4.2)$$

Definition 4.3 *For all $M = \mathcal{M}(\epsilon, N, V)$ and all $q \in Q$, if $\overline{M}(q, \phi) = 1$, we write $M, q \models \phi$ and say that ϕ is true at q on M. If $M, q \models \phi$ for all $q \in Q$ we write $M \models \phi$ and say that ϕ is true on M, while if $\mathcal{M}(\epsilon, N, V) \models \phi$ for all N, V, we write $\models_\epsilon \phi$ and say that ϕ is valid under ϵ.*

For an example, consider again the AF N from Fig. 3. Assume someone claims the following: "if your beliefs are based on N it should be at least as hard to believe that you must reject an argument as it is to disbelieve that you should accept it". Quite a mouthful, but also meaningful, as it expresses absence of a certain kind of normative conflict about what meta-arguments to accept. In terms of $\mathcal{L}_\blacklozenge$, we can represent the claim as follows: $\blacklozenge(p \rightarrow p) \rightarrow (\blacksquare\lozenge\neg p \rightarrow \blacklozenge\neg p)$, for all $p \in \Pi$. It is not hard to see that it holds for N. This follows, in particular, from the fact that all atoms that can be assigned $\frac{1}{2}$ by an admissible labeling can be assigned 0 by some other such labeling. However, a stronger principle, making the same claim about *formulas* rather than atoms, fails on N. In particular, we do not have $\blacklozenge(p \rightarrow p) \rightarrow (\blacksquare\lozenge\phi \rightarrow \blacklozenge\phi)$ on any serial model induced by N. This is witnessed by the formula $\phi = \neg p \vee \neg p'$, since any admissible assignment must assign $\frac{1}{2}$ to at least one of p, p', meaning that while $\lozenge\phi$ evaluates to 1 in every state corresponding to an admissible labeling, the formula $\neg\phi$ evaluates to $\frac{1}{2}$ in all such states (hence is believed to be harder to accept).

In fact, we can prove a general result about this kind of normative conflict in argumentation assessment.

Proposition 4.4 *For all ϵ, N, V, we have that if N is finite then $\mathcal{M}(\epsilon, N, V) \models \blacksquare\Diamond\phi \rightarrow$ $\blacklozenge\phi$ if, and only if, for all $q \in Q$, $R(q) \cap s(\epsilon) \neq \emptyset$*

Proof. To prove ($\Rightarrow$) we assume towards contradiction that $R(q)$ contains no q' such that $V(q')$ is boolean-valued (meaning, in particular, that no q' corresponds to a stable set in N). Since N is finite it follows that $\Pi(N)$ is finite as well. Hence the formula $\bigwedge_{p \in \Pi(N)}\{p \vee \neg p\}$ is in $\mathcal{L}_\blacklozenge$ and it evaluates to $\frac{1}{2}$ at all $q' \in R(q)$. It follows that $\blacksquare\Diamond \bigwedge_{p \in \Pi(N)}\{p \vee \neg p\}$ evaluates to 1 at q while $\blacklozenge \bigwedge_{p \in \Pi(N)}\{p \vee \neg p\}$ evaluates to $\frac{1}{2}$, contradicting the assumption that $\blacksquare\Diamond\phi \rightarrow \blacklozenge\phi$ is true at q for all ϕ. For ($\Leftarrow$), let $q' \in R(q)$ be such that $V(q')$ is a stable labeling for N. Notice first that $\blacksquare\Diamond\phi$ can not evaluate to $\frac{1}{2}$ at q since $\Diamond\phi$ is always boolean-valued. Moreover, the case when it evaluates to 0 is trivial. So assume it evaluates to 1. Then it follows that $\Diamond\phi$ evaluates to 1 at q'. Since $V(q')$ is boolean-valued, we conclude that ϕ evaluates to 1 as well, so q' witnesses to the fact that $\blacklozenge\phi$ evaluates to 1 at q.

This result is only an example of the potential for making interesting use of modal reasoning about argumentation semantics.[20] In future work we would like to explore characterizations such as these in more depth. However, the most obvious meta-logical question raised by Definition 4.1 is the question of finding a sound and complete reasoning system. This question can be approached by checking if every three-valued $KD45$ model admits a modally equivalent model that is induced by an AF. If this can be established, modulo some argumentation semantics, it follows that the validities of modal reasoning about AFs under this semantics coincide with those of regular three-valued $KD45$.

Preliminary work suggests that such a result holds for the preferred and semi-stable semantics for argumentation. It seems, in particular, that under preferred and semi-stable semantics we can induce any set of three-valued labellings with finite domain using an appropriately constructed AF. We plan to work out the implications of this for modal reasoning about AFs in a future paper, where we will also consider the matter of completeness and canonicity with respect to other argumentation semantics.

5 Conclusion

We started from the study of truth and went on to establish connections to argumentation and belief, through formal equivalences between models used to study these notions. The link to kernel theory and Łukasiewicz logic was emphasized, and we made use of the latter to provide axiomatizations of the skeptical validities arising from formal argumentation. We then proposed a modal extension, where credulous and skeptical forms of reasoning are captured as dual modalities. The semantics was provided by a special

---

[20] We remark that Proposition 4.4 does not hold for infinite AFs. Consider for instance the AF $N = \bigcup_{i \in \mathbb{N}}\{(p_i, p_i), (q_i, p_i), (r_i, q_i), (q_i, r_i), (z, q_i), (r_i, z)\}$. It is not hard to verify that for all finite subsets $P \subseteq \Pi$ there is an admissible labeling for N, c_P, that is boolean-valued on P. Let Q be the set of all finite subsets of Π and consider the Kripke model $M = \mathcal{M}(\epsilon, N, V)$ with V defined by $V(q) = \mathsf{c}_q$ for all $q \in Q$. It is easy to verify that $\blacksquare\Diamond\phi \rightarrow \blacklozenge\phi$ is true on M, even though N does not admit any stable labeling.

class of three-valued Kripke frames, those that can be induced from AFs using an argu-mentation semantics. We conjectured that the classes obtained under the preferred and semi-stable semantics are canonical for three-valued $KD45$ models.

In general, we think that the connections addressed in this paper can serve to motivate further work in all the fields we addressed. We think there is much to be gained from keeping formal links in mind, also if one feels that the underlying phenomena under consideration require different conceptual frameworks.[21] However, we think striking similarities at the formal level might also suggest deeper theoretical connections, and that this possibility should be explored further.

We are particularly keen on philosophical assessment of the formal link established between truth, argumentation and three-valued belief. It seems likely to us that it can inspire new philosophical ideas concerning the nature of these notions. To what extent are they mutually dependent? How are they related at a high level of abstraction? More concretely: Is it always possible for the truth to prevail in an argument? Should it be? Can false belief be distinguished from true belief on the basis of assessing arguments? Does this hold if "true" is replaced by "rational"? The formal connections mapped out in this paper naturally raise questions such as these, and we think they should be addressed. It seems, moreover, that we have identified a versatile formal framework for doing so.

References

1. Aczel, P.: Non-wellfounded sets. Technical Report 14, CSLI (1988)
2. Arieli, O., Caminada, M.W.A.: A QBF-based formalization of abstract argumentation se-mantics. Journal of Applied Logic 11(2), 229–252 (2013)
3. Barwise, J., Moss, L.: Vicious Circles: On the Mathematics of Non-Wellfounded Phenomena. CSLI, Stanford (1996)
4. Beall, J.C.: Revenge of the Liar: New Essays on the Paradox. Oxford University Press (2007)
5. Bezem, M., Grabmayer, C., Walicki, M.: Expressive power of digraph solvability. Ann. Pure Appl. Logic 163(3), 200–213 (2012)
6. Béziau, J.-Y.: A sequent calculus for Łukasiewicz's three-valued logic based on Suszko's bivalent semantics. Bulletin of the Section of Logic 28(2), 89–97 (1998)
7. Boros, E., Gurvich, V.: Perfect graphs, kernels and cooperative games. Discrete Mathemat-ics 306, 2336–2354 (2006)
8. Caminada, M.: On the issue of reinstatement in argumentation. In: Fisher, M., van der Hoek, W., Konev, B., Lisitsa, A. (eds.) JELIA 2006. LNCS (LNAI), vol. 4160, pp. 111–123. Springer, Heidelberg (2006)
9. Caminada, M.W.A., Gabbay, D.M.: A logical account of formal argumentation. Studia Log-ica 93(2-3), 109–145 (2009)

---

[21] It is worth noting, for instance, that some results from formal argumentation (such as the exis-tence of stable labellings for AFs without odd cycles [16]) are immediate corollaries of much older results from kernel theory ([37]). In the search for new results, it seems prudent to look across disciplinary boundaries to assess whether they have in fact already been established, or follow as trivial corollaries from existing theorems.

10. Caminada, M.W.A.: Semi-stable semantics. In: Proceedings of the 2006 Conference on Computational Models of Argument: Proceedings of COMMA 2006, pp. 121–130. IOS Press, Amsterdam (2006)
11. Cook, R.: Patterns of paradox. The Journal of Symbolic Logic 69(3), 767–774 (2004)
12. Dimopoulos, Y., Torres, A.: Graph theoretical structures in logic programs and default theories. Theoretical Computer Science 170(1-2), 209–244 (1996)
13. Doutre, S.: Autour de la sématique préférée des systèmes d'argumentation. PhD thesis, Université Paul Sabatier, Toulouse (2002)
14. Duchet, P.: Graphes noyau-parfaits, II. Annals of Discrete Mathematics 9, 93–101 (1980)
15. Duchet, P., Meyniel, H.: Une généralisation du théorème de Richardson sur l'existence de noyaux dans les graphes orientés. Discrete Mathematics 43(1), 21–27 (1983)
16. Dung, P.M.: On the acceptability of arguments and its fundamental role in nonmonotonic reasoning, logic programming and n-person games. Artificial Intelligence 77, 321–357 (1995)
17. Dyrkolbotn, S., Walicki, M.: Kernels in digraphs that are not kernel perfect. Discrete Mathematics 312(16), 2498–2505 (2012)
18. Dyrkolbotn, S., Walicki, M.: Propositional discourse logic. Synthese 191(5), 863–899 (2014)
19. Gabbay, D.: Modal provability foundations for argumentation networks. Studia Logica 93(2-3), 181–198 (2009)
20. Galeana-Sánchez, H., Guevara, M.-K.: Some sufficient conditions for the existence of kernels in infinite digraphs. Discrete Mathematics 309(11), 3680–3693 (2009); 7th International Colloquium on Graph Theory (ICGT) (2005)
21. Galeana-Sánchez, H., Neumann-Lara, V.: On kernels and semikernels of digraphs. Discrete Mathematics 48(1), 67–76 (1984)
22. Gottlob, G.: Complexity results for nonmonotonic logics. Journal of Logic and Computation 2(3), 397–425 (1992)
23. Grossi, D.: Argumentation in the view of modal logic. In: McBurney, P., Rahwan, I., Parsons, S. (eds.) ArgMAS 2010. LNCS (LNAI), vol. 6614, pp. 190–208. Springer, Heidelberg (2011)
24. Grossi, D.: On the logic of argumentation theory. In: van der Hoek, W., Kaminka, G.A., Lespérance, Y., Luck, M., Sen, S. (eds.) AAMAS, pp. 409–416. IFAAMAS (2010)
25. Kripke, S.: Outline of a theory of truth. The Journal of Philosophy 72(19), 690–716 (1975)
26. Łukasiewicz, J.: Selected works, edited by L. Borkowski. Studies in Logic and the Foundations of Mathematics. North Holland, Amsterdam (1970)
27. Mercier, H., Sperber, D.: Why do humans reason? Arguments for an argumentative theory. Behavioral and Brain Sciences 34, 57–74 (2011)
28. Milner, E.C., Woodrow, R.E.: On directed graphs with an independent covering set. Graphs and Combinatorics 5, 363–369 (1989)
29. Minari, P.: A note on Łukasiewicz's three-valued logic. Annali del Dipartimento di Filosofia dell'Universitá di Firenze 8(1) (2002)
30. Neumann-Lara, V.: Seminúcleos de una digráfica. Technical report, Anales del Instituto de Matemáticas II, Universidad Nacional Autónoma México (1971)
31. Pollock, J.L.: How to reason defeasibly. Artif. Intell. 57(1), 1–42 (1992)
32. Priest, G.: The logic of paradox. Journal of Philosophical Logic 8, 219–241 (1979)
33. Quine, W.V.: The ways of paradox and other essays. Random House, New York (1966)
34. Rabern, L., Rabern, B., Macauley, M.: Dangerous reference graphs and semantic paradoxes. Journal of Philosophical Logic 42(5), 727–765 (2013)
35. Rahwan, I., Simari, G.R. (eds.): Argumentation in artificial intelligence. Springer (2009)
36. Reiter, R.: A logic for default reasoning. Artif. Intell. 13(1-2), 81–132 (1980)
37. Richardson, M.: On weakly ordered systems. Bulletin of the American Mathematical Society 52, 113–116 (1946)
38. Richardson, M.: Solutions of irreflexive relations. The Annals of Mathematics, Second Series 58(3), 573–590 (1953)

39. Tarski, A.: The concept of truth in formalised languages. In: Corcoran, J. (ed.) Logic, Semantics, Metamathematics, papers from 1923 to 1938, Hackett Publishing Company (1983) (translation of the Polish original from 1933)
40. von Neumann, J., Morgenstern, O.: Theory of Games and Economic Behavior. Princeton University Press (1944, 1947)
41. Walicki, M., Dyrkolbotn, S.: Finding kernels or solving SAT. Journal of Discrete Algorithms 10, 146–164 (2012)
42. Wu, Y., Caminada, M.W.A., Gabbay, D.M.: Complete extensions in argumentation coincide with 3-valued stable models in logic programming. Studia Logica 93(2-3), 383–403 (2009)
43. Yablo, S.: Paradox without self-reference. Analysis 53(4), 251–252 (1993)

The Impact of Including Model Update Operators in Modal Logics

Raul Fervari

FaMAF, Universidad Nacional de Córdoba & CONICET, Argentina
fervari@famaf.unc.edu.ar

Abstract. In this paper we discuss ideas about *dynamic modal logics*. Modal logics are appropriate to describe properties of relational structures, and several operators have been already introduced to describe dynamic properties of such structures. However, we are interested in those operators which can modify models during the evaluation of a formula. First, we introduce different dynamic operators to clarify which of them are interesting for us. Then we focus on operators which modify the accessibility relation of relational models, and we show some expressivity results.

Keywords: modal logics, model updates, bisimulation, complexity.

1 What Kind of Dynamic Logics?

Modal logics [8,9] extend classical logics with operators that represent the modal character of some situation, for instance, necessity, possibility, knowledge, belief or permissions, just to name a few. In particular, the Basic Modal Logic ($\mathcal{ML}$) is an extension of propositional logic with a new operator which can describe the structural properties of a relational model. Formally:

Definition 1 (Syntax). *Let* PROP *be an infinite, countable set of propositional symbols. The set* FORM *of* $\mathcal{ML}$ *formulas over* PROP *is defined as:*

$$\text{FORM} ::= \bot \mid p \mid \neg\varphi \mid \varphi \wedge \psi \mid \Diamond\varphi,$$

where $p \in$ PROP *and* $\varphi, \psi \in$ FORM. *We use* $\Box\varphi$ *as a shorthand for* $\neg\Diamond\neg\varphi$, *while* $\top$ *and* $\varphi \vee \psi$ *are defined as usual.*

Definition 2 (Semantics). *A model* $\mathcal{M}$ *is a triple* $\mathcal{M} = \langle W, R, V \rangle$, *where* W *is a non-empty set;* $R \subseteq W \times W$ *is the accessibility relation; and* $V :$ PROP $\rightarrow$ $\mathcal{P}(W)$ *is a valuation. Let* w *be a state in* $\mathcal{M}$, *the pair* $(\mathcal{M}, w)$ *is called a pointed model; we will usually drop parentheses and write* $\mathcal{M}, w$. *Given a pointed model* $\mathcal{M}, w$ *and a formula* φ *we say that* $\mathcal{M}, w$ *satisfies* φ ($\mathcal{M}, w \models \varphi$) *when*

$$
\begin{aligned}
&\mathcal{M}, w \models \bot && never \\
&\mathcal{M}, w \models p && \textit{iff } w \in V(p) \\
&\mathcal{M}, w \models \neg\varphi && \textit{iff } \mathcal{M}, w \not\models \varphi \\
&\mathcal{M}, w \models \varphi \wedge \psi && \textit{iff } \mathcal{M}, w \models \varphi \textit{ and } \mathcal{M}, w \models \psi \\
&\mathcal{M}, w \models \Diamond\varphi && \textit{iff } \textit{for some } v \in W \textit{ s.t. } (w,v) \in R, \mathcal{M}, v \models \varphi.
\end{aligned}
$$

φ *is satisfiable if for some pointed model* $\mathcal{M}, w$ *we have* $\mathcal{M}, w \models \varphi$.

M. Colinet et al. (Eds.): ESSLLI 2012/2013, LNCS 8607, pp. 91–108, 2014.
© Springer-Verlag Berlin Heidelberg 2014

As shown in Definition 2, modal logics describe characteristics of relational structures. Given a pointed model, the $\Diamond$ operator moves the evaluation of the formula in its scope to some successor of the evaluation point. In this way, it is possible to describe the model by traversing its structure. But these are *static* characteristics of the structure, i.e. properties never change after the application of certain operations. If we want to describe *dynamic aspects* of a given situation, e.g. how the relations between a set of elements *evolve* through time or through the application of certain operations, the use of modal logics (or actually, any logic with classical semantics) becomes less clear. We can always resort to modeling the whole space of possible evolutions of the system as a graph, but this soon becomes unwieldy. It would be more elegant to use truly dynamic modal logics with operators that can mimic the changes that the structure will undergo.

We should take some care here, because some modal operators have been devised in the past to model dynamic phenomena, but not in the sense we just mentioned. One example is *Propositional Dynamic Logic* (**PDL**) [16,12,14]. This logic is a formal system for reasoning about programs. Originally, it was designed to formalize correctness specifications and prove that those specifications correspond to a particular program. **PDL** is a modal logic that contains an infinite number of modalities $\langle\pi\rangle$, where each π corresponds to a *program*. The interpretation of $\langle\pi\rangle\varphi$ is that *"some terminating execution of π from the current state leads to a state where the property φ holds"*. The structure of a program is defined inductively from a set of basic programs $\{a, b, c, \ldots\}$ as:

- **Choice:** if π and π' are programs, then $\pi \cup \pi'$ is a program which executes non-deterministically π or π'.
- **Composition:** if π and π' are programs, then $\pi; \pi'$ is a program which executes first π and then π'.
- **Iteration:** if π is a program, π^* is the program that executes a finite number (possibly zero) of times π.
- **Test:** if φ is a formula, then φ? is a program. It tests whether φ holds, and if so, continues; if not, it fails.

The expressive power of **PDL** is high (notice that it goes beyond first-order logic, as it can express the reflexive-transitive closure of a relation), and **PDL** can express some interesting properties. For example the formula

$$\langle(\varphi?; a)^*; (\neg\varphi)?\rangle\psi$$

represents that the program "**while** φ **do** a" ends in a state satisfying ψ (the program inside the modality executes a a finite, but not specified number of times after checking that φ holds, and after finishing the loop $\neg\varphi$ must holds. This captures exactly the behaviour of a while loop).

Clearly, the language gives us a practical way to deal with the notion of state and change, but this is a weak notion of dynamic behaviour. Formulas do not change the model, they only formalize program executions. We are more interested in operations that can change the model while we are evaluating a formula, i.e., *model update operators*. We will see in the next section, various concrete examples of this kind of logics.

1.1   Some Examples of Dynamic Modal Logics

A typical example when we think in logics that can change the model are *Dynamic Epistemic Logics* [23]. This is a family of logics that are used to reason about knowledge and belief, with operators that let us change such knowledge or belief by communicating some information. The Epistemic Logic $\mathcal{EL}$ is an extension of Propositional Logic with the knowledge operator K_a, where a is an agent name. K_a has the same semantics of $\Box$ but in a multiagent framework: edges of models are labeled by agent names, and each K_a is interpreted on the accessibility relation labeled by a. $K_a\varphi$ is interpreted as "the agent a knows that φ is the case". This logic only represents static information, but there are different extensions to model information exchange among the agents, which involves a dynamic behaviour.

Public Announcement Logic ($\mathcal{PAL}$) was introduced in [20] (first published in 1989), as an extension of $\mathcal{EL}$ with the operator $[!\varphi]$ which communicates some common information to the agents ($\langle!\psi\rangle\varphi$ is a shorthand for $\neg[!\psi]\neg\varphi$.) The formula $[!\psi]\varphi$ is read as "after ψ is (truthfully) announced, φ is the case". The formula ψ is revealed to all the agents (the announcement is public), then φ is evaluated. Announcements are represented by removing the access to states of the model where the announced fact does not hold. We introduce the formal semantics of $\mathcal{PAL}$:

$$\mathcal{M}, w \models [!\psi]\varphi \text{ iff } \mathcal{M}, w \models \psi \text{ implies } \mathcal{M}_{|\psi}, w \models \varphi,$$

where $\mathcal{M}_{|\psi} = \langle W', R', V'\rangle$ is defined as follows:

$$
\begin{aligned}
W' &= \{w \in W \mid \mathcal{M}, w \models \psi\} \\
R'_a &= R_a \cap (W' \times W') \\
V'(p) &= V(p) \cap W'.
\end{aligned}
$$

After making an announcement, the model is transformed to a new one and evaluation of the rest of the formula continues in the new model. Agents cannot access anymore information which contradicts the announcement: the knowledge of the agents has changed. Notice that the propositional information contained in states (the valuation) does not change. The only information affected is the knowledge that the agents have of this information.

Another family of model update logics is memory logics [4,19]. The semantics of these logics is specified on models that come equipped with a set of states called the *memory*. The simplest memory logic includes a modality $\widehat{r}$ that *stores* the current point of evaluation into memory, and a modality $\widehat{k}$ that verifies whether the current state of evaluation has been memorized. The memory can be seen as a special proposition symbol whose extension grows whenever the $\widehat{r}$ modality is used. In contrast with public announcements, the basic memory logic *expands* the model with an ever increasing set of memorized elements.

Definition 3 (Syntax of Memory Logics). *Given a set* PROP, *the set* FORM *of formulas of* $\mathcal{ML}(\textcircled{r}, \textcircled{k})$ *over* PROP *is defined as:*

$$\text{FORM} ::= \perp \mid p \mid \textcircled{k} \mid \neg\varphi \mid \varphi \wedge \psi \mid \Diamond\varphi \mid \textcircled{r}\varphi,$$

where $p \in$ PROP *and* $\varphi, \psi \in$ FORM.

Given a set PROP, *the set* FORM *of formulas of* $\mathcal{ML}(\langle\!\langle r \rangle\!\rangle, \textcircled{k})$ *over* PROP *is defined as:*

$$\text{FORM} ::= \perp \mid p \mid \textcircled{k} \mid \neg\varphi \mid \varphi \wedge \psi \mid \langle\!\langle r \rangle\!\rangle\varphi,$$

where $p \in$ PROP *and* $\varphi, \psi \in$ FORM.

We turn now to semantics. Models of memory logics are modal models, but with an extra set where we store the elements that we visited.

Definition 4 (Semantics of Memory Logics). *A model* $\mathcal{M} = \langle W, R, V, S \rangle$ *is an extension of an Kripke model with a memory* $S \subseteq W$. *Let* w *be a state in* $\mathcal{M}$, *we inductively define the notion of satisfiability of a formula as:*

$$\langle W, R, V, S \rangle, w \models \textcircled{k} \quad \textit{iff } w \in S$$
$$\langle W, R, V, S \rangle, w \models \textcircled{r}\varphi \quad \textit{iff } \langle W, R, V, S \cup \{w\} \rangle, w \models \varphi$$
$$\langle W, R, V, S \rangle, w \models \langle\!\langle r \rangle\!\rangle\varphi \textit{ iff } \langle W, R, V, S \rangle, w \models \textcircled{r}\Diamond\varphi.$$

A formula φ *of* $\mathcal{ML}(\textcircled{r}, \textcircled{k})$ *or* $\mathcal{ML}(\langle\!\langle r \rangle\!\rangle, \textcircled{k})$ *is* satisfiable *if there exists a model* $\langle W, R, V, \emptyset \rangle$ *such that* $\langle W, R, V, \emptyset \rangle, w \models \varphi$.

In the definition of satisfaction, the empty initial memory ensures that no point of the model satisfies the unary predicate $\textcircled{k}$ unless a formula $\textcircled{r}\varphi$ or $\langle\!\langle r \rangle\!\rangle\varphi$ has previously been evaluated there. The memory logic $\mathcal{ML}(\langle\!\langle r \rangle\!\rangle, \textcircled{k})$ does not have the $\Diamond$ operator, and its expressive power is strictly weaker than $\mathcal{ML}(\textcircled{r}, \textcircled{k})$ [19,5]. However, in both cases we have a logic that is strictly more expressive than the basic modal logic $\mathcal{ML}$. We show this result with a simple example.

Example 5. Given a pointed model $\langle W, R, V, \emptyset \rangle, w$, the $\mathcal{ML}(\langle\!\langle r \rangle\!\rangle, \textcircled{k})$-formula $\langle\!\langle r \rangle\!\rangle\textcircled{k}$ is satisfiable only if w is reflexive. The $\langle\!\langle r \rangle\!\rangle$ operator remembers the current element but at the same time looks for a successor. In this case, such successor has to be in the memory, but w is the only one belonging to the memory (remember that we started with the empty memory). Then, the formula is satisfiable if only if w is his own successor. The same effect can be captured with the $\mathcal{ML}(\textcircled{r}, \textcircled{k})$-formula $\textcircled{r}\Diamond\textcircled{k}$.

Memory logics will not only result interesting as an example of model update operator, but the logic $\mathcal{ML}(\textcircled{r}, \textcircled{k})$ will be useful to prove the undecidability of some other logics. The idea is taking advantage of the model update operators to simulate the capability of memorizing elements. Then we encode the satisfiability problem of some dynamic logics into the undecidable satisfiability problem of $\mathcal{ML}(\textcircled{r}, \textcircled{k})$.

Notice that all the operators introduced in this section have something in common: they all can be defined in terms of an *update function* on the models. For instance, public announcements can be represented by an update function which takes a model and some announcement, and removes all the states which do not hold such announcement. The semantics of Ⓡ can be seen as a function which adds elements to the memory. We are interested in this kind of operators, that let us transform a model during the evaluation of a formula. We introduced several examples, all of them thought in a determined context. This is the main difference with the work in this article. We are not interested in the application of dynamic operators to model a particular problem, we want to explore the impact of including dynamic operators (in particular, model update operators) in modal logics. When we use this kind of operators with a particular purpose, we can ignore some details about the behaviour of the resulting logics. By investigating dynamic operators from a theoretical point of view, we can study in detail the intrinsic properties of these operators.

As we mentioned, it is possible to modify a relational model in different ways. For instance, it is possible to remove elements of the domain ($\mathcal{PAL}$), change the valuation of the model ($\mathcal{ML}($Ⓡ$, $Ⓚ$)$ and $\mathcal{ML}(\langle\!\langle r \rangle\!\rangle, $Ⓚ$))$) and change the *accessibility relation*. We are particularly interested in this last family of operators, that we called *Relation-Changing Operators*. This is not a new idea: van Benthem introduced the *Sabotage Operator* which deletes arbitrarily edges in the model [22], as an example of a relation-changing operator used to model changes in the scenario of a two-player game. In the epistemic logics field, *Arrow Updates* [15] were introduced to encode dynamic epistemic logics. In [7] some relation-changing operators have been introduced as data structure modifiers. In the next section, we will introduce some other examples of relation-changing operators that will be discussed in the rest of this article.

2   Relation-Changing Operators

We will introduce some relation-changing operators that have been previously investigated in [1,2,3,11]. We will compare the results obtained by adding dynamic operators to modal logics (most of them, included in the publications we mentioned). We will consider relation-changing operators, and the examples we introduced in the previous section.

In this article, we only discuss the single addition to $\mathcal{ML}$ of the local version of some relation-changing operators, i.e., operators that perform modifications from the evaluation point. We will introduce $\langle sw \rangle$, an operator that swaps around edges; $\langle sb \rangle$, a local version of van Benthem's sabotage operator; and $\langle br \rangle$, which adds new edges from the evaluation point to an unaccessible point. Let us formally define the syntax of these relation-changing modal logics.

Definition 6 (Syntax). *Let* PROP *be a countable, infinite set of propositional symbols. Then the set* FORM *of formulas over* PROP *is defined as:*

$$\text{FORM} ::= \bot \mid p \mid \neg\varphi \mid \varphi \wedge \psi \mid \blacklozenge\varphi,$$

where $p \in$ PROP, $\blacklozenge \in \{\Diamond, \langle sw \rangle, \langle sb \rangle, \langle br \rangle\}$ and $\varphi, \psi \in$ FORM. *Other operators are defined as usual. In particular,* $\blacksquare\varphi$ *is defined as* $\neg\blacklozenge\neg\varphi$.

We call $\mathcal{ML}(\blacklozenge)$ the extension of $\mathcal{ML}$ allowing also the $\blacklozenge$ operator, for $\blacklozenge \in \{\langle sw \rangle, \langle sb \rangle, \langle br \rangle\}$.

Formulas of $\mathcal{ML}(\langle sb \rangle)$, $\mathcal{ML}(\langle sw \rangle)$ and $\mathcal{ML}(\langle br \rangle)$ are evaluated in standard relational models, and the meaning of the operators of the basic modal logic is unchanged. When we evaluate formulas containing relation-changing operators, we will need to keep track of the edges that have been modified. To that end, let us define precisely the models that we will use. In the rest of this thesis we will use wv as a shorthand for $\{(w, v)\}$ or (w, v). Context will always disambiguate the intended use.

Definition 7 (Models and Model Variants). *A model $\mathcal{M}$ is a triple $\mathcal{M} = \langle W, R, V \rangle$, where W is a non-empty set whose elements are called points or states; $R \subseteq W \times W$ is the accessibility relation; and $V :$ PROP $\to \mathcal{P}(W)$ is a valuation.*

Given a model $\mathcal{M} = \langle W, R, V \rangle$, we define the following notations for model variants:

(sabotaging) $\mathcal{M}_S^- = \langle W, R_S^-, V \rangle$, with $R_S^- = R \backslash S$, $S \subseteq R$.
(swapping) $\mathcal{M}_S^* = \langle W, R_S^*, V \rangle$, with $R_S^* = (R \backslash S^{-1}) \cup S$, $S \subseteq R^{-1}$.
(bridging) $\mathcal{M}_B^+ = \langle W, R_B^+, V \rangle$, with $R_B^+ = R \cup B$, $B \subseteq (W \times W) \backslash R$.

Let w be a state in $\mathcal{M}$, the pair $(\mathcal{M}, w)$ is called a pointed model; we will usually drop parenthesis and call $\mathcal{M}, w$ a pointed model.

Let us introduce the formal semantics of the new operators.

Definition 8 (Semantics). *Given a pointed model $\mathcal{M}, w$ and a formula φ we say that $\mathcal{M}, w$ satisfies φ, and write $\mathcal{M}, w \models \varphi$, when*

$$
\begin{aligned}
&\mathcal{M}, w \models \bot && never \\
&\mathcal{M}, w \models p && \textit{iff } w \in V(p) \\
&\mathcal{M}, w \models \neg\varphi && \textit{iff } \mathcal{M}, w \not\models \varphi \\
&\mathcal{M}, w \models \varphi \wedge \psi && \textit{iff } \mathcal{M}, w \models \varphi \text{ and } \mathcal{M}, w \models \psi \\
&\mathcal{M}, w \models \Diamond\varphi && \textit{iff } \textit{for some } v \in W \textit{ s.t. } (w, v) \in R, \mathcal{M}, v \models \varphi \\
&\mathcal{M}, w \models \langle sb \rangle\varphi && \textit{iff } \textit{for some } v \in W \textit{ s.t. } (w, v) \in R, \mathcal{M}_{wv}^-, v \models \varphi \\
&\mathcal{M}, w \models \langle sw \rangle\varphi && \textit{iff } \textit{for some } v \in W \textit{ s.t. } (w, v) \in R, \mathcal{M}_{vw}^*, v \models \varphi \\
&\mathcal{M}, w \models \langle br \rangle\varphi && \textit{iff } \textit{for some } v \in W \textit{ s.t. } (w, v) \notin R, \mathcal{M}_{wv}^+, v \models \varphi.
\end{aligned}
$$

φ is satisfiable if for some pointed model $\mathcal{M}, w$ we have $\mathcal{M}, w \models \varphi$.

We will discuss the impact of considering these operators and some of those introduced in Section 1.

3   Bisimulations

Bisimulations are an important tool to investigate the expressive power of the languages. In most modal logics, bisimulations are binary relations linking elements of the domains that have the same atomic information, and preserving the relational structure of the model [8]. This is the case for $\mathcal{ML}$:

Definition 9 ($\mathcal{ML}$-Bisimulations). *Let $\mathcal{M} = \langle W, R, V \rangle$, $\mathcal{M}' = \langle W', R', V' \rangle$ be two models. A non empty relation $Z \subseteq W \times W'$ is an $\mathcal{ML}$-bisimulation if it satisfies the following conditions. If wZw' then*

(atomic harmony) *for all $p \in$ PROP, $w \in V(p)$ iff $w' \in V'(p)$;*
(zig) *if $(w, v) \in R$ then for some v', $(w', v') \in R'$ and vZv';*
(zag) *if $(w', v') \in R'$ then for some v, $(w, v) \in R$ and vZv'.*

Given two pointed models $\mathcal{M}, w$ and $\mathcal{M}', w'$ we say that they are $\mathcal{ML}$-bisimilar and we write $\mathcal{M}, w \leftrightarrow_{\mathcal{ML}} \mathcal{M}', w'$ if there is an $\mathcal{ML}$-bisimulation Z such that wZw'.

In general, when we want to express differences between two models in a particular language $\mathcal{L}$, we do it by defining an $\mathcal{L}$-formula which is satisfiable in one of them and is not satisfiable in the other. On the other hand, if we want to show that some language cannot distinguish between two models, we need specific tools to capture the expressivity of the language. As we mentioned, $\mathcal{ML}$-bisimulations relate elements in models that have the same atomic and structural information. This is exactly what we can characterize using $\mathcal{ML}$, then it looks like bisimulations are the appropriate tool to compare models. Thanks to the next theorem, we can say that if there is a bisimulation between two pointed models then they satisfy the same formulas.

Theorem 10 (Invariance for Bisimulations). *Let $\mathcal{M} = \langle W, R, V \rangle$, $\mathcal{M}' = \langle W', R', V' \rangle$ be two models, $w \in W$ and $w' \in W'$. If there is an $\mathcal{ML}$-bisimulation Z between $\mathcal{M}, w$ and $\mathcal{M}', w'$ such that wZw' then for any formula $\varphi \in \mathcal{ML}$, $\mathcal{M}, w \models \varphi$ iff $\mathcal{M}', w' \models \varphi$.*

In a few words, the existence of an $\mathcal{ML}$-bisimulation between two models indicates that they are modally equivalent. Let us see an example.

Example 11. These two models that cannot be distinguished by any formula of $\mathcal{ML}$. Dotted lines represent the bisimulation Z.

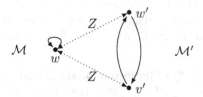

For the Public Announcement Logic $\mathcal{PAL}$ introduced in Section 1 this is also the case: if two models are bisimilar according to Definition 9 then they satisfy the same $\mathcal{PAL}$-formulas. $\mathcal{PAL}$ has the same expressive power as $\mathcal{ML}$ [23]. The translation is not straightforward (the resulting $\mathcal{ML}$-formula can be exponentially larger than the original $\mathcal{PAL}$-formula) but it can be done via reduction axioms. However, this translation is not possible for all the dynamic logics we are discussing in this article. When we increase the expressivity of the language, the definition of bisimulation we introduced before is useless. In such cases, as we pointed in previous works, we need to include additional conditions to capture the expressivity of the logic.

Definition 12 ($\mathcal{ML}(\blacklozenge)$-Bisimulations). *Let* $\mathcal{M}=\langle W, R, V\rangle$, $\mathcal{M}'=\langle W', R', V'\rangle$ *be two models. A non empty relation* $Z \subseteq (W \times \mathcal{P}(W^2)) \times (W' \times \mathcal{P}(W'^2))$ *is a* $\mathcal{ML}(\blacklozenge)$-*bisimulation if it satisfies the conditions* atomic harmony, zig *and* zag *below, and the corresponding conditions for the operators that the considered logic contains. If* $(w, S)Z(w', S')$ *then*

(atomic harmony) *for all* $p \in \mathsf{PROP}$, $w \in V(p)$ *iff* $w' \in V'(p)$;
(zig) *if* $(w, v) \in S$ *then for some* v', $(w', v') \in S'$ *and* $(v, S)Z(v', S')$;
(zag) *if* $(w', v') \in S'$ *then for some* v, $(w, v) \in S$ *and* $(v, S)Z(v', S')$;
($\langle\mathsf{sb}\rangle$-zig) *if* $(w, v) \in S$ *then for some* v', $(w', v') \in S'$ *and* $(v, S_{vw}^-)Z(v'S_{v'w'}^{'-})$;
($\langle\mathsf{sb}\rangle$-zag) *if* $(w', v') \in S'$ *then for some* v, $(w, v) \in S$ *and* $(v, S_{wv}^-)Z(v'S_{w'v'}^{'-})$;
($\langle\mathsf{sw}\rangle$-zig) *if* $(w, v) \in S$ *then for some* v', $(w', v') \in S'$ *and* $(v, S_{vw}^*)Z(v'S_{v'w'}^{'*})$;
($\langle\mathsf{sw}\rangle$-zag) *if* $(w', v') \in S'$ *then for some* v, $(w, v) \in S$ *and* $(v, S_{vw}^*)Z(v'S_{v'w'}^{'*})$;
($\langle\mathsf{br}\rangle$-zig) *if* $(w, v) \notin S$, *there is* $v' \in W'$ *s.t.* $(w', v') \notin S'$ *and* $(v, S_{wv}^+)Z(v', S_{w'v'}^{'+})$;
($\langle\mathsf{br}\rangle$-zag) *if* $(w', v') \notin S'$, *there is* $v \in W$ *s.t.* $(w, v) \notin S$ *and* $(v, S_{wv}^+)Z(v', S_{w'v'}^{'+})$.

Given two pointed models $\mathcal{M}, w$ *and* $\mathcal{M}', w'$ *we say that they are* $\mathcal{ML}(\blacklozenge)$-*bisimilar* $(\mathcal{M}, w \leftrightarrow_{\mathcal{ML}(\blacklozenge)} \mathcal{M}', w')$ *if there is a* $\mathcal{ML}(\blacklozenge)$-*bisimulation* Z *such that* $(w, R)Z(w, R')$ *where* R *and* R' *are respectively the relations of* $\mathcal{M}$ *and* $\mathcal{M}'$.

Notice that bisimulations for relation-changing modal logics relate current states and current accessibility relations of the models. Depending on which operator we are considering, different zig/zag conditions are added. Zig and zag for $\mathcal{ML}$-bisimulations are the correspondent conditions to capture $\Diamond$: they talk about the successors of the current state. Conditions for relation-changing modal logics are the same, but also keeping track of the modifications already done, and changing the relation according to the semantics of the operators. For instance, $\langle\mathsf{sb}\rangle$-zig/zag establish that there are successors of the current states that are related, and delete the edges that connect them. For $\langle\mathsf{sw}\rangle$ is the same but swapping edges instead deleting. Conditions for $\langle\mathsf{br}\rangle$ require that there exist unreachable points from the current states, and put edges to them in the accessibility relation.

As we have showed for $\mathcal{ML}$, bisimulations are important to distinguish when two models are equal for those languages. The next theorem establishes that two bisimilar models are not distinguishable for any formula of the corresponding language.

Theorem 13 (Invariance for Bisimulations). *Let* $\mathcal{M} = \langle W, R, V\rangle$, $\mathcal{M}' = \langle W', R', V'\rangle$ *be two models,* $w \in W$, $w' \in W'$, *and let* $S \subseteq W^2, S' \subseteq W'^2$. *If there is a* $\mathcal{ML}(\blacklozenge)$-*bisimulation* Z *between* $\mathcal{M}, w$ *and* $\mathcal{M}', w'$ *such that* $(w, S)Z(w', S')$ *then for any formula* $\varphi \in \mathcal{ML}(\blacklozenge)$, $\langle W, S, V\rangle, w \models \varphi$ *iff* $\langle W', S', V'\rangle, w' \models \varphi$.

Proof. We will see the case for $\mathcal{ML}(\langle\mathsf{sw}\rangle)$. The proof is by structural induction on $\mathcal{ML}(\langle\mathsf{sw}\rangle)$-formulas. The base case holds by (atomic harmony), and the $\wedge$ and $\neg$ cases are trivial.

$\varphi = \Diamond\psi$: Suppose $\langle W, S, V\rangle, w \models \Diamond\psi$. Then there is v in W s.t. $(w, v) \in S$ and $\langle W, S, V\rangle, v \models \psi$. By (zig) we have v' in W' such that $w'S'v'$ and $(v, S)Z(v', S')$. By I.H., $\langle W', S', V'\rangle', v' \models \psi$ and by definition $\langle W', S', V'\rangle, w' \models \Diamond\psi$. For the other direction use (zag).

$\varphi = \langle \mathsf{sw} \rangle \psi$: For the left to the right direction suppose $\langle W, S, V \rangle, w \models \langle \mathsf{sw} \rangle \psi$. Then there is v in W s.t. $(w, v) \in S$ and $\langle W, S^*_{vw}, V \rangle, v \models \psi$. By ($\langle \mathsf{sw} \rangle$-zig) we have v' in W' s.t. $(w', v') \in S'$ and $(v, S^*_{vw}) Z (v', S'^*_{v'w'})$. By I.H., $\langle W', S'^*_{v'w'}, V' \rangle, v' \models \psi$ and by definition $\langle W', S', V' \rangle, w' \models \langle \mathsf{sw} \rangle \psi$. For the other direction use ($\langle \mathsf{sw} \rangle$-zag).

$\square$

Example 14. The two models below are $\mathcal{ML}(\langle \mathsf{sw} \rangle)$-bisimilar. The simplest way to check this is to recast the notion of $\mathcal{ML}(\langle \mathsf{sw} \rangle)$-bisimulation as an Ehrenfeucht-Fraïssé game as the one used for $\mathcal{ML}$, but where Spoiler can also swap arrows when moving from a node to an adjacent node. It is clear that Duplicator has a winning strategy.

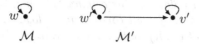

Example 15. There is no $\mathcal{ML}(\langle \mathsf{sw} \rangle)$-bisimulation between the models below. Indeed the formula $\langle \mathsf{sw} \rangle \Diamond \Box \bot$ is satisfied in $\mathcal{M}', w'$ and not in $\mathcal{M}, w$. Notice that the models are $\mathcal{ML}$-bisimilar.

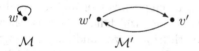

As we have seen, the first difference between $\mathcal{PAL}$ and relation-changing modal logics is the definition of bisimulation. For $\mathcal{PAL}$ it suffices with the conditions defined for $\mathcal{ML}$, but for the relation-changing modal logics we are discussing in this article we need to define new conditions which capture the new behaviour. As we showed in [1,3,11], it is natural given that these relation-changing operators increase the expressive power of $\mathcal{ML}$.

We can use bisimulations to compare the logics among them, and also with others. Definition 16 formalizes how we compare the expressive power of two logics.

Definition 16 ($\mathcal{L} \leq \mathcal{L}'$). *We say that $\mathcal{L}'$ is at least as expressive as $\mathcal{L}$ (notation $\mathcal{L} \leq \mathcal{L}'$) if there is a function* Tr *between formulas of $\mathcal{L}$ and $\mathcal{L}'$ such that for every model $\mathcal{M}$ and every formula φ of $\mathcal{L}$ we have that*

$$\mathcal{M} \models_{\mathcal{L}} \varphi \text{ iff } \mathcal{M} \models_{\mathcal{L}'} \mathsf{Tr}(\varphi).$$

$\mathcal{M}$ *is seen as a model of $\mathcal{L}$ on the left and as a model of $\mathcal{L}'$ on the right, and we use in each case the appropriate semantic relation $\models_{\mathcal{L}}$ or $\models_{\mathcal{L}'}$ as required.*

We say that $\mathcal{L}$ and $\mathcal{L}'$ are incomparable *(notation $\mathcal{L} \neq \mathcal{L}'$) if $\mathcal{L} \not\leq \mathcal{L}'$ and $\mathcal{L}' \not\leq \mathcal{L}$.*

We say that $\mathcal{L}'$ is strictly more expressive *than $\mathcal{L}$ (notation $\mathcal{L} < \mathcal{L}'$) if $\mathcal{L} \leq \mathcal{L}'$ but not $\mathcal{L}' \leq \mathcal{L}$.*

The $\leq$ relation indicates that we can embed one language into another. To do this, we need an equivalence preserving translation from the first language to the second one. Its strict version is $<$, that indicates that the second language can express strictly more than the first one. Incomparability relation says than any of the two languages cannot be embedded in the other, i.e., they are able to say different things. These definitions will be used next, when we compare the expressive power of relation-changing modal logics of Definition 6.

The comparisons have been already investigated in [1,11], establishing that relation-changing modal logics are all incomparable among them. Some cases are easy to check, but there are others in which we need more complex structures to distinguish two languages.

Lemma 17. *For every pair of pointed models $\mathcal{M}, w$ and $\mathcal{M}', w'$ in Figure 1, and for all corresponding formulas φ of the column "Distinct by", we have $\mathcal{M}, w \not\models \varphi$ and $\mathcal{M}', w' \models \varphi$. Moreover, for all corresponding logics $\mathcal{L}$ of the column "Bisimilar for", we have that (w, R) and (w', R') are in an $\mathcal{L}$-bisimulation, where R and R' are the accessibility relations of $\mathcal{M}$ and $\mathcal{M}'$ respectively.*

Proof. We will check the conditions to show that the two models in first row are bisimilar for $\mathcal{ML}(\langle sb \rangle)$ and for $\mathcal{ML}(\langle sw \rangle)$. Clearly all the states agree propositionally (their valuations are empty). For zig and zag conditions, we need to check if both have bisimilar successors, which holds because there are not successors at all. The same happens with $\langle sb \rangle$-zig/zag and $\langle sw \rangle$-zig/zag: the lack of successors makes the conditions true. Now we can prove that $\mathcal{ML}(\langle br \rangle) \not\leq \mathcal{ML}(\langle sb \rangle)$ and $\mathcal{ML}(\langle br \rangle) \not\leq \mathcal{ML}(\langle sw \rangle)$. We have to check now that there is a $\mathcal{ML}(\langle br \rangle)$-formula that distinguishes the two models. The $\mathcal{ML}(\langle br \rangle)$-formula $\langle br \rangle \langle br \rangle \top$ holds at $\mathcal{M}', w'$ but not at $\mathcal{M}, w$. Checking $\langle br \rangle$-zag, it fails starting from w' and finding two states to reach with a new edge, while starting from w we can just reach one.

The models in the second row are $\mathcal{ML}(\langle br \rangle)$-bisimilar because no new edges can be added, and we checked that they are also $\mathcal{ML}(\langle sw \rangle)$-bisimilar in Example 15. In the third row, the given models are bisimilar for $\mathcal{ML}(\langle sb \rangle)$ because they are bisimilar for $\mathcal{ML}$ and they are acyclic. In the fourth row, both models are $\mathcal{ML}(\langle br \rangle)$-bisimilar since they are infinite, hence one can add as many links as needed to points that are modally bisimilar. $\square$

Corollary 18. *For all $\blacklozenge_1, \blacklozenge_2 \in \{\langle sb \rangle, \langle sw \rangle, \langle br \rangle\}$ such that $\blacklozenge_1 \neq \blacklozenge_2$, we have $\mathcal{ML}(\blacklozenge_1) \neq \mathcal{ML}(\blacklozenge_2)$.*

We have proved in [1,11] that adding some relation-changing operators to the basic modal logic we increase its expressive power, and according to the results we just showed, each logic allows to express different things. We have seen in this section that standard tools in modal logics such as bisimulations can be adapted for logics with relation-changing operators.

$\mathcal{M}$	$\mathcal{M}'$	Distinct by	Bisimilar for
		$\langle\mathsf{br}\rangle\langle\mathsf{br}\rangle\top$	$\mathcal{ML}(\langle\mathsf{sb}\rangle)$ $\mathcal{ML}(\langle\mathsf{sw}\rangle)$
		$\langle\mathsf{sb}\rangle\Diamond\top$	$\mathcal{ML}(\langle\mathsf{sw}\rangle)$ $\mathcal{ML}(\langle\mathsf{br}\rangle)$
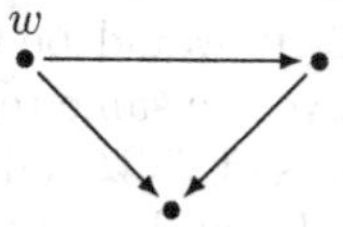	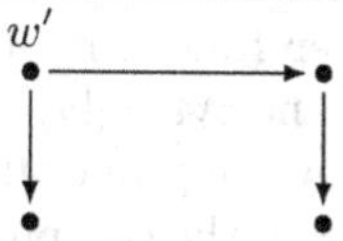	$\langle\mathsf{sw}\rangle\Diamond\Diamond\Diamond\Box\bot$	$\mathcal{ML}(\langle\mathsf{sb}\rangle)$
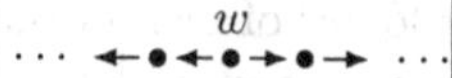		$\langle\mathsf{sw}\rangle\Diamond\Box\bot$	$\mathcal{ML}(\langle\mathsf{br}\rangle)$

Fig. 1. Bisimilar models and distinguishing formulas

4   Computational Behaviour

When we need to choose a logic to model a particular problem, First-Order Logic $\mathcal{FOL}$ [10] comes immediately to our mind. $\mathcal{FOL}$ is a nice language, very powerful and well-known for everyone who studied mathematics and/or computer science, but it has some undesirable properties. For instance, its satisfiability problem is undecidable, and model checking is PSPACE-complete. However, $\mathcal{FOL}$ is still used because it is appropriate to model many different problems. On the other hand, there are weaker languages with a better computational behaviour that we can use, such as modal logics. For any problem that requires describing structural properties of a graph, modal logics can be a good choice. The satisfiability problem for $\mathcal{ML}$ is PSPACE-complete, and its model checking problem is in P.

These two languages have very different properties (more expressive power in $\mathcal{FOL}$, better computational properties in $\mathcal{ML}$), and each of them is still appropriate in determined situations. Let us see what happens when we add model update operators to modal logics. We will analyze if by adding the kind of operators that are appropriate to model dynamic situations, we preserve the good properties of $\mathcal{ML}$, or the increasing of the expressivity leads them closer to $\mathcal{FOL}$.

Let us start by discussing the case that, so far, resulted easier to be analyzed: the public announcement logic $\mathcal{PAL}$. We mentioned that this logic has the same expressive power than $\mathcal{ML}$ but there are certain properties that can be expressed exponentially more succinct in $\mathcal{PAL}$ than in $\mathcal{ML}$. Despite this succinctness, the computational complexity of these two logics coincides [18,13]. In this case, adding dynamic behaviour we keep the properties.

On the other hand, memory logics have a more complex behaviour. We know that $\mathcal{ML}$ is a proper fragment of $\mathcal{FOL}$ [9] with good computational properties. Memory logics are also a proper fragment of $\mathcal{FOL}$ [19], but unfortunately, the good properties of modal logics are not preserved. Adding to the language the capability of remember visited elements we move closer to $\mathcal{FOL}$ than to $\mathcal{ML}$. The satisfiability problem for $\mathcal{ML}(\langle\!\langle r \rangle\!\rangle, \textcircled{k})$ is decidable but the one of $\mathcal{ML}(\textcircled{r}, \textcircled{k})$ is not, and its model checking problem is PSPACE-complete [4,6,19].

For relation-changing operators, we have similar results. In [3] we provided a translation from $\mathcal{ML}(\langle sw \rangle)$ to two sorted $\mathcal{FOL}$ by unraveling all the possible model transformations that can be done using $\langle sw \rangle$. Sorts are convenient for such translation, but it is possible to translate it to unsorted $\mathcal{FOL}$, then we can conclude that $\mathcal{ML}(\langle sw \rangle)$ is a proper fragment of $\mathcal{FOL}$. It would be easy apply a similar argument for $\mathcal{ML}(\langle sb \rangle)$ and $\mathcal{ML}(\langle br \rangle)$ to prove the same result. Such as for $\mathcal{ML}(\textcircled{r}, \textcircled{k})$, even though they are proper fragments of $\mathcal{FOL}$, adding relation-changing modal operators increases the computational complexity of the logics. We have proved that the model checking task for logics of Definition 6 is PSPACE-complete [1,11] (such as for $\mathcal{FOL}$). Also, in [3] we proved in detail that the satisfiability problem for $\mathcal{ML}(\langle sw \rangle)$ is undecidable, and in [17,21] the same was showed for a global version of the sabotage operator. This results give us an idea about the computational behaviour of this kind of operators. In the next section we will use similar arguments as the used for $\mathcal{ML}(\langle sw \rangle)$ to prove that the satisfiability problem for $\mathcal{ML}(\langle sb \rangle)$ is undecidable.

4.1 Undecidability of $\mathcal{ML}(\langle sb \rangle)$

We will prove that the satisfiability problem for $\mathcal{ML}(\langle sb \rangle)$ is undecidable. This result has been proved together with Mauricio Martel[1] and appears in [11]. First, we provide a translation from formulas of this logic to formulas of the memory logic $\mathcal{ML}(\textcircled{r}, \textcircled{k})$. In order to simulate the behaviour of the operators $\textcircled{r}$ and $\textcircled{k}$ without having a memory in the model, we impose constraints on the models where we evaluate the translated formula. Then we prove that a $\mathcal{ML}(\textcircled{r}, \textcircled{k})$-formula is satisfiable if and only if, the translation of such formula (in addition to the constraints we define) is satisfiable.

Definition 19. *Let $s \in$ PROP, we define Conds as the conjunction of the following formulas:*

$$
\begin{array}{ll}
(1) & s \wedge \Box\neg s \wedge \Box\Diamond s \\
(2) & \Box\Box(s \rightarrow \neg\Diamond s) \\
(3) & [\mathsf{sb}][\mathsf{sb}](s \rightarrow \Box\Diamond s) \\
(4) & \Box[\mathsf{sb}](s \rightarrow \Diamond\neg\Diamond s) \\
(5) & \Box\Box(\neg s \rightarrow \Diamond(s \wedge \neg\Diamond s)) \\
(6) & \Box[\mathsf{sb}](\neg s \rightarrow [\mathsf{sb}](s \rightarrow \Box\Box(\neg s \rightarrow \Diamond s))) \\
(7) & \Box\Box(\neg s \rightarrow [\mathsf{sb}](s \rightarrow \Diamond\Diamond(\neg s \wedge \neg\Diamond s))) \\
(Spy) & \Box\Box(\neg s \rightarrow [\mathsf{sb}](s \rightarrow \Diamond\neg\Diamond s)).
\end{array}
$$

Let us call s (for *spy point*) a node satisfying *Conds* in an arbitrary model. Then, the point s satisfies the propositional symbol s, and is related with all the states of the connected component of the model in the two directions. Formula (1) ensures that the propositional symbol s is satisfied at the evaluation point,

---

[1] Master student at Universidad Nacional de Río Cuarto, Argentina.

and is not satisfied in any successor. It also says that all the successors can see an s-state. (2) ensures that all the s-states that are accessible in two steps from the evaluation point, has no successors satisfying s. (3) ensures that after deleting two edges and reaching an s-state, the property that all the successors can see an s-state is maintained. The formula (4) establishes that for all the successors, after deleting an edge an reaching an s-state, there is a successor which cannot see any s-state (it was the only successor satisfying s). (5) says that reaching some state in two steps that does not satisfy s, there is always an s-state which is reachable and has no successors satisfying s. (6) ensures that after eliminating the edge from a $\neg s$-state (which is no longer accessible from the evaluation point in two steps) to an s-state, the remaining $\neg s$-states still have an edge pointing to some state satisfying s. (7) ensures that all the states reachable in two steps (which do not satisfy s) have only one successor labeled by s. Finally, (Spy) establishes that states that are accessible in two steps are also accessible in one step.

Next, we will see an example showing how we will use $Conds$. The idea is to pick an $\mathcal{ML}(\widehat{\mathbb{r}}, \widehat{\mathbb{k}})$ model, and add a spy point to satisfy $Conds$. A model where $\mathcal{M}, s \models Conds$ is illustrated below:

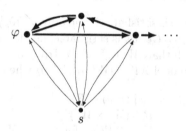

In this picture, the thick points and lines represent the model of the initial memory logic formula that can be extracted from the whole model. We introduce some properties of the models satisfying $Conds$, that will be useful in the equisatisfiability proof.

Proposition 20. Let $\mathcal{M} = \langle W, R, V \rangle$ be a model, $w \in W$. If $\mathcal{M}, w \models Conds$, then the following properties hold:

1. w is the only state in $\mathcal{M}$ that satisfies s in the connected component generated by w.
2. For all states $v \in W$ such that $v \neq w$, we have that if $(w, v) \in R$ then $(v, w) \in R$, and if $(w, v) \in R^*$ then $(v, w) \in R$ (w is a spy point).

Proposition 20 enumerates the main properties of the spy point: it is the only spy point in the connected component, and each time that there is an outgoing edge to some state of the model, there is also an edge coming back.

Now we introduce the translation from $\mathcal{ML}(\widehat{\mathbb{r}}, \widehat{\mathbb{k}})$-formulas to $\mathcal{ML}(\langle \mathsf{sb} \rangle)$-formulas.

Definition 21. *Let φ be an $\mathcal{ML}(\textcircled{r}, \textcircled{k})$-formula that does not contain the propositional symbol s. We define* $\mathsf{Tr}(\varphi) = \lozenge(\varphi)'$, *where ( )' is defined as follows:*

$$
\begin{aligned}
(p)' &= p \quad \textit{for } p \in \mathsf{PROP} \textit{ appearing in } \varphi \\
(\textcircled{k})' &= \neg\lozenge s \\
(\neg\psi)' &= \neg(\psi)' \\
(\psi \wedge \chi)' &= (\psi)' \wedge (\chi)' \\
(\lozenge\psi)' &= \lozenge(\neg s \wedge (\psi)') \\
(\textcircled{r}\psi)' &= (\lozenge s \to \langle\mathsf{sb}\rangle(s \wedge \langle\mathsf{sb}\rangle(\neg\lozenge s \wedge (\psi)'))) \wedge (\neg\lozenge s \to (\psi)').
\end{aligned}
$$

Boolean and modal cases are obvious. $\textcircled{r}$ is represented by removing the edges from the spy point to the state we want to memorize and from this state to the spy point. Notice how the translation behaves: if the point has already been memorized ($\neg\lozenge s$), then nothing needs to be done and translation continues; otherwise ($\lozenge s$), we make s inaccessible using $\langle\mathsf{sb}\rangle$ and we also delete the arrow from s to the current point. $\textcircled{k}$ is represented by checking whether there is an edge pointing to the spy point or not.

Theorem 22. *Let φ be a formula of $\mathcal{ML}(\textcircled{r}, \textcircled{k})$ that does not contain the propositional symbol s. Then, φ and $\mathsf{Tr}(\varphi) \wedge Conds$ are equisatisfiable.*

Proof. We will prove that φ is satisfiable if and only if $\mathsf{Tr}(\varphi) \wedge Conds$ is satisfiable.

($\Leftarrow$) Suppose that $\mathsf{Tr}(\varphi) \wedge Conds$ is satisfiable, i.e., there exists a model $\mathcal{M} = \langle W, R, V \rangle$, and $s \in W$ such that $\langle W, R, V \rangle, s \models \mathsf{Tr}(\varphi)$ and $\langle W, R, V \rangle, s \models Conds$. Then we can define the model $\mathcal{M}' = \langle W', R', V', \emptyset \rangle$ where

$$
\begin{aligned}
W' &= \{v \mid (s, v) \in R\} \\
R' &= R \cap (W' \times W') \\
V'(p) &= V(p) \cap W' \quad \text{for } p \in \mathsf{PROP}.
\end{aligned}
$$

Let $w' \in W'$ be a state s.t $(s, w') \in R$ and $\langle W, R, V \rangle, w' \models (\varphi)'$ (because $\langle W, R, V \rangle, s \models \lozenge(\varphi)'$). We will prove

$$
\langle W', R', V', M' \rangle, v \models \psi \quad \text{iff} \quad \langle W, R(M'), V \rangle, v \models (\psi)',
$$

where $v \in W'$, $M' \subseteq W'$, $\psi \in \mathsf{FORM}$, and $R(M') = R \backslash \{(s, t), (t, s) \mid t \in M'\}$. In particular, when $M' = \emptyset$ we have that $\langle W', R', V', \emptyset \rangle, w' \models \varphi$ iff $\langle W, R, V \rangle, w' \models (\varphi)'$.

Then we do structural induction on ψ. We have two base cases:

$\boldsymbol{\psi = p}$: Suppose that $\langle W', R', V', M' \rangle, v \models p$. By $\models$ we have $v \in V'(p)$, and this is equivalent to $v \in V(p) \cap W'$ by definition of V'. Because $v \in V(p)$, by $\models$ we have $\langle W, R(M'), V \rangle, v \models p$, and by definition of ( )' this is equivalent to $\langle W, R(M'), V \rangle, v \models (p)'$.

$\boldsymbol{\psi = \textcircled{k}}$: Suppose that $\langle W', R', V', M' \rangle, v \models \textcircled{k}$. By $\models$ we have $v \in M'$, and by Proposition 20 and definition of $R(M')$ we have $(v, s) \notin R(M')$ and $\langle W, R(M'), V \rangle, s \models s$. Then by $\models$ $\langle W, R(M'), V \rangle, v \models \neg\lozenge s$, and by definition of ( )' this is equivalent to $\langle W, R(M'), V \rangle, v \models (\textcircled{k})'$.

Now we prove inductive cases.

$\psi = \neg\phi$: Suppose $\langle W', R', V', M'\rangle, v \models \neg\phi$. By ($\models$), $\langle W', R', V', M'\rangle, v \not\models \phi$. By I.H., we have $\langle W, R(M'), V\rangle, v \not\models (\phi)'$, iff $\langle W, R(M'), V\rangle, v \models \neg(\phi)'$. Then, by definition of ( )', $\langle W, R(M'), V\rangle, v \models (\neg\phi)'$.

$\psi = \phi \wedge \chi$: Suppose $\langle W', R', V', M'\rangle, v \models \phi \wedge \chi$. By $\models$, $\langle W', R', V', M'\rangle, v \models \phi$ and $\langle W', R', V', M\rangle, v \models \chi$. By I.H. we have $\langle W, R(M'), V\rangle, v \models (\phi)'$ and $\langle W, R(M'), V\rangle, v \models (\chi)'$. Then we have $\langle W, R(M'), V\rangle, v \models (\phi)' \wedge (\chi)'$. Then by definition of ( )', $\langle W, R(M'), V\rangle, v \models (\phi \wedge \chi)'$.

$\psi = \Diamond\phi$: Suppose $\langle W, R(M'), V\rangle, v \models (\Diamond\phi)'$. By definition of ( )' we have $\langle W, R(M'), V\rangle, v \models \Diamond(\neg s \wedge (\phi)')$. By $\models$, there is $v' \in W$ s.t. $(v, v') \in R(M')$ and $\langle W, R(M'), V\rangle, v' \models \neg s \wedge (\phi)'$. Then we have $\langle W, R(M'), V\rangle, v' \models \neg s$ and $\langle W, R(M'), V\rangle, v' \models (\phi)'$. By I.H., $\langle W', R', V', M'\rangle, v' \models \phi$, hence by $\models$ and Proposition 20, we have $\langle W', R', V', M'\rangle, v \models \Diamond\phi$.

$\psi = \textcircled{r}\phi$: Suppose $\langle W, R(M'), V\rangle, v \models (\textcircled{r}\phi)'$. By definition of ( )' and $\models$,

$$\langle W, R(M'), V\rangle, v \models \Diamond s \rightarrow \langle \mathsf{sb}\rangle(s \wedge \langle \mathsf{sb}\rangle(\neg\Diamond s \wedge (\phi)')) \text{ and}$$
$$\langle W, R(M'), V\rangle, v \models \neg\Diamond s \rightarrow (\phi)'.$$

We will prove each conjunct separately. First, suppose $\langle W, R(M'), V\rangle, v \models \Diamond s$. Then $(v, s) \in R(M')$ (by Proposition 20). We want to prove that

$$\langle W, R(M'), V\rangle, v \models \langle \mathsf{sb}\rangle(s \wedge \langle \mathsf{sb}\rangle(\neg\Diamond s \wedge (\phi)')).$$

By assumption we know $(v, s) \in R(M')$ then $(s, v) \in R(M')$, because in $R(M')$ we always delete pairs in the two directions and by Proposition 20. Then we only need to prove that $\langle W, R(M')^-_{vs}, V\rangle, s \models s \wedge \langle \mathsf{sb}\rangle(\neg\Diamond s \wedge (\phi)')$. It is trivial that $\langle W, (R(M'))^-_{vs}, V\rangle, s \models s$. Let us see $\langle W, (R(M')^-_{vs}, V\rangle, s \models \langle \mathsf{sb}\rangle(\neg\Diamond s \wedge (\phi)')$. Because $(s, v) \in (R(M'))^-_{vs}$, we only need to prove that $\langle W, R(M')^-_{vs,sv}, V\rangle, v \models \neg\Diamond s \wedge (\phi)'$. First conjunct is trivial because $(v, s) \notin R(M')^-_{vs,sv}$.
On the other hand, we know that for all t, $(t, s) \notin R(M')$ iff $(s, t) \notin R(M')$. Then by I.H., $\langle W, R(M')^-_{vs,sv}, V\rangle, v \models (\phi)'$ iff $\langle W', R', V', M \cup \{v\}\rangle, v \models \phi$. Hence, by $\models$ we have $\langle W', R', V', M'\rangle, v \models \textcircled{r}\phi$.
Now suppose the other case, $\langle W, R(M'), V\rangle, v \models \neg\Diamond s$. By Proposition 20, we know $(v, s) \notin R(M')$. By definition of $R(M')$, we have $(s, v) \notin R(M')$. Then $v \in M'$, and by I.H. we have $\langle W', R', V', M'\rangle, v \models \textcircled{r}\phi$.

($\Rightarrow$) Suppose that φ is satisfiable, i.e., there exists a model $\mathcal{M} = \langle W, R, V, \emptyset\rangle$ and $w \in W$ such that $\langle W, R, V, \emptyset\rangle, w \models \varphi$.
Let s be a state that does not belong to W. Then we can define the model $\mathcal{M}' = \langle W', R', V'\rangle$ as follows:

$$\begin{aligned}
W' &= W \cup \{s\} \\
R' &= R \cup \{(s, w) \mid w \in W\} \cup \{(w, s) \mid w \in W\} \\
V'(p) &= V(p) \quad \text{for } p \in \mathsf{PROP} \text{ appearing in } \varphi \\
V'(s) &= \{s\}.
\end{aligned}$$

By construction of $\mathcal{M}'$, it is easy to check that $\mathcal{M}', s \models Conds$. Then we can verify that

$$\langle W, R, V, M \rangle, w \models \varphi \quad \text{iff} \quad \langle W', R'(M), V' \rangle, s \models \mathsf{Tr}(\varphi),$$

where $R'(M)$ is defined as for the ($\Leftarrow$) direction of the proof.

Again we need to do structural induction. Boolean cases are easy, and it is also the case for ⓚ. If $\langle W, R, V, M \rangle, w \models \Diamond\psi$, then by construction of $\mathcal{M}'$ it is clear that $w \notin V'(s)$ and $\langle W', R'(M), V' \rangle, v \models (\psi)'$. If $\langle W, R, V, M \rangle, w \models ⓡ\psi$, we can delete the edges (w, s) and (s, w) to simulate the storing of w in the memory (if those pairs are not in R' means $w \in M$) and continue by evaluating the rest of the translation $(\ )'$ (steps are similar than for the ($\Leftarrow$) direction of the proof). □

From the previous theorem, we immediately get:

Theorem 23. *The satisfiability problem of $\mathcal{ML}(\langle\mathsf{sb}\rangle)$ is undecidable.*

We showed that $\mathcal{ML}(\langle\mathsf{sb}\rangle)$ behaves in the same way as $\mathcal{ML}(\langle\mathsf{sw}\rangle)$ with respect to the satisfiability problem. For the $\mathcal{ML}(\langle\mathsf{br}\rangle)$ case, similar constructions have been done in [11]. With the relation-changing operators that we introduced we can simulate memory logics operators. The idea is to use the capability of adding, swapping and deleting edges to remember points of the model. In general, by defining any operator with this ability we increase the expressivity of the language to an undecidable one.

5   Conclusions

We have discussed several cases of modal logics with operators that let us modify a model. The *Public Announcement Logic* $\mathcal{PAL}$ [20], incorporates an operator that removes all the states of the model which do not satisfy a determined formula (the *announcement*). Adding this new operator to the basic modal logic does not affect its behaviour, because public announcements can be represented by modal formulas (which are possibly exponentially larger), and even the notion of bisimulation remains unchanged. On the other hand, *Memory Logics* [4,19] are extensions of $\mathcal{ML}$ that come equipped with operators to store states in a memory and to consult if the current state belongs to the memory. This new behaviour can be captured by adding a new propositional symbol, and changing its extension when we want to remember some state. Hence, memory logics can be seen as a model update logic with the ability of modify the valuation of the models. In this case, a new notion of bisimulation has to be defined, and the computational complexity blows up with respect to $\mathcal{ML}$.

In order to further explore the spectrum of possible modifications that can be done to a relational model, we discussed in this article *Relation-Changing Modal Logics*. Some other operators that change the accessibility relation of the models have been investigated in the past, such as van Benthem's sabotage logic [22], and some *Epistemic Logics* [7,15]. However, our goal was to investigate different relation-changing operators from a theoretical point of view to study the effects of using this kind of operators. Local Sabotage, Swap and Bridge were introduced before in [1,3,11]. In this paper we provide an analysis of their properties and compare them with other kind of model updates. We learned that it is possible to adapt the notion of bisimulation to capture their behaviour, and we obtained some experience working with relation-changing logics that can help us in the future, for instance, to find decidable fragments.

Acknowledgments. This work was partially supported by grants ANPCyT-PICT-2008-306, ANPCyT-PIC-2010-688, the FP7-PEOPLE-2011-IRSES Project "Mobility between Europe and Argentina applying Logics to Systems" (MEALS) and the Laboratoire Internationale Associé "INFINIS".

Thanks to Mauricio Martel for working with me on the undecidability results.

References

1. Areces, C., Fervari, R., Hoffmann, G.: Moving arrows and four model checking results. In: Ong, L., de Queiroz, R. (eds.) WoLLIC 2012. LNCS, vol. 7456, pp. 142–153. Springer, Heidelberg (2012)
2. Areces, C., Fervari, R., Hoffmann, G.: Tableaux for relation-changing modal logics. In: Fontaine, P., Ringeissen, C., Schmidt, R.A. (eds.) FroCoS 2013. LNCS (LNAI), vol. 8152, pp. 263–278. Springer, Heidelberg (2013)
3. Areces, C., Fervari, R., Hoffmann, G.: Swap logic. Logic Journal of the IGPL 22(2), 309–332 (2014)
4. Areces, C., Figueira, D., Figueira, S., Mera, S.: Expressive power and decidability for memory logics. In: Hodges, W., de Queiroz, R. (eds.) Logic, Language, Information and Computation. LNCS (LNAI), vol. 5110, pp. 56–68. Springer, Heidelberg (2008)
5. Areces, C., Figueira, D., Figueira, S., Mera, S.: The expressive power of memory logics. The Review of Symbolic Logic 4(2), 290–318 (2011)
6. Areces, C., Figueira, D., Gorín, D., Mera, S.: Tableaux and model checking for memory logics. In: Giese, M., Waaler, A. (eds.) TABLEAUX 2009. LNCS (LNAI), vol. 5607, pp. 47–61. Springer, Heidelberg (2009)
7. Aucher, G., Balbiani, P., Fariñas Del Cerro, L., Herzig, A.: Global and local graph modifiers. Electronic Notes in Theoretical Computer Science (ENTCS), Special issue Proceedings of the 5th Workshop on Methods for Modalities (M4M5 2007) 231, 293–307 (2009)
8. Blackburn, P., de Rijke, M., Venema, Y.: Modal Logic. Cambridge Tracts in Theoretical Comp. Scie, vol. 53. Cambridge University Press (2001)
9. Blackburn, P., van Benthem, J.: Modal logic: A semantic perspective. In: Handbook of Modal Logic. Elsevier, North-Holland (2006)
10. Enderton, H.: A mathematical introduction to logic. Academic Press (1972)
11. Fervari, R.: Relation-Changing Modal Logics. PhD thesis, Facultad de Matemática Astronomía y Física, Universidad Nacional de Córdoba, Córdoba, Argentina (March 2014)
12. Fischer, M., Ladner, R.: Propositional dynamic logic of regular programs. J. Comput. Syst. Sci. 18(2), 194–211 (1979)

13. French, T., van der Hoek, W., Iliev, P., Kooi, B.: On the succinctness of some modal logics. Artificial Intelligence 197, 56–85 (2013)
14. Harel, D.: Dynamic logic. In: Gabbay, D., Guenthner, F. (eds.) Handbook of Philosophical Logic. Vol. II. Synthese Library, vol. 165, pp. 497–604. D. Reidel Publishing Co., Dordrecht (1984); Extensions of classical logic
15. Kooi, B., Renne, B.: Arrow update logic. Review of Symbolic Logic 4(4), 536–559 (2011)
16. Ladner, R.: The computational complexity of provability in systems of modal propositional logic. SIAM J. Comput. 6(3), 467–480 (1977)
17. Löding, C., Rohde, P.: Solving the sabotage game is PSPACE-hard. In: Rovan, B., Vojtáš, P. (eds.) MFCS 2003. LNCS, vol. 2747, pp. 531–540. Springer, Heidelberg (2003)
18. Lutz, C.: Complexity and succinctness of public announcement logic. In: Nakashima, H., Wellman, M.P., Weiss, G., Stone, P. (eds.) AAMAS, pp. 137–143. ACM (2006)
19. Mera, S.: Modal Memory Logics. PhD thesis, Univ. de Buenos Aires and UFR STMIA - Ecole Doctorale IAEM Lorraine Dép. de Form. Doct. en Informat. (2009)
20. Plaza, J.: Logics of public communications. Synthese 158(2), 165–179 (2007)
21. Rohde, P.: On games and logics over dynamically changing structures. PhD thesis, RWTH Aachen (2006)
22. van Benthem, J.: An essay on sabotage and obstruction. In: Hutter, D., Stephan, W. (eds.) Mechanizing Mathematical Reasoning. LNCS (LNAI), vol. 2605, pp. 268–276. Springer, Heidelberg (2005)
23. van Ditmarsch, H., van der Hoek, W., Kooi, B.: Dynamic Epistemic Logic. Kluwer (2007)

How Arbitrary Are Arbitrary Public Announcements?

Louwe Bouke Kuijer

University of Groningen, The Netherlands

Abstract. Public announcements are used in dynamic epistemic logic to model certain kinds of information change. A formula $\langle \psi \rangle \, \varphi$ represents the statement that after ψ is publicly announced φ will be the case.

Sometimes we want to reason about whether it is possible for φ to become true after some announcement. In order to do this an arbitrary public announcement operator $\Diamond$ can be added to an epistemic logic with public announcements. Ideally a formula $\Diamond\varphi$ would hold if and only if there is a formula ψ such that $\langle \psi \rangle \, \varphi$. However, in order to avoid circularity the $\Diamond$ operator can only quantify over those ψ that are $\Diamond$-free. So $\Diamond\varphi$ holds if and only if there is a $\Diamond$-free ψ such that $\langle \psi \rangle \, \varphi$.

As a result it does not follow immediately from the definition that $\langle \psi \rangle \, \varphi$ implies $\Diamond\varphi$ if ψ contains a $\Diamond$. But the implication may still hold in some cases. In this paper I show that on finite models $\langle \psi \rangle \, \varphi$ implies $\Diamond\varphi$ for every ψ, and that on finitely branching models $\langle \psi \rangle \, \varphi$ implies $\Diamond\varphi$ for every ψ if φ is $\Diamond$-free. Finally I also show that there are φ and ψ such that $\langle \psi \rangle \, \varphi$ does not imply $\Diamond\varphi$ even on a finitely branching model.

1 Introduction

In epistemic logic we can reason about basic facts (represented by propositional variables) and about knowledge of different agents (represented by one operator K_a per agent). A commonly used example in epistemic logic is that of a simple card game. Suppose two agents a and b are playing a game where they each hold one card, and they know their own card but not the other's card. Then if a holds a queen (and we use the propositional variable q to represent this basic fact) the formulas (i) $K_a q$, (ii) $q \wedge \neg K_b q$ and (iii) $K_a \neg K_b q$ represent the (true) statements that (i) a knows that she holds a queen, (ii) a holds a queen but b does not know this, and (iii) a knows that b does not know that she holds a queen.

In such a basic epistemic logic we cannot however express information change. For example, we cannot reason about what would happen if a were to show her card to b in basic epistemic logic. If we want to reason about information change we need to use a *dynamic* epistemic logic. There are many different kinds of dynamic epistemic logic, see for example [1] for an overview. One of the most common ways to turn a (static) epistemic logic into a dynamic epistemic logic is to add *public announcements* [2,3] to the logic. A public announcement is a binary operator of the form $\langle \psi \rangle \, \varphi$. The formula $\langle \psi \rangle \, \varphi$ is true if φ will hold after ψ is announced *truthfully* and *publicly*.

M. Colinet et al. (Eds.): ESSLLI 2012/2013, LNCS 8607, pp. 109–123, 2014.
© Springer-Verlag Berlin Heidelberg 2014

Using public announcements we can reason about what would happen if a were to show her card to b; the showing of a card can be considered an announcement of the card that a holds. The statement that after a shows her card b knows what card a holds is therefore represented by the (true) formula $\langle q \rangle K_b q$. One thing to note about the formula $\langle q \rangle K_b q$ is that after q is announced agent b knows that q, so the announcing of q is a way for b to get to know q.

However, not all formulas can be learned in such a way. Consider the formula $q \wedge \neg K_b q$, representing a holding a queen and b not knowing this. This formula was introduced in [4] as a formula that can never be known by b even if it is true. Since $q \wedge \neg K_b q$ can never be known by b there is also no announcement such that b will know $q \wedge \neg K_b q$ after the announcement. So not only is it impossible for b to get to know the truth of $q \wedge \neg K_b q$ by announcing $q \wedge \neg K_b q$, there is no formula ψ such that $\langle \psi \rangle K_b(q \wedge \neg K_b q)$.

This last property, whether for a given φ there exists a ψ such that $\langle \psi \rangle K_b \varphi$ requires us to quantify over all formulas. We can of course do this quantification meta-logically, but epistemic logic with public announcements does not allow us to perform this quantification inside the logic. This is unfortunate, as this means we cannot use public announcements to reason about whether it is possible to get to know something. A solution proposed in [5,6] is to add one more operator $\Diamond$, representing *arbitrary public announcements*.

Such arbitrary public announcements can be useful when considering problems of knowability, but also in more practical scenarios such as in cryptography where it is important to know whether it is possible to make a public statement such that agent b learns the content p of a message but another agent e does not, so whether $\Diamond(K_b p \wedge \neg K_e p)$.

We would like to define $\Diamond$ in such a way that $\Diamond \varphi$ holds if and only if there is an announcement ψ such that $\langle \psi \rangle \varphi$ holds. There is a technical problem with this kind of definition, however. If we allow the announcement ψ to be *any* formula the evaluation of $\Diamond \varphi$ would become circular. After all, in order to know whether $\Diamond \varphi$ holds we would have to check whether $\langle \Diamond \varphi \rangle \varphi$ holds. But in order to know whether $\langle \Diamond \varphi \rangle \varphi$ holds we would among other things have to know whether $\Diamond \varphi$ is a truthful announcement, so whether $\Diamond \varphi$ holds.

This circularity is removed in [5,6] by restricting ψ to formulas that do not themselves contain $\Diamond$ operators. So $\Diamond \varphi$ holds if and only if there is a $\Diamond$-free formula ψ such that $\langle \psi \rangle \varphi$. Unfortunately this means that the announcements in an arbitrary announcement operator are not in fact entirely arbitrary. But while the definition of $\Diamond$ cannot allow completely arbitrary announcements it might be possible to get entirely arbitrary announcements as an "emergent property". Suppose that whenever there is a ψ containing $\Diamond$ such that $\langle \psi \rangle \varphi$ there would always also be a ψ' that is $\Diamond$-free such that $\langle \psi' \rangle \varphi$. Then $\langle \psi \rangle \varphi$ would imply $\Diamond \varphi$, even if ψ happens to contain a $\Diamond$.

A different way of phrasing this is to ask whether $\langle \psi \rangle \varphi \rightarrow \Diamond \varphi$ is valid for every ψ. It was shown in [5] that the implication is valid if there is only a single agent. In this paper I show that if there are multiple agents the validity of the implication depends on the class of models we use to evaluate the logic on and on φ. If we use only finite models then $\langle\psi\rangle\,\varphi \to \Diamond\varphi$ is valid. If we allow finitely branching infinite models then $\langle\psi\rangle\,\varphi \to \Diamond\varphi$ is valid for every ψ and every $\Diamond$-free φ. But if we allow models that are infinitely branching or if we do not restrict to $\Diamond$-free φ then there are φ and ψ such that $\langle\psi\rangle\,\varphi \to \Diamond\varphi$ is not valid.

In Section 2 I give some definitions needed to formulate and prove the results. Then in Section 3 I show that for finite models $\langle\psi\rangle\,\varphi \to \Diamond\varphi$ is valid. In Section 4.1 I prove that for finitely branching models $\langle\psi\rangle\,\varphi \to \Diamond\varphi$ is valid if φ is $\Diamond$-free. In Section 4.2 I construct ψ and $\Diamond$-free φ such that $\langle\psi\rangle\,\varphi \to \Diamond\varphi$ is not valid on infinitely branching models. Finally, in Section 4.3 I construct φ and ψ containing $\Diamond$ such that $\langle\psi\rangle\,\varphi \to \Diamond\varphi$ is not valid on finitely branching models.

2 Definitions

Let us start by defining *arbitrary public announcement logic* $\mathcal{L}_{\mathrm{APAL}}$ and the $\Diamond$-free fragment *public announcement logic* $\mathcal{L}_{\mathrm{PAL}}$ of $\mathcal{L}_{\mathrm{APAL}}$. Let us fix a countably infinite set $\mathcal{P}$ of propositional variables and a finite set $\mathcal{A}$ of agents. The language of $\mathcal{L}_{\mathrm{APAL}}$ is then defined as follows.

Definition 1. *The formulas of* $\mathcal{L}_{\mathrm{APAL}}$ *are given by*

$$\varphi ::= p \mid \neg\varphi \mid (\varphi \vee \varphi) \mid K_a\varphi \mid \langle\varphi\rangle\,\varphi \mid \Diamond\varphi$$

where p ranges over $\mathcal{P}$ and a ranges over $\mathcal{A}$.

Definition 2. *The logic* $\mathcal{L}_{\mathrm{PAL}}$ *is the $\Diamond$-free fragment of* $\mathcal{L}_{\mathrm{APAL}}$.

Parentheses are omitted where this should not cause confusion and $\wedge, \to, \leftrightarrow, \bigvee$ and $\bigwedge$ are used in the usual way as abbreviations. Furthermore, $\hat{K}_a$, $[\varphi]$ and $\Box$ are used as abbreviations for $\neg K_a\neg$, $\neg\,\langle\varphi\rangle\,\neg$ and $\neg\Diamond\neg$ respectively. Integer superscripts are used to indicate multiple copies of an operator, so K_a^3 stands for $K_aK_aK_a$. Finally, if B is a set of agents then K_B stands for $\bigwedge_{a\in B} K_a$ and $\hat{K}_B$ for $\bigvee_{a\in B} K_a$.

The intended reading of the non-boolean operators is as follows:

- $K_a\varphi$ is read as "agent a knows that φ",
- $\langle\psi\rangle\,\varphi$ is read as "after it is publicly announced that ψ is the case φ holds",
- $\Diamond\varphi$ is read as "there is a $\Diamond$-free announcement ψ such that $\langle\psi\rangle\,\varphi$ holds".

Since $\mathcal{L}_{\mathrm{APAL}}$ and $\mathcal{L}_{\mathrm{PAL}}$ are epistemic logics they are usually considered over the class of S5 models. We will follow this tradition, but it should be noted that none of the proofs in this paper depend on the special properties of S5 models. So all the results presented here also hold over the class of K models.

Definition 3. *A model $\mathcal{M}$ is a triple (W, R, v) where W is a set of worlds, $R : \mathcal{A} \to \wp(W \times W)$ assigns to each agent an equivalence relation on W and $v : \mathcal{P} \to \wp(W)$ is a valuation function that assigns an extension to each propositional variable.*

A model $\mathcal{M} = (W, R, v)$ is said to be finitely branching *if for each $w \in W$ and each $a \in \mathcal{A}$ the set $\{w' \mid (w, w') \in R(a)\}$ is finite. A model $\mathcal{M} = (W, R, v)$ is said to be* finite *if W is a finite set.*

The semantics for most operators of $\mathcal{L}_{APAL}$ are as usual. For the only unusual operator $\Diamond$ it should be noted that it quantifies not over the formulas of $\mathcal{L}_{APAL}$ but over the formulas of $\mathcal{L}_{PAL}$.

Definition 4. *Given a model* $\mathcal{M} = (W, R, v)$, *a world* w *of* $\mathcal{M}$ *and* φ, ψ *formulas of* $\mathcal{L}_{APAL}$ *the satisfaction relation* $\models$ *is given by*

$$\mathcal{M}, w \models p \quad \Leftrightarrow w \in v(p)$$
$$\mathcal{M}, w \models \neg\varphi \quad \Leftrightarrow \mathcal{M}, w \not\models \varphi$$
$$\mathcal{M}, w \models \varphi \vee \psi \Leftrightarrow \mathcal{M}, w \models \varphi \text{ or } \mathcal{M}, w \models \psi$$
$$\mathcal{M}, w \models K_a\varphi \Leftrightarrow \mathcal{M}, w' \models \varphi \text{ for all } w' \in W \text{ such that } (w, w') \in R(a)$$
$$\mathcal{M}, w \models \langle\varphi\rangle\psi \Leftrightarrow \mathcal{M}, w \models \varphi \text{ and } \mathcal{M}_\varphi, w \models \psi$$
$$\mathcal{M}, w \models \Diamond\varphi \quad \Leftrightarrow \text{ there is a } \mathcal{L}_{PAL} \text{ formula } \psi \text{ such that } \mathcal{M}, w \models \langle\psi\rangle\varphi$$

with $\mathcal{M}_\varphi = (W_\varphi, R_\varphi, v_\varphi)$ *where* $W_\varphi = \{w \in W \mid \mathcal{M}, w \models \varphi\}$ *and* R_φ *and* v_φ *are the restrictions of* R *and* v *to* W_φ.

We write $\mathcal{M} \models \varphi$ *if* $\mathcal{M}, w \models \varphi$ *for every* $w \in W$ *and* $\models \varphi$ *if* $\mathcal{M} \models \varphi$ *for every model* $\mathcal{M}$. *Furthermore, we write* $\models_{br} \varphi$ *if* $\mathcal{M} \models \varphi$ *for every finitely branching model* $\mathcal{M}$ *and* $\models_{fin} \varphi$ *if* $\mathcal{M} \models \varphi$ *for every finite model* $\mathcal{M}$.

3   APAL on Finite Models

With the definitions out of the way I can show that $\models_{fin} \langle\psi\rangle\varphi \rightarrow \Diamond\varphi$ for all $\mathcal{L}_{APAL}$ formulas ψ. This is not a very surprising result; in a finite model we can replace any $\Diamond$ operator by the announcement of a disjunction of $\mathcal{L}_{PAL}$ formulas, one for each world where the $\Diamond$ is replaced by the "chosen announcement" for that world.

Lemma 1. *Fix a finite model* $\mathcal{M} = (W, R, v)$ *and a* $\mathcal{L}_{APAL}$ *formula* φ. *Then there is a* $\mathcal{L}_{PAL}$ *formula* ψ *such that* $\mathcal{M} \models \varphi \leftrightarrow \psi$.

Proof. By induction on the construction of φ. The lemma trivially holds if φ is atomic. Suppose then as induction hypothesis that φ is not atomic, and that the lemma holds for all finite models and all subformulas of φ.

The formula φ is not atomic, so it is of one of the following forms:

1. $\varphi = \neg\varphi'$,
2. $\varphi = \varphi' \vee \varphi''$,
3. $\varphi = K_a\varphi'$,
4. $\varphi = \langle\varphi''\rangle\varphi'$ or
5. $\varphi = \Diamond\varphi'$.

By the induction hypothesis there is a $\mathcal{L}_{PAL}$ formula ψ' such that $\mathcal{M} \models \varphi' \leftrightarrow \psi'$ and, if applicable, a $\mathcal{L}_{PAL}$ formula ψ'' such that $\mathcal{M} \models \varphi'' \leftrightarrow \psi''$. So if we take ψ to be $\neg\psi'$, $\psi' \vee \psi''$ or $K_a\psi'$ then we have $\mathcal{M} \models \varphi \leftrightarrow \psi$ in the first, second or third case respectively.

Let us then consider fourth case. By the induction hypothesis there are $\mathcal{L}_{PAL}$ formulas ψ'' such that $\mathcal{M} \models \varphi'' \leftrightarrow \psi''$ and ψ' such that $\mathcal{M}_{\varphi''} \models \varphi' \leftrightarrow \psi'$. This implies that $\mathcal{M} \models \varphi \leftrightarrow \langle\psi''\rangle\psi'$.

Let us then consider the fifth case, $\varphi = \Diamond\varphi'$. Let W' be the extension of φ, so $W' := \{w \in W \mid \mathcal{M}, w \models \Diamond\varphi'\}$. For each $w_i \in W'$ we have $\mathcal{M}, w_i \models \Diamond\varphi'$, so there is a $\mathcal{L}_{\mathrm{PAL}}$ formula φ_i'' such that $\mathcal{M}, w_i \models \langle\varphi_i''\rangle \varphi'$. By the induction hypothesis there is a $\mathcal{L}_{\mathrm{PAL}}$ formula ψ_i' such that $\mathcal{M}_{\varphi_i''} \models \psi_i' \leftrightarrow \varphi'$. We therefore have $\mathcal{M} \models \langle\varphi_i''\rangle \psi_i' \leftrightarrow \langle\varphi_i''\rangle \varphi'$.

Now let $\psi := \bigvee_{w_i \in W'} \langle\varphi_i''\rangle \psi_i'$. This is a $\mathcal{L}_{\mathrm{PAL}}$ formula, since all its subformulas are $\mathcal{L}_{\mathrm{PAL}}$ formulas and W' is a finite set. Furthermore, for each $w_i \in W'$ we have $\mathcal{M}, w_i \models \psi$.

Suppose now towards a contradiction that for some $w' \in W \setminus W'$ we have $\mathcal{M}, w' \models \psi$. Then one of the disjuncts of ψ holds in w', so for some $w_i \in W'$ we have $\mathcal{M}, w' \models \langle\varphi_i''\rangle \psi_i'$. Then we also have $\mathcal{M}, w' \models \langle\varphi_i''\rangle \varphi'$, since $\mathcal{M} \models \langle\varphi_i''\rangle \psi_i' \leftrightarrow \langle\varphi_i''\rangle \varphi'$. But φ_i'' is a $\mathcal{L}_{\mathrm{PAL}}$ formula so this implies that $\mathcal{M}, w' \models \Diamond\varphi'$. This contradicts w' being an element of $W \setminus W'$, so we must have $\mathcal{M}, w' \not\models \psi$.

This shows that $\mathcal{M} \models \varphi \leftrightarrow \psi$, which completes the induction step and thereby the proof. □

It now follows immediately that $\models_{\mathrm{fin}} \langle\psi\rangle \varphi \to \Diamond\varphi$.

Theorem 1. *For every $\mathcal{L}_{\mathrm{APAL}}$ formulas φ and χ we have $\models_{\mathrm{fin}} \langle\varphi\rangle \chi \to \Diamond\chi$.*

Proof. Fix any $\mathcal{L}_{\mathrm{APAL}}$ formulas φ and χ, and any finite model $\mathcal{M}$. Then by Lemma 1 there is a $\mathcal{L}_{\mathrm{PAL}}$ formula ψ such that $\mathcal{M} \models \varphi \leftrightarrow \psi$. This implies that $\mathcal{M} \models \langle\varphi\rangle \chi \leftrightarrow \langle\psi\rangle \chi$. But ψ is a $\mathcal{L}_{\mathrm{PAL}}$ formula so $\models \langle\psi\rangle \chi \to \Diamond\chi$ and therefore $\mathcal{M} \models \langle\varphi\rangle \chi \to \Diamond\chi$. Since this is true for any finite model $\mathcal{M}$ this implies that $\models_{\mathrm{fin}} \langle\varphi\rangle \chi \to \Diamond\chi$. □

4   APAL on Infinite Models

On infinite models we cannot use the method that worked for finite models, $\bigvee_{w_i \in W'} \langle\varphi_i''\rangle \psi_i'$ is in general not a formula on infinite models since W' may be infinite. Here I show that no other method can work; there are infinite models $\mathcal{M}$, worlds w of $\mathcal{M}$ and $\mathcal{L}_{\mathrm{APAL}}$ formulas φ and ψ such that $\mathcal{M}, w \models \langle\psi\rangle \varphi \wedge \neg\Diamond\varphi$.

Like the result for the finite case this should not surprise us. What is somewhat surprising however is that the result extends to finitely branching models; there are φ and ψ such that $\not\models_{\mathrm{br}} \langle\psi\rangle \varphi \to \Diamond\varphi$. To see why it is unexpected that $\not\models_{\mathrm{br}} \langle\psi\rangle \varphi \to \Diamond\varphi$ consider the following. Fix any finitely branching model $\mathcal{M}$ and any world w of $\mathcal{M}$. We cannot guarantee the existence of a $\mathcal{L}_{\mathrm{PAL}}$ formula ψ' such that $\mathcal{M} \models \psi \leftrightarrow \psi'$, but since $\mathcal{M}$ is finitely branching we can for any $n \in \mathbb{N}$ guarantee the existence of a $\mathcal{L}_{\mathrm{PAL}}$ formula ψ'' such that $\mathcal{M}, w' \models \psi \leftrightarrow \psi''$ for every world w' that is reachable within n steps from w.

The language of $\mathcal{L}_{\mathrm{APAL}}$ does not contain common knowledge, so it would at first glance seem like such a ψ'' that is equivalent to ψ up to a given distance might be sufficient to make φ have the same value after both announcements. If φ does not contain any $\Diamond$ operators then this does indeed work, for any $\mathcal{L}_{\mathrm{APAL}}$ formula ψ and any $\mathcal{L}_{\mathrm{PAL}}$ formula φ we have $\models_{\mathrm{br}} \langle\psi\rangle \varphi \to \Diamond\varphi$. But a $\Diamond$ operator (or more precisely: a $\Box$ operator) can make a formula depend on worlds that are arbitrarily far away in such a way that in certain models no finite approximation ψ'' of ψ will suffice.

I first show that for $\Diamond$-free φ we have $\models_{br} \langle \psi \rangle \varphi \to \Diamond \varphi$, then that there are ψ and $\Diamond$-free φ such that $\not\models \langle \psi \rangle \varphi \to \Diamond \varphi$ and finally that for some φ that do contain $\Diamond$ we have $\not\models_{br} \langle \psi \rangle \varphi \to \Diamond \varphi$. This order of proofs is chosen for reasons of clarity of exposition; the proof that $\not\models_{br} \langle \psi \rangle \varphi \to \Diamond \varphi$ uses more complicated variants on some of the same techniques that are used in the proof of $\not\models \langle \psi \rangle \varphi \to \Diamond \varphi$.

4.1   Validity of $\langle \psi \rangle \varphi \to \Diamond \varphi$ for $\Diamond$-free φ

Before proving that $\models_{br} \langle \psi \rangle \varphi \to \Diamond \varphi$ we need one auxiliary lemma.

Lemma 2. *Let $\mathcal{M}$ be any finitely branching model and w_1, w_2 two worlds of $\mathcal{M}$. Then there is a $\mathcal{L}_{APAL}$ formula that distinguishes between $\mathcal{M}, w_1$ and $\mathcal{M}, w_2$ if and only if there is a $\mathcal{L}_{PAL}$ formula that distinguishes between them.*

Proof. If there is a $\mathcal{L}_{PAL}$ formula ψ' that distinguishes between two worlds then there is also a $\mathcal{L}_{APAL}$ formula ψ that distinguishes between the two worlds, namely $\psi = \psi'$. Left to show is that if $\mathcal{L}_{APAL}$ can distinguish between two worlds then so can $\mathcal{L}_{PAL}$.

The formulas of $\mathcal{L}_{APAL}$ are invariant under bisimulation (see [6]), so if a $\mathcal{L}_{APAL}$ formula distinguishes between $\mathcal{M}, w_1$ and $\mathcal{M}, w_2$ then $\mathcal{M}, w_1$ and $\mathcal{M}, w_2$ are not bisimilar. On finitely branching models worlds are bisimilar if and only if they are indistinguishable by basic modal logic (see for example [1]). So since $\mathcal{M}, w_1$ and $\mathcal{M}, w_2$ are not bisimilar they can be distinguished by a $\mathcal{L}_{PAL}$ formula.   □

Lemma 2 also holds for models that are not finitely branching, but that requires a more complicated proof and we only need the result for finitely branching models.[1]

Lemma 3. *Let ψ be any $\mathcal{L}_{APAL}$ formula and let φ be any $\mathcal{L}_{PAL}$ formula. Then $\models_{br} \langle \psi \rangle \varphi \to \Diamond \varphi$.*

Proof. Fix any finitely branching model $\mathcal{M}$ and any world w of $\mathcal{M}$. It was shown in [2] that every $\mathcal{L}_{PAL}$ formula is equivalent to a $\mathcal{L}_{PAL}$ formula that does not contain any public announcements. Let φ' be the announcement-free formula equivalent to φ, and let n be the maximum nesting depth of K operators in φ'. Then the truth of φ'—and therefore also φ—on $\mathcal{M}, w$ does not depend on changes to worlds that are not reachable from w in at most n steps.

Let W' be the set of worlds that are reachable from w in at most n steps, and let $W_1 := \{w' \in W' \mid \mathcal{M}, w' \models \psi\}$ and $W_2 := W' \setminus W_1$. Then for each $w_i \in W_1$ and $w_j \in W_2$ the formula ψ distinguishes $\mathcal{M}, w_i$ from $\mathcal{M}, w_j$, so by Lemma 2 there is also a $\mathcal{L}_{PAL}$ formula that distinguishes the two worlds. Let

---

[1] For an idea of why Lemma 2 also holds for infinitely branching models consider the case where $\mathcal{M}, w \models \Diamond \varphi$ and $\mathcal{M}, w' \not\models \Diamond \varphi$. Then there is a ψ such that $\mathcal{M}, w \models \langle \psi \rangle \varphi$ and in particular $\mathcal{M}, w' \not\models \langle \psi \rangle \varphi$ so the formula $\langle \psi \rangle \varphi$ distinguishes the two worlds as well. This can be extended to any formula containing a $\Diamond$ operator.

$\psi'_{i,j}$ be this distinguishing $\mathcal{L}_{\text{PAL}}$ formula and assume without loss of generality that $\mathcal{M}, w_i \models \psi'_{i,j}$ and $\mathcal{M}, w_j \not\models \psi'_{i,j}$.

For $w_i \in W_1$ let $\psi'_i := \bigwedge_{w_j \in W_2} \psi'_{i,j}$. Then $\mathcal{M}, w_i \models \psi'_i$ and $\mathcal{M}, w_j \not\models \psi'_i$ for each $w_j \in W_2$. Finally, let $\psi' := \bigvee_{w_i \in W_1} \psi'_i$. This ψ' satisfies $\mathcal{M}, w_i \models \psi'$ for each $w_i \in W_1$ and $\mathcal{M}, w_j \not\models \psi'$ for each $w_j \in W_2$.

As such, the models $\mathcal{M}_\psi$ and $\mathcal{M}_{\psi'}$ only differ in worlds that are not reachable from w within n steps, so $\mathcal{M}, w \models \langle\psi\rangle\,\varphi \leftrightarrow \langle\psi'\rangle\,\varphi$. Because ψ' is a $\mathcal{L}_{\text{PAL}}$ formula this implies that $\mathcal{M}, w \models \langle\psi\rangle\,\varphi \to \Diamond\varphi$. The model $\mathcal{M}$ and world w were chosen as any finitely branching model and any world of that model, so we have $\models_{\text{br}}$ $\langle\psi\rangle\,\varphi \to \Diamond\varphi$. $\qquad\square$

4.2 Invalidity of $\langle\psi\rangle\,\varphi \to \Diamond\varphi$ on Infinitely Branching Models

If we do not restrict ourselves to finite or finitely branching models there are φ and ψ such that $\langle\psi\rangle\,\varphi \to \Diamond\varphi$ is not valid. Let

$$\varphi_1 := \hat{K}_c p \wedge K_c(r \to \hat{K}_d \neg r) \wedge K_c((p \wedge \neg r) \to \hat{K}_e r),$$
$$\varphi_2 := K_c(\neg q \to (\hat{K}_f(\neg \hat{K}_c q \wedge K_a p) \wedge \hat{K}_f \neg K_a p)),$$
$$\varphi := \varphi_1 \wedge \varphi_2$$
$$\psi := p \vee q \vee K_a \neg \Diamond(\hat{K}_b K_a p \wedge \hat{K}_b \neg K_a p).$$

Furthermore, let $\mathcal{M}$ be the model shown in Figure 1 and let $\mathcal{M}_n$ for $n \in \mathbb{N}$ be the submodels indicated in Figure 1.

We want to show that $\mathcal{M}, w \not\models \langle\psi\rangle\,\varphi \to \Diamond\varphi$. This requires us to show that $\mathcal{M}, w \models \langle\psi\rangle\,\varphi$ and that $\mathcal{M}, w \not\models \Diamond\varphi$. In order to prove that $\mathcal{M}, w \not\models \Diamond\varphi$ we have to demonstrate that if $\mathcal{M}, w \models \langle\psi'\rangle\,\varphi$ then ψ' contains a $\Diamond$ operator. The subformula φ_1 is constructed in such a way that if $\mathcal{M}, w \models \langle\psi'\rangle\,\varphi$ then the update $\langle\psi'\rangle$ retains an infinite number of worlds. The subformula φ_2 guarantees that if $\mathcal{M}, w \models \langle\psi'\rangle\,\varphi$ and $\langle\psi'\rangle$ retains an infinite number of worlds then ψ' must perform an infinite number of different updates, which cannot be done without a $\Diamond$ operator. But before looking at the details of the proof that $\mathcal{M}, w \not\models \Diamond\varphi$ let us start by proving the simpler part of the statement, namely that $\mathcal{M}, w \models \langle\psi\rangle\,\varphi$.

Lemma 4. *We have $\mathcal{M}, w \models \langle\psi\rangle\,\varphi$.*

Proof. To show is that $\mathcal{M}_\psi \models \varphi$, so let us look at which worlds are retained by $\langle\psi\rangle$. The disjuncts p and q of ψ guarantee that any world in the leftmost three columns is retained.

The worlds in the fourth column from the left satisfy neither p nor q though, so they are retained only if they satisfy $K_a \neg \Diamond(\hat{K}_b K_a p \wedge \hat{K}_b \neg K_a p)$. These worlds themselves always satisfy $\neg\Diamond(\hat{K}_b K_a p \wedge \hat{K}_b \neg K_a p)$; there is no update that would let them satisfy $\hat{K}_b K_a p$ because every b-reachable world satisfies $\neg p$.

So the worlds in the fourth column are retained if and only if the p world to the left of them (which they are a-connected with) satisfies $\neg\Diamond(\hat{K}_b K_a p \wedge \hat{K}_b \neg K_a p)$.

Now we reach the difference between the rows of a submodel $\mathcal{M}_n$. Consider the p world in the bottom row of $\mathcal{M}_n$ for any n. The only world b-reachable from

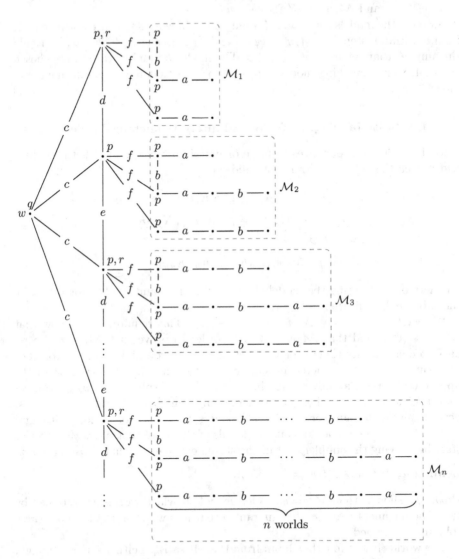

Fig. 1. The model $\mathcal{M}$. Some accessibility arrows are not drawn.

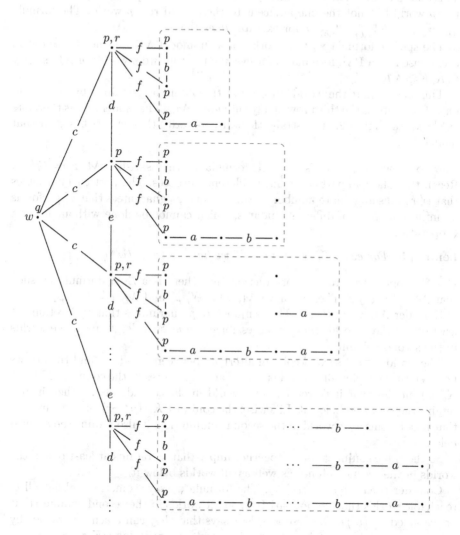

Fig. 2. The model $\mathcal{M}_\psi$. Some accessibility arrows are not drawn.

this world is itself, so there is no update that can make the world satisfy $\hat{K}_b K_a p \wedge \hat{K}_b \neg K_a p$. So the p world in the bottom row satisfies $\neg \Diamond (\hat{K}_b K_a p \wedge \hat{K}_b \neg K_a p)$.

Now consider one of the p worlds in the top two rows of $\mathcal{M}_n$. These two worlds can be distinguished from each other because their "tails" are of different lengths. This allows us to create an update χ_n that removes the $\neg p$ world adjacent to the top p world but not the one adjacent to the second row p world. The formula $\chi_n := \neg p \to \hat{K}^{n-1}_{\{a,b\}} K^n_{\{a,b\}} \neg p$ for example does this.

The specific formula χ_n that works for a submodel $\mathcal{M}_n$ depends on n, but in every case it is a PAL formula so for every n the top two p worlds of $\mathcal{M}_n$ satisfy $\Diamond (\hat{K}_b K_a p \wedge \hat{K}_b \neg K_a p)$.

This means that the worlds in the fourth column are retained by $\langle \psi \rangle$ if and only if they are in the third row of any submodel $\mathcal{M}_n$. The model $\mathcal{M}_\psi$ is therefore as shown in Figure 2. It is straightforward to verify that w satisfies φ in that model.   □

Now to show that there is no PAL formula ψ' that satisfies $\mathcal{M}, w \models \langle \psi' \rangle \varphi$. Recall that the two parts of φ have different purposes. The part φ_1 guarantees that ψ' retains an infinite number of worlds while φ_2 guarantees that ψ' performs an infinite number of different updates, which cannot be done without using a $\Diamond$ operator.

Lemma 5. *For every $\mathcal{L}_{\text{PAL}}$ formula ψ' we have $\mathcal{M}, w \not\models \langle \psi' \rangle \varphi$.*

Proof. Suppose towards a contradiction that there is a $\mathcal{L}_{\text{PAL}}$ formula ψ' such that $\mathcal{M}, w \models \langle \psi' \rangle \varphi$. Then we have $\mathcal{M}, w \models \langle \psi' \rangle \varphi_1$ and $\mathcal{M}, w \models \langle \psi' \rangle \varphi_2$.

Consider $\mathcal{M}, w \models \langle \psi' \rangle \varphi_1$. The conjunct $\hat{K}_c p$ guarantees that $\langle \psi' \rangle$ retains at least one of the p worlds that are accessible from w, so at least one of the worlds in the second column.

The worlds in the second column alternate between r and $\neg r$, and the arrows between those worlds alternate between d and e. As a result the conjunct $K_c(r \to \hat{K}_d \neg r)$ implies that if ψ' retains an r world in the second column then it also retains the $\neg r$ world below it. Likewise, the conjunct $K_c((p \wedge \neg r) \to \hat{K}_e r)$ implies that if ψ' retains a $\neg r$ world in the second column then it also retains the r world below it.

So the three conjuncts of φ_1 together imply that ψ' retains at least one of the worlds in the second column as well as all worlds below it.

Consider then $\mathcal{M}, w \models \langle \psi' \rangle \varphi_2$. The formula φ_2 says something about all c-reachable worlds that do not satisfy q, so all worlds in the second column (that are retained by $\langle \psi' \rangle$). Of these worlds it says that they can reach two worlds by using f, one world satisfying $\neg \hat{K}_c q \wedge K_a p$ and one satisfying $\neg K_a p$.

The worlds in the first two columns all satisfy $\hat{K}_c q$ and $K_a p$ so these two f-reachable worlds must be in the third column. If the n-th world of the second column is retained by $\langle \psi' \rangle$ there must therefore be two p worlds retained in $\mathcal{M}_n$. Furthermore, one of those worlds in $\mathcal{M}_n$ must be adjacent to a $\neg p$ world that is retained while the other must not be adjacent to a retained $\neg p$ world.

One of the $\neg p$ worlds in the second column of $\mathcal{M}_n$ (so the fourth column of $\mathcal{M}$) must be retained and one must not be retained, so in particular ψ' must distinguish between two of those worlds. But the only way to distinguish between those worlds is to use the fact that one "tail" is shorter than the others, and doing this requires a formula with K-depth at least $2n - 2$.

The K-depth of ψ' is fixed and finite, so there is some $N \in \mathbb{N}$ such that for every $n \geq N$ the formula ψ' cannot distinguish between the worlds in the second column of $\mathcal{M}_n$. Putting all of the above together, we get that ψ':

- must retain all worlds in the second column below a certain point,
- must distinguish between two worlds in the second column of $\mathcal{M}_n$ if the n-th world of the second column is retained and
- cannot distinguish between the worlds in the second column of $\mathcal{M}_n$ for all n greater than some number N.

This is a contradiction, so our initial assumption that such a ψ' exists must be false, which proves the lemma. □

The theorem now follows easily.

Theorem 2. *There are a $\mathcal{L}_{\text{PAL}}$ formula φ and a $\mathcal{L}_{\text{APAL}}$ formula ψ such that $\not\models \langle \psi \rangle \varphi \to \Diamond \varphi$.*

Proof. Let $\mathcal{M}, w, \varphi$ and ψ be as defined above. Then $\mathcal{M}, w \models \langle \psi \rangle \varphi$ by Lemma 4. Furthermore, by Lemma 5 we know that $\mathcal{M}, w \not\models \langle \psi' \rangle \varphi$ for every $\mathcal{L}_{\text{PAL}}$ formula ψ' so we have $\mathcal{M}, w \not\models \Diamond \varphi$. This implies that $\mathcal{M}, w \not\models \langle \psi \rangle \varphi \to \Diamond \varphi$ and so that $\not\models \langle \psi \rangle \varphi \to \Diamond \varphi$. □

4.3 Invalidity of $\langle \psi \rangle \varphi \to \Diamond \varphi$ on Finitely Branching Models

Now to show that $\not\models_{\text{br}} \langle \psi \rangle \varphi \to \Diamond \varphi$. The method used to show this is very similar to the method used to show that $\not\models \langle \psi \rangle \varphi \to \Diamond \varphi$. We use φ to force ψ to retain an infinite number of worlds in a pointed model $(\mathcal{N}, w)$. Additionally we force ψ to distinguish between infinitely many pairs of worlds, and we let the difference between the two worlds in a pair get further and further away.

Unfortunately, forcing ψ to retain an infinite number of worlds is much more complicated in a finitely branching frame, so we need more complex formulas and models. Let $\mathcal{N}$ be the model shown in Figure 3 and let

$$\psi := (\neg p \land \hat{K}_b(p \land \hat{K}_a(q \lor r))) \to \Diamond(\hat{K}_a K_b p \land \hat{K}_a(p \land \neg K_b p)),$$
$$\varphi_1 := (q \lor r) \to (\hat{K}_a K_b p \land \hat{K}_a(p \land \neg K_b p)),$$
$$\varphi_2 := (q \to \neg \hat{K}_c \hat{K}_a \hat{K}_b \hat{K}_c r) \land (r \to \neg \hat{K}_c \hat{K}_a \hat{K}_b \hat{K}_c q),$$
$$\varphi := \langle \varphi_1 \rangle (\hat{K}_a \hat{K}_b \hat{K}_c q \land \hat{K}_a \hat{K}_b \hat{K}_c r \land \langle \varphi_2 \rangle \Box \neg(\hat{K}_a K_b p \land \hat{K}_a(p \land \neg K_b p))).$$

Note the recurring a-triangles with two p worlds in the model and the recurring subformula $\hat{K}_a K_b p \land \hat{K}_a(p \land \neg K_b p)$. These subformulas have the property that they hold in the $\neg p$ world of such a triangle if and only if for one of the p worlds in the triangle a b-reachable $\neg p$ world is retained but for the other it is not.

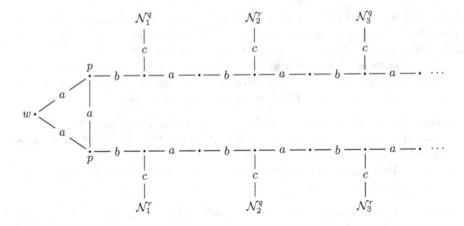

Fig. 3. The model $\mathcal{N}$. Reflexive arrows are not drawn. The submodels $\mathcal{N}_n^x$ for $n \in \mathbb{N}$, $x \in \{q,r\}$ are shown in Figure 4.

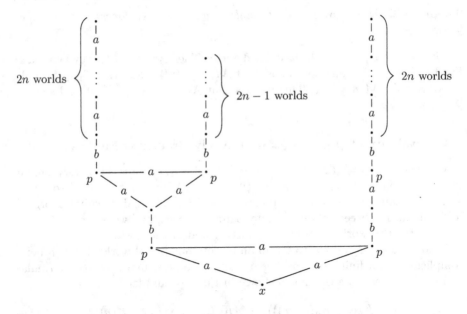

Fig. 4. The submodel $\mathcal{N}_n^x$ for $x \in \{q,r\}$ and $n \in \mathbb{N}_{>0}$. The origin world that connects it to $\mathcal{N}$ is the world satisfying x.

Lemma 6. *We have* $\mathcal{N}, w \models \langle \psi \rangle \, \varphi$.

Proof. Let us consider the update $\langle \psi \rangle$. It places the conditions on $\neg p \wedge \hat{K}_b(p \wedge \hat{K}_a(q \vee r))$ worlds that they must satisfy $\Diamond(\hat{K}_a K_b p \wedge \hat{K}_a(p \wedge \neg K_b p))$. The $\neg p \wedge \hat{K}_b(p \wedge \hat{K}_a(q \vee r))$ worlds are exactly those that are in the third line from the bottom in $\mathcal{N}_n^x$ submodels. Furthermore, of the two such worlds in a submodel

$\mathcal{N}_n^x$ the left one satisfies $\Diamond(\hat{K}_a K_b p \wedge \hat{K}_a(p \wedge \neg K_b p))$, and the right one does not.[2] The updated submodel $\mathcal{N}_{n\,\psi}^x$ is therefore as shown in Figure 5. (The worlds of $\mathcal{N}$ that are not in one of the submodels $\mathcal{N}_n^x$ are all retained by the update so nothing changes there.)

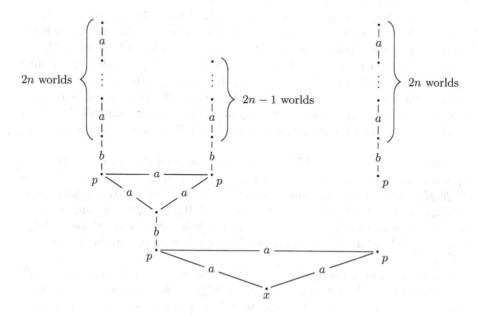

Fig. 5. The submodel $\mathcal{N}_{n\,\psi}^x$ for $x \in \{q, r\}$ and $n \in \mathbb{N}_{>0}$.

After the update $\langle\psi\rangle$ the formula $\hat{K}_a K_b p \wedge \hat{K}_a(p \wedge \neg K_b p)$ therefore holds in the origin world of each submodel $\mathcal{N}_n^x$. Since q and r only hold in the origin worlds of these submodels the update $\langle\varphi_1\rangle = \langle(q \vee r) \to (\hat{K}_a K_b p \wedge \hat{K}_a(p \wedge \neg K_b p))\rangle$ does nothing if executed immediately after $\langle\psi\rangle$. We therefore have $\mathcal{N}, w \models \langle\psi\rangle\langle\varphi_1\rangle$ $(\hat{K}_a \hat{K}_b \hat{K}_c q \wedge \hat{K}_a \hat{K}_b \hat{K}_c r)$.

Finally consider the third update $\langle\varphi_2\rangle$. It places conditions on $q \vee r$ worlds; q worlds must satisfy $\neg\hat{K}_c \hat{K}_a \hat{K}_b \hat{K}_c r$ and r worlds must satisfy $\neg\hat{K}_c \hat{K}_a \hat{K}_b \hat{K}_c q$. After the other updates there are no q or r worlds that satisfy this condition.

As such the result of applying the three updates $\langle\psi\rangle\langle\varphi_1\rangle\langle\varphi_2\rangle$ removes the origin worlds of all $\mathcal{N}_n^x$ submodels. In the resulting model the two p worlds that are a-reachable form w are indistinguishable, so $\mathcal{N}, w \models \langle\psi\rangle\langle\varphi_1\rangle\langle\varphi_2\rangle \square\neg(\hat{K}_a K_b p \wedge \hat{K}_a(p \wedge \neg K_b p))$. Together with the previous result $\mathcal{N}, w \models \langle\psi\rangle\langle\varphi_1\rangle(\hat{K}_a \hat{K}_b \hat{K}_c q \wedge \hat{K}_a \hat{K}_b \hat{K}_c r)$ this shows that $\mathcal{N}, w \models \langle\psi\rangle\varphi$. $\square$

---

[2] Announcements that make $\hat{K}_a K_b p \wedge \hat{K}_a(p \wedge \neg K_b p)$ true in the leftmost world in the third row do so by removing one of the $\neg p$ worlds in the fifth row but not the other. This can be done because there are formulas that differentiate between a "tail" of $2n$ worlds and a "tail" of $2n - 1$ worlds, as in the infinitely branching case.

Lemma 7. *For every $\mathcal{L}_{\text{PAL}}$ formula ψ' we have $\mathcal{N}, w \not\models \langle \psi' \rangle \varphi$.*

Proof. Suppose towards a contradiction that there is a $\mathcal{L}_{\text{PAL}}$ formula ψ' such that $\mathcal{N}, w \models \langle \psi' \rangle \varphi$. Then after the updates $\langle \psi' \rangle \langle \varphi_1 \rangle$ the formula $\hat{K}_a \hat{K}_b \hat{K}_c q \wedge \hat{K}_a \hat{K}_b \hat{K}_c r$ must hold in w. The origin worlds of $\mathcal{N}_1^q$ and $\mathcal{N}_1^r$ and the paths to those worlds must therefore be retained by $\langle \psi' \rangle \langle \varphi_1 \rangle$.

But after those two updates it must also hold in w that $\langle \varphi_2 \rangle \Box \neg (\hat{K}_a K_b p \wedge \hat{K}_a (p \wedge \neg K_b p))$, so after the update $\langle \varphi_2 \rangle$ the two worlds that are b-reachable from the p worlds that are a-reachable from w must be indistinguishable. In particular this means that neither the origin world of $\mathcal{N}_1^q$ nor that of $\mathcal{N}_1^r$ may be reachable, as otherwise $\hat{K}_c q$ or $\hat{K}_c r$ would distinguish the worlds.

Since the update $\langle \varphi_2 \rangle$ only removes $q \vee r$ worlds this implies that the origin worlds of $\mathcal{N}_1^q$ and $\mathcal{N}_1^r$ must satisfy $\neg \varphi_2$ after the first two updates. But then $\hat{K}_c \hat{K}_a \hat{K}_b \hat{K}_c r$ must hold in the origin of $\mathcal{N}_1^q$ and $\hat{K}_c \hat{K}_a \hat{K}_b \hat{K}_c q$ in the origin of $\mathcal{N}_1^r$.

But then the origins of $\mathcal{N}_2^q$ and $\mathcal{N}_2^r$ must be reachable after the first two updates. But these two origin worlds must also be removed by $\langle \varphi_2 \rangle$ as otherwise $\hat{K}_a \hat{K}_b \hat{K}_c q$ would distinguish the two worlds that must be indistinguishable. But then the origins of $\mathcal{N}_3^q$ and $\mathcal{N}_3^r$ must be retained. Repeating the argument shows that if the origins of $\mathcal{N}_n^q$ and $\mathcal{N}_n^r$ remain reachable then so do those of $\mathcal{N}_{n+1}^q$ and $\mathcal{N}_{n+1}^r$. Therefore, the updates $\langle \psi' \rangle \langle \varphi_1 \rangle$ must leave the origin of every $\mathcal{N}_n^x$ submodel reachable.

But then consider the update $\langle \varphi_1 \rangle$. This update retains the origin of a $\mathcal{N}_n^x$ submodel if and only if it satisfies $\hat{K}_a K_b p \wedge \hat{K}_a (p \wedge \neg K_b p)$. This implies that for each $n \in \mathbb{N}$ and $x \in \{q, r\}$ the update $\langle \psi' \rangle$ must retain one of the worlds on the third row of the submodel but not the other. However, in $\mathcal{N}_n^x$ these worlds are indistinguishable up to depth $2n$, so a $\mathcal{L}_{\text{PAL}}$ formula must contain at least $2n + 1$ iterations of a K-operator to distinguish them. There is therefore no single formula in $\mathcal{L}_{\text{PAL}}$ that distinguishes the two worlds for every submodel. This contradicts the assumption that such a ψ' exists. □

The theorem now follows easily.

Theorem 3. *There are $\mathcal{L}_{\text{APAL}}$ formulas φ, ψ such that $\not\models_{\text{br}} \langle \psi \rangle \varphi \to \Diamond \varphi$.*

Proof. For the $\mathcal{L}_{\text{APAL}}$ formulas φ, ψ, finitely branching model $\mathcal{N}$ and world w of $\mathcal{N}$ as defined above we have $\mathcal{N}, w \models \langle \psi \rangle \varphi$ by Lemma 6 and $\mathcal{N}, w \not\models \langle \psi' \rangle \varphi$ for every $\mathcal{L}_{\text{PAL}}$ formula ψ' by Lemma 7. This implies that $\mathcal{N}, w \not\models \langle \psi \rangle \varphi \to \Diamond \varphi$ so $\not\models_{\text{br}} \langle \psi \rangle \varphi \to \Diamond \varphi$. □

5 Conclusion and Further Research

I showed that for any $\mathcal{L}_{\text{APAL}}$ formula φ and ψ we have $\models_{\text{fin}} \langle \psi \rangle \varphi \to \Diamond \varphi$ and that for any $\mathcal{L}_{\text{PAL}}$ formula φ and any $\mathcal{L}_{\text{APAL}}$ formula ψ we also have $\models_{\text{br}} \langle \psi \rangle \varphi \to \Diamond \varphi$. Additionally, I showed that there are $\mathcal{L}_{\text{APAL}}$ formulas φ and ψ such that $\not\models_{\text{br}} \langle \psi \rangle \varphi \to \Diamond \varphi$ and that there are a $\mathcal{L}_{\text{PAL}}$ formula φ and a $\mathcal{L}_{\text{APAL}}$ formula ψ such that $\not\models \langle \psi \rangle \varphi \to \Diamond \varphi$.

The operator $\Diamond$ therefore only represents a truly arbitrary public announcement on finite models. There are scenarios that can be modeled in finite models and where arbitrary public announcements are useful, such as the cryptography example mentioned in the introduction. The message p for which we want to know whether $\Diamond(K_b p \wedge \neg K_e p)$ is generally taken from a finite set of possible messages which allows for a finite model to be used.

However, not all interesting scenarios allow for finite modeling, so it seems like an interesting topic for further research whether semantics for a different arbitrary public announcement operator $\blacklozenge$ can be found such that for any $\mathcal{L}_{PAL+\blacklozenge}$ formulas φ, ψ we have $\models \langle \psi \rangle \varphi \rightarrow \blacklozenge \varphi$. One possibility that might work is an infinite hierarchy of $\Diamond_i$ operators, where each $\Diamond_i$ quantifies over all formulas that use only $\Diamond_j$ with $j < i$. I conjecture that if we then define $\blacklozenge \varphi$ as $\bigvee_{i \in \mathbb{N}} \Diamond_i \varphi$ we have $\models \langle \psi \rangle \varphi \rightarrow \blacklozenge \varphi$.

References

1. van Ditmarsch, H., van der Hoek, W., Kooi, B.: Dynamic Epistemic Logic. Springer (2007)
2. Plaza, J.: Logics of public communication. In: Emrich, M., Phifer, M., Hadzikadic, M., Ras, Z. (eds.) Proceedings of the Fourth International Symposium on Methodologies for Intelligent Systems, Poster Session Program, pp. 201–216. Oak Ridge National Laboratory, ORNL/DSRD-24 (1989); Reprinted as [7]
3. Baltag, A., Moss, L., Solecki, S.: The logic of public announcements, common knowledge, and private suspicions. In: Gilboa, I. (ed.) Proceedings of the 7th Conference on Theoretical Aspects of Rationality and Knowledge, pp. 43–56. Morgan Kaufmann Publishers Inc. (1998)
4. Fitch, F.: A logical analysis of some value concepts. The Journal of Symbolic Logic 28, 135–142 (1963)
5. Balbiani, P., Baltag, A., van Ditmarsch, H., Herzig, A., Hoshi, T., de Lima, T.: What can we achieve by arbitrary announcements? A dynamic take on Fitch's knowability. In: Samet, D. (ed.) Proceedings of the 11th Conference on Theoretical Aspects of Rationality and Knowledge, pp. 42–51 (2007)
6. Balbiani, P., Baltag, A., van Ditmarsch, H., Herzig, A., Hoshi, T., de Lima, T.: 'Knowable' as 'Known after an Announcement'. The Review of Symbolic Logic 1, 305–334 (2008)
7. Plaza, J.: Logics of public communication. Synthese 158, 165–179 (2007)

A Quantitative Measure of Relevance
Based on Kelly Gambling Theory

Mathias Winther Madsen

Institute for Logic, Language, and Computation,
University of Amsterdam, The Netherlands
mathias.winther@gmail.com

Abstract. This paper proposes a quantitative measure relevance which can quantify the difference between useful and useless facts. This measure evaluates sources of information according to how they affect the expected logarithmic utility of an agent. A number of reasons are given why this is often preferable to a naive value-of-information approach, and some properties and interpretations of the concept are presented, including a result about the relation between relevant information and Shannon information. Lastly, a number of illustrative examples of relevance measurements are discussed, including random number generation and job market signaling.

Defining a good concept of relevance is a key problem in all disciplines that theorize about information, including information retrieval [3], epistemology [5], and the pragmatics of natural languages [12].

Shannon information theory [10] provides an interesting quantification of the notion of information, but no tools for distinguishing useless from useful facts. The microeconomic concept of value-of-information can formalize this concept in terms of expected gains [1], but this notion is not easily combined with information theory, and is largely unable to exploit its tools and insights.

In this paper, I propose a framework that integrates information theory more natively with utility theory and thus tackles these problems. Specifically, I draw on John Kelly's application of information theory to gambling situations [7]. Kelly showed that when we take logarithmic capital growth as our measure of real utility, information theory can integrate seamlessly with the classical calculus of expectations. My approach here is to turn this idea on its head and base a novel notion of information on the concept of utility.

The resulting measure coincides with Shannon information when all information can be converted into a strategy improvements. However, when the environment provides sources of both useful and useless information, the concept explains and quantifies the difference, and thus suggests a novel notion of value-of-information.

1 Doubling Rates and Kelly Gambling

The ideas in this paper are based on some observations about the relationship about logarithmic information measures and good gambling strategies [7]. I will

M. Colinet et al. (Eds.): ESSLLI 2012/2013, LNCS 8607, pp. 124–141, 2014.
© Springer-Verlag Berlin Heidelberg 2014

thus start by giving a largely self-contained discussion of some background concepts.

My presentation loosely follows that of Cover and Thomas [4, ch. 6], and readers who already know this material should recognize everything up until section 1.4, in which I sketch the generalization which drives the rest of the argument. Throughout the paper, I assume a small amount of prior familiarity with the basic concepts of information theory, such as entropy.

1.1 Growth Rates and Degenerate Gambling

In many gambling situations, people evaluate a strategy in terms of its effect on the **growth rate** of their capital, that is,

$$R = \frac{\text{Posterior capital}}{\text{Prior capital}}.$$

As an example, consider a horse race with n horses with winning probabilities $p_1, p_2, \ldots, p_n$, and suppose that a bookmaker pays the odds $o_1, o_2, \ldots, o_n$ for these horses. So for example, if you bet everything on horse 1, you get your money back o_1 times if it wins and 0 times if it loses. If you split your capital into the n piles $b_1, b_2, \ldots, b_n$ and bet those on horses $1, 2, \ldots n$, your payoff is $b_i o_i$ when horse i wins.

If we normalize the initial bankroll to 1, the **expected growth rate** associated with such a betting scheme b is thus

$$E[R] = \sum_i p_i \left(b_i o_i \right)$$

This is a linear function of the betting scheme, so it always has a maximum in one of the corner points of the simplex of capital distributions. In other words, you can achieve the maximal expected growth rate by betting your whole capital on a horse with a maximal growth factor $p_i o_i$. I will call this betting scheme **degenerate gambling**, in analogy with degenerate probability distributions.

There are some reasons to disprefer degenerate gambling, however. Since it will suggest that you bet all of your capital on a single horse, it will usually also entail a substantial probability of losing all of your money. This has some highly unfortunate consequences in situations involving repeated investments and reinvestments: After k runs of the game, your initial capital of 1 will have grown or shrunk by a factor of

$$R_1 \cdot R_2 \cdot R_3 \cdots R_k.$$

If each of these factors have a positive probability of vanishing, the whole product will tend to 0 with probability 1. In other words, if you keep exposing yourself to the risk of bankruptcy, it will eventually happen with probability 1.

1.2 Doubling Rates

The growth of an initial stock of capital is described by a long product whose factors are random variables. This suggests that the quality of a betting scheme should be measured by a statistic which interacts just as nicely with products as expectations interact with sums.

A candidate for such a statistic is the logarithm of the growth rate,

$$W = \log R.$$

When the logarithm is base two, this quantity is called the **doubling rate** of the capital, in analogy with the half-life of a radioactive material. W measures how many times your capital is expected to double in a single game, and $1/W$ measures the average waiting time before your capital is doubled once.

A betting scheme which maximizes W is in many ways preferable to one that maximizes R. Since $2^W = R$, the capital after k horse races is

$$R_1 \cdot R_2 \cdot R_3 \cdots R_k = 2^{W_1 + W_2 + W_3 + \cdots + W_k}.$$

The exponent in this expression is a sum of random variables. If the runs of the game are independent and identically distributed, and the gambling scheme is fixed, then the string of doubling rates will also be independent and identically distributed.

The weak law of large numbers thus applies, and we can conclude that the sum is very close to its expected value with high probability. Thus,

$$W_1 + W_2 + \cdots + W_k \approx kE[W],$$

where $E[W]$ is the **expected doubling rate**

$$E[W] = \sum_i p_i \log(b_i o_i).$$

This statistic can be both positive and negative, depending on whether the game is favorable or unfavorable. In the long run, the evolution of a stock of capital in a horse race can thus be considered as an exponential function $2^{kE[W]}$, and if you invest the capital according to the scheme which maximizes $E[W]$, you will achieve the fastest exponential growth that the game allows for (cf. Fig. 1).

1.3 Proportional Gambling in the Horse Race

What can we say about the strategy that optimizes the doubling rate? To answer this question, let us rewrite the odds o_i of the horse race in the form $o_i = c/r_i$, where c is a constant chosen so that $\sum_i r_i = 1$. For instance, the odds $o = 2, 4, 4$ would be transformed into $r = 1/2, 1/4, 1/4$. (Horses with odds 0 will never be part of an optimal solution and can be discarded from the analysis.)

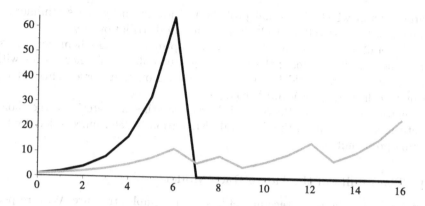

Fig. 1. Accumulated capital after 16 rounds of betting on the outcome of a bent coin flip with bias $\theta = 3/4$ and even odds. The black graph shows the performance of the strategy which maximizes $E[R]$, that is, betting the entire capital on the most likely event; this leads to a short life of explosive growth and then bankruptcy. The gray graph illustrates the strategy which maximizes $E[W]$, as explained in the next section; this strategy leads to a moderate but consistent exponential growth.

Having changed representation in this way, we can rewrite the expected doubling rate as

$$E[W] = \sum_i p_i \log(b_i o_i)$$

$$= \sum_i p_i \log\left(b_i \times \frac{c}{r_i}\right)$$

$$= \sum_i p_i \log\left(\frac{1}{r_i}\right) - \sum_i p_i \log\left(\frac{1}{b_i}\right) + \log c$$

$$= \sum_i p_i \log\left(\frac{p_i}{r_i}\right) - \sum_i p_i \log\left(\frac{p_i}{b_i}\right) + \log c$$

Considering the shape of the last two expressions, it becomes apparent that the bookmaker and the gambler are in a completely symmetric situation when $c = 1$: Both are trying to express an approximation of the winning probabilities of the horses by the means of the two distributions they control, namely, the bets and the odds. Whoever has the better approximation will, in the long run, make money at the expense of the other. If $c < 1$, the game is unfairly skewed in favor of the house so that the bookmaker can make money even when the gambler has a better approximation of the underlying distribution.

A compact representation of these statements is that

$$E[W] = D(p \,\|\, r) + D(p \,\|\, b) + \log c,$$

where $D(s \,\|\, t)$ is the **Kullback-Leibler divergence** from s to t [8]. This divergence is a measure of the error a probability distribution t will make in an environment in which the actual probabilities are given by s. Its minimum is obtained at $D(s \,\|\, s) = 0$, and $D(s \,\|\, t) > 0$ for all distributions $t \neq s$.

As either of these representations show, the optimal betting scheme for a horse race is the one that matches the underlying probabilities: If a horse wins with probability 1/3, you should bet 1/3 of you capital on that horse. This betting scheme is called **proportional betting**.

Somewhat surprisingly, the odds of the race are thus immaterial to the choice of betting scheme. The optimal capital distribution is determined solely by the winning probabilities.

1.4   The Limits of the Horse Race Model

As it turns out, the horse race model has a very simple structure: With respect to long-term growth, the optimal strategy is simply to translate your subjective probability distribution into a distribution of capital. As a consequence, updates of your subjective probabilities translate directly into updates of your capital distribution. This gives rise to a tight connection between the collection of information and increases in doubling rates.

However, this correspondence relies on a number of assumptions that are particular to the horse race model, such as the assumption that the situation involves only one random variable, and that the gambler can distribute his wealth onto the horses without any restrictions. In more complex and more realistic situations, an agent's representation of the environment may contain more variables and fewer feasible strategies. In such a situation, not all messages will afford the same opportunities for capital growth.

In fact, even a random variable which strongly affects an agent's payoff might not contain any useful information. If you don't own an umbrella, it might not be worth anything for you to know whether it rains or not, even if it changes your utility drastically. Thus, the announcement of a fact about the world does not always translate into a possibility for strategy improvement.

The suggestion I want to make in this paper is that we take the notions of utility and strategy as primitives and derive a notion of relevance from those. This contrasts with Shannon information theory, which defines information independently of the agents using that information. It also contrasts with arithmetic value-of-information in using a logarithmic target statistic, rather than the nominal size of the capital.

Both of these assumptions lead to a number of unique features which are illustrated by several examples in section 3. However, in the next section, I will give a few more formal definitions and discuss some of the properties of the concept of relevant information.

2   Expected Relevant Information and Relevance Rates

Relevant information is a notion of information defined in terms of utility. The notion of utility itself only makes sense in the context of agents faced with choices, so I first need to define a notion of a decision problem.

Definition 1. *A **decision problem** $D = (S, \Omega, p, u)$ consists of*

- *a strategy set S;*
- *a sample space Ω;*
- *a probability measure $p : \Omega \to \mathbb{R}$;*
- *a utility function $u : S \times \Omega \to \mathbb{R}$.*

*When u is bounded and non-negative, we further define the **(expected) doubling rate** of the strategy s as*

$$W(s) = \int p(x) \log u(s, x) \, dx$$

$W^* = \sup_s W(s)$ *is the **optimal (expected) doubling rate** of the decision problem.*

By convention, we set $W = -\infty$ if $R = 0$ with positive probability. A risk of bankruptcy will thus outweigh potential gains of any finite magnitude.

Definition 2. *Let a decision problem $D = (S, \Omega, p, u)$ be given as above, and let Y be a random variable. Then the **posterior decision problem given the event** $Y = y$ is $D' = (S, \Omega, p', u)$, where $p'(x) = p(x \mid y)$. The **amount of relevant information** in $Y = y$ is the increase in optimal doubling rate that the announcement of $Y = y$ leads to,*

$$K(y) = \sup_{s \in S} W'(s) - \sup_{s \in S} W(s),$$

where W' is the doubling rate in D'. Further,

$$K(Y) = E_y[K(Y = y)] = \int p(y) \, K(Y = y) \, dy$$

*is the **expected amount of relevant information** contained in Y.*

Theorem 1. *Expected relevant information is non-negative.*

Proof. With respect to the marginal distribution of Y, we have the expectations

$$E_y \left[\sup_{s \in S} W'(s) \right] = \int p(y) \left(\sup_{s \in S} \int p(x \mid y) \log u(s, x) \, dx \right) dy$$

$$\geq \sup_{s \in S} \left(\int p(y) p(x \mid y) \log u(s, x) \, dx \, dy \right)$$

$$= \sup_{s \in S} \left(\int p(x) \log u(s, x) \, dx \right)$$

$$= \sup_{s \in S} W(s)$$

$$= E_y \left[\sup_{s \in S} W(s) \right],$$

where the last equality follows from the fact that W does not depend on Y. The expected posterior doubling rate is thus higher than the expected prior, and so

$$E_y\left[K(Y)\right] = E_y\left[\sup_{s\in S} W'(s) - \sup_{s\in S} W(s)\right] = E_y\left[\sup_{s\in S} W'(s)\right] - E_y\left[\sup_{s\in S} W(s)\right]$$

is non-negative.

This proposition closely mirrors the well-known fact that Shannon information content is non-negative on average. So although bad news may occasionally represent a setback, information cannot hurt you on average. Notice also that the proof can be read as saying that an irrationally risk-averse agent can secure an unchanged average doubling rate by ignoring all incoming information.

Theorem 2. $1 - 2^{-K(Y)}$ *is the greatest taxation rate that an agent can accept, without expected loss, in exchange for learning the value of Y.*

Proof. Let $D = (S, \Omega, p, u)$ be the original decision problem, and let its prior and posterior doubling rates be W and W', respectively. Accepting a taxation rate of f in exchange for the ability to observe the value of Y will modify this problem so that the utility function in the posterior decision problem is downscaled by a factor of $1 - f$.

Let $K'(Y)$ denote the expected amount of relevant information contained in Y in this modified problem. We then have

$$K'(Y) = \int p(y)\left(\int p(x|y)\, \log\left((1-f)u(s,x)\right)\, dx\right) dy$$

$$= \int p(y)\left(\int p(x|y)\, \log u(s,x)\, dx\right) dy \; + \; \log(1-f)$$

$$= K(Y) + \log(1-f).$$

The agent can thus expect an on-average loss in the modified problem if and only if $K(Y) + \log(1 - f) < 0$. Solving this inequality for f gives the desired result.

To distinguish relevant information from Shannon information in the usual sense, we further define a concept of "raw" information:

Definition 3. *Let p be a probability measure on Ω, and let X be the random variable whose values are the sample points $\omega \in \Omega$. For any random variable Y, the expected amount of* **raw information** *contained in Y is then*

$$G(Y) = I(X;Y) = H(X) - H(X\,|\,Y).$$

Here $I(X;Y)$ is the mutual information between X and Y, and $H(X)$ and $H(X\,|\,Y)$ are the unconditional and conditional entropies of X, respectively [4,9].

Raw information is thus defined by comparing a variable Y to the unique state variable $X(\omega) = \omega$, that is, the maximally specific random variable. As a consequence, raw information is not a measure of dependence between two random variables in particular, but rather a measure of global decrease in uncertainty. Any source of uncertainty in your environment is thus a potential source of raw information, but not necessarily of relevant information.

These two measures of information suggest a natural measure of relevance:

Definition 4. *Let $D = (S, \Omega, p, u)$ be a decision problem, and Y a random variable on Ω. Then the* **relevance rate** *of Y is $K(Y)/G(Y)$.*

The relevance rate of a random variable can be both larger than and smaller than 1. However, the following theorem shows that the two coincide when an agent can bet with fair odds on the outcome of any random event whatsoever:

Theorem 3. *Let $D = (S, \Omega, p, u)$ be a decision problem in which the strategy space is the set of probability distributions on Ω, and Y is a random variable on Ω. Suppose further that the utility function u has the form $u(s, x) = s(x)v(x)$ for some non-negative, real-valued function v. Then $K(Y) = G(Y)$.*

Proof. This observation is due to Kelly [7]. We prove it by noting that a utility function of the form $u(s, x) = s(x)v(x)$ leads to a doubling rate of the form

$$
W(s) = \int p(x) \log s(x)v(x)\, dx
$$

$$
= \int p(x) \log \left(\frac{p(x)}{o(x)}\right) dx - \int p(x) \log \left(\frac{p(x)}{s(x)}\right) dx
$$

$$
= D(p \,\|\, o) - D(p \,\|\, s),
$$

where $o = 1/v$ can be interpreted as the odds, and s as the bets.

As a consequence of Jensen's inequality, the unique minimum of $D(p \,\|\, s)$ is $s = p$. The doubling rate is thus maximized by proportional betting ($s = p$), regardless of the probability environment and the odds. The optimal doubling rate under the distribution p is thus

$$
\sup_{s \in S} W(s) = D(p \,\|\, o) - D(p \,\|\, p)
$$

$$
= D(p \,\|\, o)
$$

$$
= \int p(x) \log \left(\frac{p(x)}{o(x)}\right) dx.
$$

Similarly, the optimal doubling rate given the condition $Y = y$ is

$$
\sup_{s \in S} W'(s) = \int p(x|y) \log \left(\frac{p(x|y)}{o(x)}\right) dx.
$$

Taking expectations with respect to Y and subtracting the prior doubling rate from the posterior, we find

$$K(Y) = E_y \left[\sup_{s \in S} W'(s) \right] - E_y \left[\sup_{s \in S} W(s) \right]$$
$$= E_y \left[\int p(x|y) \log p(x|y) \, dx \right] - E_y \left[\int p(x) \log p(x) \, dx \right]$$
$$= -H(X|Y) - (-H(X))$$
$$= G(X).$$

This establishes the desired equality.

3   Examples of Relevance Measurements

Having now introduced the notion of relevance, I would to like present a series of examples that illustrate how and where the concept can be used.

In the following five subsections, I will thus sketch five different gambling situations and analyze their informational properties. The first three cases are intended as toy examples illustrating the mechanics of relevant information, while the last two hint in the direction of more realistic applications.

3.1   A Horse Race with Pocket

Consider a horse race with winning probabilities $p = (2/3, 1/3)$ and odds $o = (2, 2)$. If you place a proportion b of your capital on the first horse in this race, your doubling rate will be

$$E[W(b)] = \frac{2}{3} \log (2b) + \frac{1}{3} \log (2(1 - b)).$$

As discussed in section 1, this doubling rate is maximized by $b^* = 2/3$, and the optimal doubling rate is then

$$E[W(b^*)] = \frac{2}{3} \log \left(\frac{2 \cdot 2}{3} \right) + \frac{1}{3} \log \left(\frac{2 \cdot 1}{3} \right) = 0.08 \text{ bits.}$$

However, this assumes that you invest your whole capital in the horse race. What does the situation look like if you could keep some of your money in your pocket rather than risking it in the game?

In such a situation, a gambling strategy would be defined by two parameters, the fraction f of capital invested in the game, and the fraction b of that capital betted on the first horse. Allowing f to take other values than 1 would change the payoff structure of the game as informally summarized in Table 1. In effect, the modified race would have a third horse which always paid one cent of winnings for one cent of bets.

In this modified game, the expected doubling rate associated with a strategy (f, b) is

$$E[W(f, b)] = \frac{2}{3} \log (1 - f + 2bf) + \frac{1}{3} \log (1 - f + 2(1 - b)f).$$

Table 1. Payoff structures for two different gambling situations: Once in which each situation comes with a single and unique source of income, and one in which there is a universal "pocket" source with a fixed rate of return.

Winner	1	2		Winner	0	1	2
Returns	2	0		Returns	$1-f$	$2f$	0
Returns	0	2		Returns	$1-f$	0	$2f$

By differentiating, we find that this doubling rate is optimal when

$$b = \frac{3f+1}{6f}.$$

It thus turns out that the doubling rate function has an optimal ridge running through the unit square. This ridge contains the old optimum $(1, 2/3)$, but also other equally good strategies like $(1/2, 5/6)$ and $(1/3, 1)$. Since the odds of the game favors the gambler, investments less than $f = 1/3$ are strictly irrational.

Note that if we somehow learned that the first horse would win the game, the expression for the expected doubling rate would reduce to

$$E\left[W(f,b)\right] = \log\left(1 + (2b-1)f\right).$$

This expression is optimized by $f = b = 1$, corresponding to the fact that one should invest as much money as possible in a rigged game. With this betting scheme, the doubling rate would be $\log 2 = 1$, consistent with the fact that this game would deterministically double your capital. The transition between the prior and posterior doubling rate function is illustrated in Figure 2.

Note that in this game, it turned out that the relevance analysis did not actually lead to a new result: Since an optimal solution existed which did not make use of the pocket, information about the winning horse had the same relevance rate in either game. However, we are now in a better position to vary more parameters, and the following examples will present some scenarios in which relevant and raw information diverge.

3.2   Guessing with Optional Investment

Suppose you can invest any fraction of your capital in a lottery defined as follows: If you can guess the four binary digits of my credit card code in a single try, you get your investment back 16-fold; otherwise, you lose it. Note that unlike the horse race, this game does not allow for arbitrary capital distributions.

Let us assume that you invest a fraction f of your capital in this lottery and keep a fraction of $1 - f$ in your pocket. Since your chance of guessing correctly is $1/2^4$, your expected doubling rate will be

$$E\left[W(f)\right] = \left(\frac{1}{2^4}\right)\log(1 - f + 16f) + \left(1 - \frac{1}{2^4}\right)\log(1 - f).$$

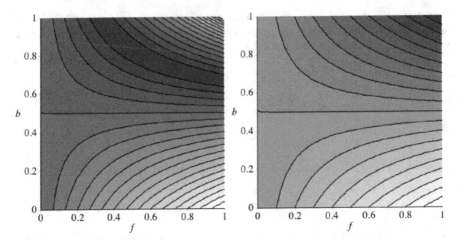

Fig. 2. Prior and posterior doubling rates for various combinations of f and b. The left graph shows the contours of W when the first horse wins with probability $p_1 = 2/3$, and the right graph shows the same for $p_1 = 1$. Darker hues correspond to higher values, but for visibility, the two graphs use different color scales.

This is a decreasing function in f, and your optimal strategy is $f^* = 0$, i.e., not betting anything.

However, suppose that you have an inside source that can supply you with some of the digits of the credit card code. For each digit you receive, your chance of guessing the code in a single attempt obviously increases; more specifically, the new expected doubling rate will be

$$E\left[W_i(f)\right] = \left(\frac{2^i}{2^4}\right)\log(1 + 15f) + \left(1 - \frac{2^i}{2^4}\right)\log(1 - f)$$

after you have received i of the four digits. As illustrated in Figure 3, this function attains its maximum on the unit interval at $f^* = 0/15, 1/15, 3/15, 7/15, 15/15$ after $i = 0, 1, 2, 3, 4$ bits of the code has been revealed.

The optimal expected doubling rates in these cases are

$$W_i^* = 0.00, 0.04, 0.26, 1.05, 4.00, \qquad \text{for } i = 0, 1, 2, 3, 4.$$

It thus turns out that the four digits you receive are not equally relevant to you. The first contains only 0.04 bits of relevant information although the revealed digit contained one bit of raw information. The second contains 0.22 bits of relevant information per bit of raw information, the third 0.79, and the fourth 2.95.

The raw and the relevant information add up to the same number, 4 bits of information, but do so at different paces. This difference in accumulation speed is only present because you are not forced to invest all your money in the lottery: If you were, all four bits would supply you with exactly one bit of relevant information, giving them a relevance rate of 1.

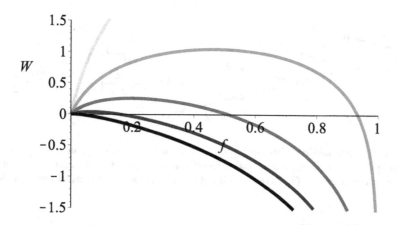

Fig. 3. Doubling rate as a function of investment level in the guessing game. Brighter gray lines represent doubling rates after more messages have been received.

3.3 Guessing with Irrelevant Side-Information

Continuing the code-guessing scenario as above, suppose now that you receive your side-information about my credit card code from an unreliable source which may abort the communication at any time. In this case, you have uncertainty about two independent variables, my actual credit card code (C), and the number digits you will receive (L).

Since receiving a digit removes uncertainty not only about C but also about L, you will, paradoxically, receive more than one bit of raw information per transmitted digit under these assumptions. The amount of relevant information you receive, however, will remain the same as in the previous scenario, since the information about L has no bearing on your guessing strategy.

To make this more concrete, suppose that just before transmitting each character, your source flips a coin to decide whether to continue or abort. This means that L takes the five values 0, 1, 2, 3, and 4 with probabilities $1/2$, $1/4$, $1/8$, $1/16$, and $1/16$, respectively (cf. Fig. 4). Excluding one of those possible outcomes at a time, beginning from the left, gives the conditional entropies shown in the table in Figure 4.

By subtracting these entropy levels from each other, we find that the four digits you receive contain $9/8$, $5/4$, $3/2$, and 2 bits of raw information, respectively. However, the optimality of a strategy in this game depends only on your chance of guessing the code in a single try. The amount of relevant information contained in the messages consequently remains as in the previous example.

This example illustrates how the addition of a variable to a model can change the information-theoretic analysis of a situation without necessarily the decision-theoretic. By my judgment, the notion of relevant information captures many intuitively important aspects of this scenario.

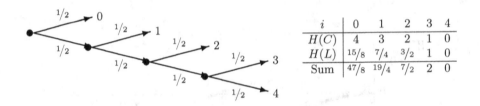

i	0	1	2	3	4
$H(C)$	4	3	2	1	0
$H(L)$	$15/8$	$7/4$	$3/2$	1	0
Sum	$47/8$	$19/4$	$7/2$	2	0

Fig. 4. Left, a probability distribution on L, the number of digits you receive from the unreliable source in example 3.3. Right, the decreasing uncertainty about the code (C) and transmission length (L) after i digits of my credit card code have been revealed.

3.4   Randomization

Suppose the two of us put down 1 cent for a game of Rock-Paper-Scissors, and that the winner gets both coins. If you play the three moves with probabilities $p = (p_1, p_2, p_3)$, and I play them with probabilities $q = (q_1, q_2, q_3)$, then your expected payoff is

$$u_1(p, q) = q_1(p_1 + 2p_2) + q_2(p_2 + 2p_3) + q_3(p_3 + 2p_1).$$

My expected payoff is $u_2 = 2 - u_1$.

This function is depends linearly on the probabilities I assign to the three moves, and it consequently has a maximum and a minimum in a corner point. Whatever your strategy is, one of my three pure strategies $(1, 0, 0)$, $(0, 1, 0)$, or $(0, 0, 1)$ is thus a best response to your strategy. Because of the zero-sum nature of the game, this best response will also minimize your expected payoff.

From your perspective, the consequence is that if you have chosen some randomized strategy $p = (p_1, p_2, p_3)$, and I have chosen a deterministic response q^* which minimizes $u_1(p, q)$, then your expected payoff is

$$u_1(p, q^*) = \min\{p_1 + 2p_2,\ p_2 + 2p_3,\ p_3 + 2p_1\},$$

and your doubling rate is the logarithm of this minimum. These quantities are optimal when you use the uniform strategy $p = (1/3, 1/3, 1/3)$.

This describes the game-theoretical aspects of this situation from an abstract, normative perspective. However, as Claude Shannon noted in the early 1950s [13], real people are in fact curiously bad at making random choices, and they invariably slip computable structure into the "random" sequences they produce. This means that a computer (or a statistician) equipped with a simple inference algorithm often outperforms humans vastly in randomization games such as Matching Pennies or Rock-Paper-Scissors.

The purpose of this example is to present a model of this limitation, and to measure how much it would change the situation if people had access to additional randomization resources like random number tables, coins, or quantum-random hardware. For clarity, I will analyze the most extreme case of this situation, namely that of a purely deterministic device that plays Rock-Paper-Scissors using a finite number of calls to a randomization oracle.

Suppose therefore that you have to play Rock-Paper-Scissors by submitting a publicly accessible program for a Turing machine. Since the program is completely deterministic, your strategy is going to be completely predictable, and your opponent can adapt perfectly to your strategy. This leads to a doubling rate of $\log \min\{0, 1, 2\} = -\infty$.

However, suppose now that your Turing machine has a module which can request a fixed number of fair coin flips per game. You can then encode these coin flips into the strategy in order to make it less predictable. The optimal way to do this is to feed the coin flips into an arithmetic decoder [9,4] which translates them into an approximately uniform distribution on $\{R, P, S\}$. The more coin flips you have, the flatter this distribution can get.

Table 2. Payoffs for increasingly randomized rock-paper-scissors strategies (left), and a graphical representation of the approximation process (right)

i	0	1	2	3	4	5	6	7	$\cdots$	∞
p_1	$1/1$	$1/2$	$2/4$	$3/8$	$6/16$	$11/32$	$22/64$	$43/128$	$\cdots$	$1/3$
p_2	$-$	$1/2$	$1/4$	$3/8$	$5/16$	$11/32$	$21/64$	$43/128$	$\cdots$	$1/3$
p_3	$-$	$-$	$1/4$	$2/8$	$5/16$	$10/32$	$21/64$	$42/128$	$\cdots$	$1/3$
u_1	0	$1/2$	$3/4$	$7/8$	$5/16$	$31/32$	$63/64$	$127/128$	$\cdots$	1

This situation is depicted in Table 2, which shows the arithmetic payoffs that you can achieve with i calls to the coin flipping module. As the table shows, the first coin flip will contain infinitely much relevant information, since it increases your arithmetic payoff from 0 to 1/2. The second contains

$$\log 3/4 - \log 1/2 = 0.59 \text{ bits of relevant information.}$$

The third, fourth, and fifth contain 0.22, 0.10, and 0.05 bits, respectively. As you add more calls to the randomization module in your program, the marginal benefit of adding another one decreases quite rapidly.

Readers familiar with Kelly gambling should note the difference between a horse race and the present model. In the horse race, a gambler chooses bets but does not control the probabilities. In the present example, the reverse is true.

Notice also that just like the horse race model implicitly measures deviations from the optimal strategy in terms of the Kullback-Leibler divergence, this model too defines a notion of distances from the uniform distributions on $\{R, P, S\}$. In this sense, the notion of relevant information as it appears here is a relatively natural generalization of entropy as a measure of how far away we are from an optimal solution (e.g., as in the context of ideal codeword lengths).

3.5 Non-cooperative Pragmatics

Following loosely the ideas from [11] and [6], suppose you regularly hire new staff from a pool of people that have taken two qualifying exams. Suppose further that the grades on these two exams, X and Y, are distributed uniformly on the set $\{1, 2, 3, \ldots, 10\}$. We can define the productivity of a hired person as units of profit per unit of salary, and we may assume that this profit rate depends on the two qualifying grades as

$$R = \frac{X + Y}{10}.$$

Hiring a person will thus in general affect your doubling rate $W = \log R$ either negatively or positively, depending on whether that person is qualified above or below a threshold of $X + Y = 10$ (cf. Fig. 5a and Table 3).

Table 3. The doubling rates associated with different combinations of grades

	1	2	3	4	5	6	7	8	9	10
10	0.14	0.26	0.38	0.49	0.58	0.68	0.77	0.85	0.93	1.00
9	0.00	0.14	0.26	0.38	0.49	0.58	0.68	0.77	0.85	0.93
8	−0.15	0.00	0.14	0.26	0.38	0.49	0.58	0.68	0.77	0.85
7	−0.32	−0.15	0.00	0.14	0.26	0.38	0.49	0.58	0.68	0.77
6	−0.51	−0.32	−0.15	0.00	0.14	0.26	0.38	0.49	0.58	0.68
5	−0.74	−0.51	−0.32	−0.15	0.00	0.14	0.26	0.38	0.49	0.58
4	−1.00	−0.74	−0.51	−0.32	−0.15	0.00	0.14	0.26	0.38	0.49
3	−1.32	−1.00	−0.74	−0.51	−0.32	−0.15	0.00	0.14	0.26	0.38
2	−1.74	−1.32	−1.00	−0.74	−0.51	−0.32	−0.15	0.00	0.14	0.26
1	−2.32	−1.74	−1.32	−1.00	−0.74	−0.51	−0.32	−0.15	0.00	0.14

However, under the distribution assumed here, it is in fact rational to hire a person in the absence of any information about that person's skill level. This holds because your average doubling rate across the whole pool of applicants, $E[W]$, is slightly larger than 0:

$$E\left[\log \frac{X + Y}{10}\right] = \sum_{i=1}^{10} \sum_{j=1}^{10} \Pr(X = i, Y = j) \times \left(\log \frac{i + j}{10}\right)$$

$$= 15.23 \text{ millibits}.$$

Hiring a randomly plucked person will thus give you an expected productivity of $E[2^W] = 2^{E[W]} = 2^{0.01523} = 1.01$ units of profit per unit of salary.

However, suppose you take a person into an interview, and that person shows you one of his or her grades. Assuming that you were shown the largest of the two grades, how much relevant information does this piece of data give you? At which grade level should you hire the applicant?

To answer this question, let $M = \max\{X, Y\}$ be the grade you were shown. The doubling rate you can expect from hiring an applicant with $M = m$ is then

$$E\left[W \mid M = m\right] = \sum_{i=1}^{10}\sum_{j=1}^{10} \Pr(X = i, Y = j \mid M = m)\log\frac{i+j}{10}.$$

By fixing m and summing up over all pairs (i, j) for which $i, j \leq m$ and $i = m$ or $j = m$ (cf. Fig. 5b), this doubling rate turns out to be negative for $m < 7$ and positive for $m \geq 7$. In other words, hiring a person whose largest grade is smaller than 7 will, on average, lead to a loss. The optimal decision in that case is thus to keep the salary in your pocket, retaining an expected doubling rate of 0.

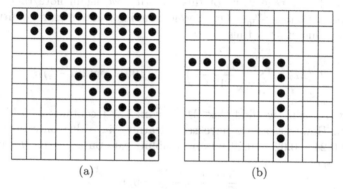

(a)   (b)

Fig. 5. (a) The applicants with positive productivity. (b) The applicants whose largest grade is $M = 7$ (see also Table 3).

So, observing $m < 7$ leads to a doubling rate of 0. On average, the doubling rate resulting from learning the value of M will thus be

$$\sum_{m=7}^{10} \Pr(M = m) \times E\left[W \mid M = m\right].$$

The probabilities in this sum are of the form

$$\Pr\left(M = m\right) = \frac{2m - 1}{100},$$

and the expected doubling rates are 0.08, 0.27, 0.44, 0.59 for $m = 7, 8, 9, 10$. A bit of computation then gives the posterior doubling rate $E\left[W'\right] = 0.24$ bits. Since the prior doubling rate was $E\left[W\right] = 0.01523 \approx 0.02$, an announcement of the value of M will on average give you

$$K(M) = 0.24 - 0.02 = 0.22 \text{ bits of relevant information.}$$

It follows that if you can observe the applicant's maximal grade before hiring him or her, your expected capital will, on average, grow by a factor of

$$R = 2^{E[W']} = 2^{0.24} = 1.18.$$

Further, you should thus be willing to trade up to $1 - 2^{-0.22} = 14.1\%$ of your future profits in return for this piece of information.

Finally, let us compute the amount of raw information contained in M. Observing the event $M = m$ narrows down the space of possible values for $X \times Y$ so that it has $2m - 1$ possible values instead of 100. Since these values are equally probable, the amount of information contained in the message $M = m$ is

$$H(X \times Y) - H(X \times Y \,|\, M = m) = \log 100 - \log(2m - 1).$$

To compute the average value of this quantity, we again note that $M = m$ has point probabilities $\frac{2m-1}{100}$. On average, the information gain resulting from learning the value of M is thus

$$G(M) = \sum_{m=1}^{10} \left(\frac{2m - 1}{100} \right) \times (\log 100 - \log(2m - 1)).$$

Computing this sum, we find that $M = \max\{X, Y\}$ contains 3.05 bits of raw information. However, as we have seen, learning its value only buys you an increase of $0.24 - 0.02 = 0.22$ bits in doubling rate on average. M thus has a relevance rate of

$$\frac{K(M)}{G(M)} = \frac{0.22}{3.05} = 0.07$$

bits of relevant information per bit of raw information.

4   Conclusion

In this paper, I have proposed a logarithmic of value-of-information measure as a quantitative elaboration of the concept of relevance.

This leads to an agent-oriented measure of relevance, as opposed to a system-oriented one [2]: It takes relevance to be a relation between events and agents rather than events and events. Because of this connection to agents and their utilities, the approach taken here forms a natural bond with ideas from decision theory, Bayesian statistics, and Shannon information theory. It thus represents a fairly conservative extension of the calculus of reasoning which is already canonical in the behavioral sciences.

This new concept of relevance may shed some new light on the ways in which dynamics of information can interact with problems of resource allocation. The examples I have given can, I believe, only be fully understood if we see the microeconomic and the information-theoretic aspects of the situation as two sides of a single coin. The notion of relevant information might be one out of several paths into such a style of analysis.

References

1. Avriel, M., Williams, A.C.: The value of information and stochastic programming. Operations Research 18(5), 947–954 (1970)
2. Borlund, P.: The concept of relevance in IR. Journal of the American Society for Information Science and Technology 54(10), 913–925 (2003)
3. Cooper, W.S.: A definition of relevance for information retrieval. Information Storage and Retrieval 7(1), 19–37 (1971)
4. Cover, T.M., Thomas, J.A.: Elements of information theory. Wiley-interscience (2006)
5. Floridi, L.: Understanding epistemic relevance. Erkenntnis 69(1), 69–92 (2008)
6. Glazer, J., Rubinstein, A.: A study in the pragmatics of persuasion: a game theoretical approach. Theoretical Economics 1(4), 395–410 (2006)
7. Kelly, J.L.: A new interpretation of information rate. IRE Transactions on Information Theory 2(3), 185–189 (1956)
8. Kullback, S., Leibler, R.A.: On information and sufficiency. The Annals of Mathematical Statistics 22(1), 79–86 (1951)
9. MacKay, D.: Information theory, inference and learning algorithms. Cambridge University Press (2003)
10. Shannon, C.E.: A mathematical theory of communication. Bell System Technical Journal 27, 379–423, 623–656 (1948)
11. Spence, M.: Job market signaling. The Quarterly Journal of Economics 87(3), 355–374 (1973)
12. Sperber, D., Wilson, D.: Relevance: Communication and Cognition. Blackwell Publishing (1995)
13. Zucker, J.-D., Meyer, C.: Apprentissage pour l'anticipation de comportements de joueurs humains dans les jeux à information complète et imparfaite: les «mind-reading machines». Revue d'intelligence Artificielle 14(3-4), 313–338 (2000)

A Generalization of Modal Frame Definability

Tin Perkov

Polytechnic of Zagreb, Croatia
`tin.perkov@tvz.hr`

Abstract. A class of Kripke frames is called modally definable if there
is a set of modal formulas such that the class consists exactly of frames
on which every formula from that set is valid, i.e. globally true under
any valuation. Here, existential definability of Kripke frame classes is
defined analogously, by demanding that each formula from a defining set
is satisfiable under any valuation. The notion of modal definability is then
generalized by combining these two. Model theoretic characterizations of
these types of definability are given.

Keywords: modal logic, model theory, modal definability.

1 Introduction

Some questions about the power of modal logic to express properties of relational structures are addressed in this paper. One way to determine the expressive power of a language is to establish a model theoretic characterization of
properties definable in that language. Such characterizations depend not only
on language, but also on a choice of semantics.

Only the Kripke semantics is considered in this paper. Even so, we have
several perspectives on the meaning of modal formulas: we distinguish between
their truth at some designated world, global truth on a model, and validity
on a frame. Because of this, model theory provides several characterizations of
modal definability, which answer to the following questions: which properties of
Kripke frames (Goldblatt-Thomason [4], see also [1]), Kripke models (de Rijke
and Sturm [3]), and pointed models (de Rijke, see [1]), are expressible in modal
logic.

Moreover, on the level of Kripke models, we can also use the notion of satisfiability, which is dual to the global truth. In [9] the notion of existential definability
of Kripke model classes (or properties) is defined as follows: a class is existen-
tially definable if there is a set of formulas such that this class consists exactly
of models in which every formula from that set is satisfiable. In [8] we combine
the usual (universal) and existential definability to obtain further generalizations
and we prove model theoretic characterizations for these types of definability.

The aim of this paper is to provide similar generalizations for the level of
Kripke frames. Since we abstract away from the effect of the valuations, frames
are the most natural semantic level for one of the basic purposes of modal logic:
to express properties of accessibility relation. So, it is of interest to get a broader
perspective on modal frame definability.

M. Colinet et al. (Eds.): ESSLLI 2012/2013, LNCS 8607, pp. 142–153, 2014.
© Springer-Verlag Berlin Heidelberg 2014

As it turns out, an appropriate notion of existential definability of Kripke frame classes demands that each formula from a defining set is satisfiable under any valuation. This is equivalent to the definability by the existential fragment of modal language enriched with the universal modality, similarly as it is on the level of models (see [9] and [8]). A generalized notion of modal definability, which is defined exactly like in the case of models, by combining universal and existential definability, also corresponds to a fragment of this language.

This paper provides two characterizations at the level of frames: one for the existential definability and one for the generalized definability. Characterizations are obtained by similar proof techniques as for the definability in the usual sense, which means that saturated models (ultraproducts and ultrafilter extensions) are deeply involved in the results. Characterization theorems are useful for obtaining non-definability results, some of which are given in Section 5. These examples also show that the conditions in the characterizations are necessary.

2   Preliminaries

For the sake of notational simplicity, only the basic modal language is considered in this paper, with the exception of few remarks concerning the universal modality.

Let Φ be a set of *propositional variables*. The syntax of the *basic propositional modal language* (BML) is given by

$$\varphi ::= p \mid \bot \mid \varphi_1 \vee \varphi_2 \mid \neg\varphi \mid \Diamond\varphi,$$

where $p \in \Phi$. We define other connectives and $\Box$ as usual. Namely, $\Box\varphi := \neg\Diamond\neg\varphi$.

The basic notions and results on the Kripke semantics are only briefly recalled here (see [1] for details if needed). A *Kripke frame* for the basic modal language is a relational structure $\mathfrak{F} = (W, R)$, where $W \neq \emptyset$ and $R \subseteq W \times W$. A *Kripke model* based on a frame $\mathfrak{F}$ is $\mathfrak{M} = (W, R, V)$, where $V : \Phi \to 2^W$ is a mapping called *valuation*. For $w \in W$, we call $(\mathfrak{M}, w)$ a *pointed model*.

The *truth* of a formula is defined locally and inductively as usual, and denoted $\mathfrak{M}, w \Vdash \varphi$. Namely, a formula of a form $\Diamond\varphi$ is *true* at $w \in W$ if $\mathfrak{M}, u \Vdash \varphi$ for some u such that Rwu. A valuation is naturally extended to all modal formulas by putting $V(\varphi) = \{w \in W : \mathfrak{M}, w \Vdash \varphi\}$.

We say that a formula is *globally true* on $\mathfrak{M}$ if it is true at every $w \in W$, and we denote this by $\mathfrak{M} \Vdash \varphi$. On the other hand, a formula is called *satisfiable* in $\mathfrak{M}$ if it is true at some $w \in W$.

If a formula φ is true at w under any valuation on a frame $\mathfrak{F}$, we write $\mathfrak{F}, w \Vdash \varphi$. We say that a formula is *valid* on a frame $\mathfrak{F}$ if we have $\mathfrak{M} \Vdash \varphi$ for any model $\mathfrak{M}$ based on $\mathfrak{F}$. This is denoted $\mathfrak{F} \Vdash \varphi$. For a set Σ of formulas we write $\mathfrak{F} \Vdash \Sigma$ if $\mathfrak{F} \Vdash \varphi$ for all $\varphi \in \Sigma$. A class $\mathcal{K}$ of Kripke frames is *modally definable* if there is Σ such that $\mathcal{K}$ consists exactly of frames on which every formula from Σ is valid, i.e. $\mathcal{K} = \{\mathfrak{F} : \mathfrak{F} \Vdash \Sigma\}$. If this is the case, we say that $\mathcal{K}$ is *defined* by Σ and denote $\mathcal{K} = \mathrm{Fr}(\Sigma)$.

Model theoretic closure conditions that are necessary and sufficient for an elementary class of frames (i.e. first-order definable property of relational structures) to be modally definable are given by the famous Goldblatt-Thomason Theorem.

Theorem (Goldblatt-Thomason [4]). *An elementary class $\mathcal{K}$ of frames is definable by a set of modal formulas if and only if $\mathcal{K}$ is closed under surjective bounded morphisms, disjoint unions and generated subframes, and reflects ultrafilter extensions.*

All of the frame constructions used in the theorem – bounded morphisms, disjoint unions, generated subframes and ultrafilter extensions – are presented briefly in Section 4 (see [1] for more details if needed). Just to be clear, we say that a class $\mathcal{K}$ *reflects* a construction if its complement $\mathcal{K}^c$, that is the class of all Kripke frames not in $\mathcal{K}$, is closed under that construction.

Now, an alternative notion of definability is proposed here as follows.

Definition 1. A class $\mathcal{K}$ of Kripke frames is called *modally $\exists$-definable* if there is a set Σ of modal formulas such that for any Kripke frame $\mathfrak{F}$ we have: $\mathfrak{F} \in \mathcal{K}$ if and only if each $\varphi \in \Sigma$, is satisfiable in $\mathfrak{M}$, for every model $\mathfrak{M}$ based on $\mathfrak{F}$. If this is the case, we denote $\mathcal{K} = \mathrm{Fr}_\exists(\Sigma)$.

The definition does not require that all formulas of Σ are satisfied at the same point – it suffices that each formula of Σ is satisfied at some point.

In the following, a notation $\mathrm{Mod}(F)$ is used for a class of structures defined by a first-order formula F. Similarly, if $\Sigma = \{\varphi\}$ is a singleton set of modal formulas, we write $\mathrm{Fr}_\exists(\varphi)$ instead of $\mathrm{Fr}_\exists(\{\varphi\})$.

Example 1. It is well-known that the formula $p \to \Diamond p$ defines reflexivity, i.e. $\mathrm{Fr}(p \to \Diamond p) = \mathrm{Mod}(\forall x R x x)$. Now, it is easy to see that $\mathrm{Fr}_\exists(p \to \Diamond p)$ is the class of all frames such that $R \neq \emptyset$, that is $\mathrm{Fr}_\exists(p \to \Diamond p) = \mathrm{Mod}(\exists x \exists y R x y)$. This class is not modally definable in the usual sense, since it is clearly not closed under generated subframes. Note that the condition $R \neq \emptyset$ is $\exists$-definable also by a simpler formula $\Diamond \top$.

Next, we define a notion which generalizes both universal and existential definability.

Definition 2. A class $\mathcal{K}$ of Kripke frames is called *modally $\forall\exists$-definable* if there is a pair (Σ_1, Σ_2) of sets of modal formulas such that for any Kripke frame $\mathfrak{F}$ we have: $\mathfrak{F} \in \mathcal{K}$ if and only if each $\varphi \in \Sigma_1$ is valid on $\mathfrak{F}$ and each $\varphi \in \Sigma_2$ is satisfiable in $\mathfrak{M}$, for any model $\mathfrak{M}$ based on $\mathfrak{F}$, i.e. $\mathcal{K} = \mathrm{Fr}(\Sigma_1) \cap \mathrm{Fr}_\exists(\Sigma_2)$.

Model theoretic characterizations of these notions are given in Section 5.

3   First and Second-Order Standard Translations

The starting point of correspondence between first-order and modal logic is the *standard translation*, a mapping that translates each modal formula φ to the first-order formula $\mathrm{ST}_x(\varphi)$, as follows:

$ST_x(p) = Px$, for each $p \in \Phi$,

$ST_x(\bot) = \bot$,

$ST_x(\neg\varphi) = \neg ST_x(\varphi)$,

$ST_x(\varphi \vee \psi) = ST_x(\varphi) \vee ST_x(\psi)$,

$ST_x(\Diamond\varphi) = \exists y(Rxy \wedge ST_y(\varphi))$.

Clearly, we have $\mathfrak{M}, w \Vdash \varphi$ if and only if $\mathfrak{M} \models ST_x(\varphi)[w]$, and $\mathfrak{M} \Vdash \varphi$ if and only if $\mathfrak{M} \models \forall x\, ST_x(\varphi)$. But, validity of a formula on a frame generally is not first-order expressible, since we need to quantify over valuations. We have a second-order standard translation, that is, $\mathfrak{F} \Vdash \varphi$ if and only if $\mathfrak{F} \models \forall P_1 \ldots \forall P_n \forall x\, ST_x(\varphi)$, where $P_1, \ldots, P_n$ are monadic second-order variables, one for each propositional variable occurring in φ. So, the notion of modal definability is equivalent to the definability by a set of second-order formulas of the form $\forall P_1 \ldots \forall P_n \forall x\, ST_x(\varphi)$. However, in many cases a formula of this type is equivalent to a first-order formula. Namely, this holds for any *Sahlqvist formula* (the definition is omitted here – see [10] or [1]), for which an equivalent first order formula is effectively computable. On the other hand, the Goldblatt-Thomason Theorem characterizes those first-order properties that are modally definable.

Now, $\exists$-definability is clearly also equivalent to the definability by a type of second-order formulas – those of the form $\forall P_1 \ldots \forall P_n \exists x\, ST_x(\varphi)$. Consider another example of a modally $\exists$-definable class.

Example 2. The condition $F = \exists x \forall y(Rxy \rightarrow \exists z Ryz)$ is not modally definable, since it is not closed under generated subframes, but it is modally $\exists$-definable by the formula $\varphi = p \rightarrow \Box\Diamond p$.

To prove this, we need to show $\mathrm{Fr}_\exists(\varphi) = \mathrm{Mod}(F)$. But $\mathfrak{F} = (W, R) \in \mathrm{Fr}_\exists(\varphi)$ if and only if $\mathfrak{F} \models \forall P \exists x(Px \rightarrow \forall y(Rxy \rightarrow \exists z(Ryz \wedge Pz)))$. So in particular, under the assignment which assigns the entire W to the second-order variable P, we get $\mathfrak{F} \models \exists x \forall y(Rxy \rightarrow \exists z Ryz)$, thus $\mathfrak{F} \in \mathrm{Mod}(F)$. The reverse inclusion is proved similarly.

Other changes of quantifiers or the order of first and second-order quantifiers would result in other types of definability, perhaps also worthy of exploring. In fact, this has already been done by Venema [12] and Hollenberg [7], who consider *negative definability*, which corresponds to second-order formulas of the form $\forall x \exists P_1 \ldots \exists P_n\, ST_x(\neg\varphi)$. The class of frames negatively defined by Σ is denoted $\mathrm{Fr}^-(\Sigma)$. It should be noted here that the definition of $\forall\exists$-definability is inspired by the analogous notion of $\pm$-definability from [7].

A general characterization of negative definability has not been obtained, and neither has been a characterization of elementary classes which are negatively definable – it even remains unknown if all negatively definable classes are in fact elementary. But, to digress a little from the main point of this paper, we easily get the following fairly broad result.

Proposition 1. *Let φ be a modal formula which has a first-order local correspondent, i.e. there is a first-order formula $F(x)$ such that for any frame $\mathfrak{F} = (W, R)$ and any $w \in W$ we have $\mathfrak{F}, w \Vdash \varphi$ if and only if $\mathfrak{F} \models F(x)[w]$. (In particular, this holds for any Sahlqvist formula.)*

Then we have $\mathrm{Fr}^-(\varphi) = \mathrm{Mod}(\forall x \neg F(x))$.

Proof. We have $\mathfrak{F} \in \mathrm{Fr}^-(\varphi)$ if and only if $\mathfrak{F} \models \forall x \exists P_1 \ldots \exists P_n \, \mathrm{ST}_x(\neg\varphi)$ if and only if $\mathfrak{F} \not\models \exists x \forall P_1 \ldots \forall P_n \, \mathrm{ST}_x(\varphi)$. But this means that there is no $w \in W$ such that $\mathfrak{F} \models \forall P_1 \ldots \forall P_n \, \mathrm{ST}_x(\varphi)[w]$. The latter holds if and only if $\mathfrak{F}, w \Vdash \varphi$, which is by assumption equivalent to $\mathfrak{F} \models F(x)[w]$. The fact that such w does not exist, is equivalent to $\mathfrak{F} \in \mathrm{Mod}(\forall x \neg F(x))$. $\qquad\square$

So for example, since $p \to \Diamond p$ locally corresponds to Rxx, we have that $p \to \Diamond p$ negatively defines irreflexivity, which is not modally definable property, since it is not preserved under surjective bounded morphisms.

4   Model-Theoretic Constructions

This section can be used, if needed, as a reference for the basic facts about the constructions used in the proofs of the characterizations. Otherwise it can be omitted.

A *bisimulation* between Kripke models $\mathfrak{M} = (W, R, V)$ and $\mathfrak{M}' = (W', R', V')$ is a relation $Z \subseteq W \times W'$ such that:

(at) if wZw' then we have: $w \in V(p)$ if and only if $w' \in V'(p)$, for all $p \in \Phi$,

(forth) if wZw' and Rwv, then there is a v' such that vZv' and $R'w'v'$,

(back) if wZw' and $R'w'v'$, then there is a v such that vZv' and Rwv.

The basic property of bisimulations is that (at) extends to all formulas: if wZw' then $\mathfrak{M}, w \Vdash \varphi$ if and only if $\mathfrak{M}', w' \Vdash \varphi$, i.e. $(\mathfrak{M}, w)$ and $(\mathfrak{M}', w')$ are modally equivalent. We get the definition of bisimulation between frames by omitting the condition (at).

A *bounded morphism* from a frame $\mathfrak{F} = (W, R)$ to $\mathfrak{F}' = (W', R')$ is a function $f : W \to W'$ such that:

(forth) Rwv implies $R'f(w)f(v)$,

(back) if $R'f(w)v'$, then there is v such that $v' = f(v)$ and Rwv.

Clearly, the graph of a bounded morphism is a bisimulation.

A *generated subframe* of $\mathfrak{F} = (W, R)$ is a frame $\mathfrak{F}' = (W', R')$ where $W' \subseteq W$ such that $w \in W'$ and Rwv implies $v \in W'$, and $R' = R \cap (W' \times W')$. A *generated submodel* of $\mathfrak{M} = (W, R, V)$ is a model based on a generated subframe, with the valuation $V'(p) = V(p) \cap W'$, for all $p \in \Phi$. It is easy to see that the global truth of a modal formula is preserved on a generated submodel.

The *disjoint union* of a family of models $\{\mathfrak{M}_i = (W_i, R_i, V_i) : i \in I\}$ is the model $\biguplus_{i \in I} \mathfrak{M}_i = (W, R, V)$ such that:

(1) $W = \bigcup_{i \in I}(W_i \times \{i\})$,

(2) $R(w, i)(v, j)$ if and only if $i = j$ and $R_i wv$,

(3) $(w, i) \in V(p)$ if and only if $w \in V_i(p)$, for all p.

It is easy to see that the disjoint union preserves the global truth of a modal formula. The definition of the disjoint union of a family of frames is obtained by omitting (3).

To define the ultraproducts and ultrafilter extensions, we need the notion of ultrafilters. An *ultrafilter* over a set $I \neq \emptyset$ is a family $U \subseteq \mathcal{P}(I)$ such that:

(1) $I \in U$,

(2) if $A, B \in U$, then $A \cap B \in U$,

(3) if $A \in U$ and $A \subseteq B \subseteq I$, then $B \in U$,

(4) for all $A \subseteq I$ we have: $A \in U$ if and only if $I \setminus A \notin U$.

The existence of ultrafilters is provided by a fact that any family of subsets which has the finite intersection property (that is, each finite intersection is non-empty) can be extended to an ultrafilter (see e.g. [2]).

Let $\{\mathfrak{M}_i = (W_i, R_i, V_i) : i \in I\}$ be a family of Kripke models and let U be an ultrafilter over I. The *ultraproduct* of this family over U is the model $\prod_U \mathfrak{M}_i = (W, R, V)$ such that:

(1) W is the set of equivalence classes f^U of the following relation defined on the Cartesian product of the family: $f \sim g$ if and only if $\{i \in I : f(i) = g(i)\} \in U$,

(2) $f^U R g^U$ if and only if $\{i \in I : f(i) R_i g(i)\} \in U$,

(3) $f^U \in V(p)$ if and only if $\{i \in I : f(i) \in V_i(p)\} \in U$, for all p.

The basic property of ultraproducts is that (3) extends to all formulas.

Proposition 2. *Let $\{\mathfrak{M}_i : i \in I\}$ be a family of Kripke models and let U be an ultrafilter over I.*

Then we have $\prod_U \mathfrak{M}_i, f^U \Vdash \varphi$ if and only if $\{i \in I : \mathfrak{M}_i, f(i) \Vdash \varphi\} \in U$, for any f^U. Furthermore, we have $\prod_U \mathfrak{M}_i \Vdash \varphi$ if and only if $\{i \in I : \mathfrak{M}_i \Vdash \varphi\} \in U$.

This is an analogue of Łoś's Fundamental Theorem on ultraproducts from the first-order model theory (see [2] for this, and [1] for the proof of the modal analogue). Łoś's Theorem also implies that an elementary class of models is closed under ultraproducts.

An ultraproduct such that $\mathfrak{M}_i = \mathfrak{M}$ for all $i \in I$ is called an *ultrapower* of $\mathfrak{M}$ and denoted $\prod_U \mathfrak{M}$. From Łoś's Theorem it follows that any ultrapower of a model is elementarily equivalent to the model, that is, the same first-order sentences are true on $\mathfrak{M}$ and $\prod_U \mathfrak{M}$. Definition of an ultraproduct of a family of frames is obtained by omitting the clause regarding valuation.

Another notion needed in the proofs of the characterizations is *modal saturation*, the modal analogue of ω-*saturation* from the classical model theory. The definition of saturation is omitted here (see e.g. [1]), since we only need some facts which it implies:

- While a bisimulation implies modal equivalence, the converse generally does not hold, but it does hold for modally saturated models. In fact, a modal equivalence between points of modally saturated models is a bisimulation.
- Any ω-saturated Kripke model is also modally saturated (see [1] for proofs of these facts).

Finally, the *ultrafilter extension* of a model $\mathfrak{M} = (W, R, V)$ is the model $ue\mathfrak{M} = (\mathrm{Uf}(W), R^{ue}, V^{ue})$, where $\mathrm{Uf}(W)$ is the set of all ultrafilters over W, $R^{ue}uv$ holds if and only if $A \in v$ implies $m_\Diamond(A) \in u$, where $m_\Diamond(A)$ denotes the set of all $w \in W$ such that Rwa for some $a \in A$, and $u \in V^{ue}(p)$ if and only if $V(p) \in u$. The basic property is that this extends to any modal formula, i.e. we have $u \in V^{ue}(\varphi)$ if and only if $V(\varphi) \in u$ (see [1]). From this it easily follows that the global truth of a modal formula is preserved on the ultrafilter extension. Another important fact is that the ultrafilter extension of a model is modally saturated (see [1]).

The ultrafilter extension of a frame $\mathfrak{F} = (W, R)$ is $\mathrm{ue}\mathfrak{F} = (\mathrm{Uf}(W), R^{\mathrm{ue}})$.

5   Characterizations

Arguments and techniques used in the proofs of the following characterizations are similar to the ones used in the proof of Goldblatt-Thomason theorem as presented in [1], so the reader might find it interesting to compare these proofs to note analogies and differences.

Theorem 1. *Let $\mathcal{K}$ be an elementary class of Kripke frames. Then $\mathcal{K}$ is modally $\exists$-definable if and only if it is closed under surjective bounded morphisms and reflects generated subframes and ultrafilter extensions.*

Proof. Let $\mathcal{K} = \mathrm{Fr}_{\exists}(\Sigma)$. Let $\mathfrak{F} = (W, R) \in \mathcal{K}$ and let f be a surjective bounded morphism from $\mathfrak{F}$ to some $\mathfrak{F}' = (W', R')$. Take any $\varphi \in \Sigma$ and any model $\mathfrak{M}' = (W', R', V')$ based on $\mathfrak{F}'$. Put $V(p) = \{w \in W : f(w) \in V'(p)\}$. Then V is a well defined valuation on $\mathfrak{F}$. Put $\mathfrak{M} = (W, R, V)$. Since $\mathfrak{F} \in \mathcal{K}$, there exists $w \in W$ such that $\mathfrak{M}, w \Vdash \varphi$. But then $\mathfrak{M}', f(w) \Vdash \varphi$. This proves that $\mathcal{K}$ is closed under surjective bounded morphisms.

To prove that $\mathcal{K}$ reflects generated subframes and ultrafilter extensions, let $\mathfrak{F} = (W, R) \notin \mathcal{K}$. This means that there is $\varphi \in \Sigma$ and a model $\mathfrak{M} = (W, R, V)$ based on $\mathfrak{F}$ such that $\mathfrak{M} \Vdash \neg\varphi$. Let $\mathfrak{F}' = (W', R')$ be a generated subframe of $\mathfrak{F}$. Define $V'(p) = V(p) \cap W'$, for all p. Then we have $\mathfrak{M}' \Vdash \neg\varphi$, which proves $\mathfrak{F}' \notin \mathcal{K}$, as desired. Also, $\mathrm{ue}\mathfrak{M}$ is a model based on the ultrafilter extension $\mathrm{ue}\mathfrak{F}$ and we have $\mathrm{ue}\mathfrak{M} \Vdash \neg\varphi$, which proves $\mathrm{ue}\mathfrak{F} \notin \mathcal{K}$.

For the converse, let $\mathcal{K}$ be an elementary class of frames that is closed under surjective bounded morphisms and reflects generated subframes and ultrafilter extensions. Denote by Σ the set of all formulas that are satisfiable in all models based on all frames in $\mathcal{K}$. Then $\mathcal{K} \subseteq \mathrm{Fr}_{\exists}(\Sigma)$ and it remains to prove the reverse inclusion.

Let $\mathfrak{F} = (W, R) \in \mathrm{Fr}_{\exists}(\Sigma)$. Let Φ be a set of propositional variables that contains a propositional variable p_A for each $A \subseteq W$. Let $\mathfrak{M} = (W, R, V)$, where $V(p_A) = A$ for all $A \subseteq W$. Denote by Δ the set of all modal formulas over Φ which are globally true on $\mathfrak{M}$. Now, for any finite $\delta \subseteq \Delta$ there is $\mathfrak{F}_{\delta} \in \mathcal{K}$ and a model $\mathfrak{M}_{\delta}$ based on $\mathfrak{F}_{\delta}$ such that $\mathfrak{M}_{\delta} \Vdash \delta$. Otherwise, since Δ is closed under conjunctions, there is $\varphi \in \Delta$ such that $\neg\varphi \in \Sigma$, thus $\neg\varphi$ is satisfiable in $\mathfrak{M}$, which contradicts $\mathfrak{M} \Vdash \Delta$.

Now, let I be the family of all finite subsets of Δ. For each $\varphi \in \Delta$, put $\hat{\varphi} = \{\delta \in I : \varphi \in \delta\}$. The family $\{\hat{\varphi} : \varphi \in \Delta\}$ clearly has the finite intersection property, so it can be extended to an ultrafilter U over I. But for all $\varphi \in \Delta$ we have $\{\delta \in I : \mathfrak{M}_{\delta} \Vdash \varphi\} \supseteq \hat{\varphi}$ and $\hat{\varphi} \in U$, thus $\{\delta \in I : \mathfrak{M}_{\delta} \Vdash \varphi\} \in U$, so the Proposition 2 implies $\prod_U \mathfrak{M}_{\delta} \Vdash \varphi$. The model $\prod_U \mathfrak{M}_{\delta}$ is based on the frame $\prod_U \mathfrak{F}_{\delta}$. Since $\mathcal{K}$ is elementary, it is also closed under ultraproducts, so

$\prod_U \mathfrak{F}_\delta \in \mathcal{K}$. It remains to prove that there is a surjective bounded morphism from some ultrapower of $\prod_U \mathfrak{F}_\delta$ to a generated subframe of $\mathfrak{ueF}$. Then the assumed properties of $\mathcal{K}$ imply that $\mathfrak{F} \in \mathcal{K}$, as desired.

Classical model theory provides us with an ω-saturated ultrapower of $\prod_U \mathfrak{M}_\delta$ (cf. [2]). Let $\mathfrak{M}_\Delta$ be such an ultrapower. We have that $\mathfrak{M}_\Delta$ is modally saturated. Also, it is elementarily equivalent to $\prod_U \mathfrak{M}_\delta$, so using standard translation we obtain $\mathfrak{M}_\Delta \Vdash \Delta$. The model $\mathfrak{M}_\Delta$ is based on a frame $\mathfrak{F}_\Delta$, which is an ultrapower of $\prod_U \mathfrak{F}_\delta$. Now define a mapping from $\mathfrak{F}_\Delta$ to $\mathfrak{ueF}$ by putting $f(w) = \{A \subseteq W : \mathfrak{M}_\Delta, w \Vdash p_A\}$.

First we need to prove that f is well-defined, i.e. that $f(w)$ is indeed an ultrafilter over W.

(1) We easily obtain $W \in f(w)$, since $p_W \in \Delta$ by the definition of V.

(2) If $A, B \in f(w)$, then $\mathfrak{M}_\Delta, w \Vdash p_A \wedge p_B$. Clearly, $\mathfrak{M} \Vdash p_A \wedge p_B \leftrightarrow p_{A \cap B}$. Thus $\mathfrak{M}_\Delta \Vdash p_A \wedge p_B \leftrightarrow p_{A \cap B}$, so $\mathfrak{M}_\Delta, w \Vdash p_{A \cap B}$, i.e. $A \cap B \in f(w)$.

(3) If $A \in f(w)$ and $A \subseteq B \subseteq W$, then from the definition of V it follows $\mathfrak{M} \Vdash p_A \rightarrow p_B$. But then also $\mathfrak{M}_\Delta \Vdash p_A \rightarrow p_B$, hence $\mathfrak{M}_\Delta, w \Vdash p_B$, so $B \in f(w)$.

(4) For all $A \subseteq W$ we have $\mathfrak{M} \Vdash p_A \leftrightarrow \neg p_{W \setminus A}$, which similarly as in the previous points implies $A \in f(w)$ if and only if $W \setminus A \notin f(w)$, as desired.

Assume for the moment that we have: $u = f(w)$ if and only if $(\mathfrak{ueM}, u)$ and $(\mathfrak{M}_\Delta, w)$ are modally equivalent. Since $\mathfrak{ueM}$ and $\mathfrak{M}_\Delta$ are modally saturated, the modal equivalence between their points is a bisimulation. So f is a bisimulation, but it is also a function, which means that it is a bounded morphism from $\mathfrak{F}_\Delta$ to $\mathfrak{ueF}$. But then the corestriction of f to its image is a surjective bounded morphism from an ultrapower of $\prod_U \mathfrak{F}_\delta$ to a generated subframe of $\mathfrak{ueF}$, which we needed.

So to conclude the proof, it remains to show that $u = f(w)$ holds if and only if $(\mathfrak{ueM}, u)$ and $(\mathfrak{M}_\Delta, w)$ are modally equivalent. Let $u = f(w)$. Then we have $\mathfrak{ueM}, u \Vdash \varphi$ if and only if $V(\varphi) \in u$, which is by the definition of f equivalent to $\mathfrak{M}_\Delta, w \Vdash p_{V(\varphi)}$. But the definition of V clearly implies $\mathfrak{M} \Vdash \varphi \leftrightarrow p_{V(\varphi)}$, so also $\mathfrak{M}_\Delta \Vdash \varphi \leftrightarrow p_{V(\varphi)}$, which provides the needed modal equivalence.

For the converse, the assumption implies that we have $\mathfrak{ueM}, u \Vdash p_A$ if and only if $\mathfrak{M}_\Delta, w \Vdash p_A$, for all $A \subseteq W$. This means that $V(p_A) = A \in u$ if and only if $A \in f(w)$, i.e. $u = f(w)$. □

In the characterization of $\forall\exists$-definability we need the following non-standard closure condition.

Definition 3. We say that a class $\mathcal{K}$ of Kripke frames is *closed under generated interframes* if the following holds:

Let $\mathfrak{F}_1$, $\mathfrak{F}$ and $\mathfrak{F}_2$ be frames such that $\mathfrak{F}_1$ is a generated subframe of $\mathfrak{F}$ and $\mathfrak{F}$ is a generated subframe of $\mathfrak{F}_2$. Then we have: if $\mathfrak{F}_1$ and $\mathfrak{F}_2$ are in $\mathcal{K}$, then $\mathfrak{F}$ is also in $\mathcal{K}$ (cf. [8] for the analogous notion for Kripke models).

Theorem 2. *Let $\mathcal{K}$ be an elementary class of Kripke frames. Then $\mathcal{K}$ is modally $\forall\exists$-definable if and only if it is closed under surjective bounded morphisms, disjoint unions and generated interframes, and reflects ultrafilter extensions.*

Proof. It is easy to show that any $\forall\exists$-definable class have the desired properties, using the same arguments as in the respective directions of the proofs of Goldblatt-Thomason theorem (see [1]) and Theorem 1.

For the converse, let $\mathcal{K}$ be an elementary class of frames that is closed under surjective bounded morphisms, disjoint unions and generated interframes, and reflects ultrafilter extensions. Let Σ_1 be the set of all formulas that are valid on all frames in $\mathcal{K}$, and let Σ_2 be the set of all formulas that are satisfiable in all models based on all frames in $\mathcal{K}$. Then $\mathcal{K} \subseteq \mathrm{Fr}(\Sigma_1) \cap \mathrm{Fr}_\exists(\Sigma_2)$ and it remains to prove the reverse inclusion.

Let $\mathfrak{F} \in \mathrm{Fr}(\Sigma_1) \cap \mathrm{Fr}_\exists(\Sigma_2)$ and let Φ be a set of propositional variables that contains p_A for each $A \subseteq W$. Let $\mathfrak{M}$ be a model based on $\mathfrak{F}$ such that $V(p_A) = A$ for all $A \subseteq W$. Let $\Delta_\forall$ be the set of all formulas over Φ which are globally true on $\mathfrak{M}$ and let $\Delta_\exists$ be the set of all formulas over Φ which are satisfiable in $\mathfrak{M}$.

Denote $D_\forall = \{\forall x\, \mathrm{ST}_x(\varphi) : \varphi \in \Delta_\forall\}$, $D_\exists = \{\exists x\, \mathrm{ST}_x(\varphi) : \varphi \in \Delta_\exists\}$, and $D = D_\forall \cup D_\exists$. It is easy to see that for all $F \in D$ there is a model $\mathfrak{M}_F$ based on some $\mathfrak{F}_F \in \mathcal{K}$ such that $\mathfrak{M}_F \models F$ (the opposite assumption easily leads to a contradiction).

Using the same arguments as in the proof of Theorem 1, we conclude that there is an ω-saturated model $\mathfrak{M}_\forall$ based on some frame $\mathfrak{F}_\forall \in \mathcal{K}$ such that $\mathfrak{M}_\forall \models D_\forall$, i.e. $\mathfrak{M}_\forall \Vdash \Delta_\forall$. We define a mapping f from $\mathfrak{F}_\forall$ to $\mathrm{ue}\mathfrak{F}$ by putting $f(w) = \{A \subseteq W : \mathfrak{M}_\forall, w \Vdash p_A\}$. In the same way as in the proof of Theorem 1, we show that f is a bounded morphism. Denote its image by $\mathfrak{F}'_\forall$. It is a generated subframe of $\mathrm{ue}\mathfrak{F}$, and since $\mathcal{K}$ is closed under surjective bounded morphisms, we have $\mathfrak{F}'_\forall \in \mathcal{K}$.

On the other hand, since $\mathcal{K}$ is closed under disjoint unions, we have that $\biguplus_{F \in D_\exists} \mathfrak{F}_F \in \mathcal{K}$, while clearly $\biguplus_{F \in D_\exists} \mathfrak{M}_F \models D_\exists$. Since $\mathcal{K}$ is elementary, it is closed under ultraproducts, so an ω-saturated ultrapower $\mathfrak{M}_\exists$ of the disjoint union $\biguplus_{F \in D_\exists} \mathfrak{M}_F$ is based on some $\mathfrak{F}_\exists \in \mathcal{K}$ and it holds $\mathfrak{M}_\exists \models D_\exists$. Hence, all formulas that are satisfiable in $\mathfrak{M}$ are also satisfiable in $\mathfrak{M}_\exists$. By contraposition, all formulas that are globally true on $\mathfrak{M}_\exists$ are also globally true on $\mathfrak{M}$, thus also on $\mathrm{ue}\mathfrak{M}$. It is not hard to show that the modal equivalence between worlds of $\mathfrak{M}_\exists$ and $\mathrm{ue}\mathfrak{M}$ is a surjective bisimulation (this follows immediately from Lemma 1 in [8]). The domain of this bisimulation is a generated submodel $\mathfrak{M}'_\exists$ of $\mathfrak{M}_\exists$.

To prove that this bisimulation is in fact a surjective bounded morphism from $\mathfrak{M}'_\exists$ to $\mathrm{ue}\mathfrak{M}$, it remains to prove that it is a function. Assume the opposite, i.e. that there is a world in $\mathfrak{M}_\exists$ which is modally equivalent to two different ultrafilters u, v in $\mathrm{ue}\mathfrak{M}$. Hence, u and v are modally equivalent, i.e. for all φ we have $V(\varphi) \in u$ if and only if $V(\varphi) \in v$. In particular, for all $A \subseteq W$ we have $V(p_A) = A \in u$ if and only if $V(p_A) = A \in v$, thus $u = v$. This proves that there is a surjective bounded morphism g from $\mathfrak{F}'_\exists$ to $\mathrm{ue}\mathfrak{F}$, where $\mathfrak{F}'_\exists$ is a generated subframe of $\mathfrak{F}_\exists$.

Let $\mathfrak{F}''_\exists$ be the frame built from $\mathrm{ue}\mathfrak{F} \uplus (\mathfrak{F}_\exists \setminus \mathfrak{F}'_\exists)$, by extending its accessibility relation with all pairs $(w, g(v))$, for w in $\mathfrak{F}_\exists \setminus \mathfrak{F}'_\exists$ and v in $\mathfrak{F}'_\exists$ such that v is accessible from w in $\mathfrak{F}_\exists$. Now, extend g to $\mathfrak{F}_\exists$ by putting $g(w) = w$ for w in $\mathfrak{F}_\exists \setminus \mathfrak{F}'_\exists$. This makes g a surjective bounded morphism from $\mathfrak{F}_\exists$ to $\mathfrak{F}''_\exists$. Since $\mathcal{K}$ is closed under surjective bounded morphisms, we have $\mathfrak{F}''_\exists \in \mathcal{K}$. Clearly, $\mathfrak{ue}\mathfrak{F}$ is a generated subframe of $\mathfrak{F}''_\exists$. We have already proved that there is $\mathfrak{F}'_\forall \in \mathcal{K}$ which is a generated subframe of $\mathfrak{ue}\mathfrak{F}$, so the closure under generated interframes implies $\mathfrak{ue}\mathfrak{F} \in \mathcal{K}$. Since $\mathcal{K}$ reflects ultrafilter extensions, it follows $\mathfrak{F} \in \mathcal{K}$. □

The following examples show that the conditions of Theorems 1 and 2, and Goldblatt-Thomason theorem, are minimal. Each example is an elementary class which satisfies all but one of the conditions of a characterization, thus showing that this condition cannot be omitted. Almost all claims are proved routinely, so most of the details are skipped.

Example 3. Irreflexivity, i.e. the class $\mathrm{Mod}(\forall x \neg Rxx)$, is not modally definable, since it is not closed under surjective bounded morphisms. It is easy to see that this class is closed under generated subframes, generated interframes, disjoint unions, and reflects ultrafilter extensions. This shows that the closure under surjective bounded morphisms cannot be omitted in Goldblatt-Thomason theorem or Theorem 2.

To show that this condition cannot be omitted from Theorem 1 either, consider the class $\mathrm{Mod}(\exists x \neg Rxx)$, i.e. the class of frames which are not reflexive. It is easy to construct an example which shows that this class is not closed under surjective bounded morphisms, but it is also not hard to show that it reflects generated subframes and ultrafilter extensions.

Example 4. The class $\mathrm{Mod}(\forall x \forall y Rxy)$ is obviously not closed under disjoint unions, but it is closed under surjective bounded morphisms, generated subframes and generated interframes, and reflects ultrafilter extensions. This proves that the closure under disjoint unions is essential in Goldblatt-Thomason theorem and Theorem 2. It is also obvious that this class does not reflect generated subframes, which means that this condition cannot be omitted in Theorem 1.

Example 5. The class $\mathrm{Mod}(\exists x \exists y Rxy)$ is not closed under generated subframes, but it satisfies all other conditions of Goldblatt-Thomason theorem.

Example 6. Let $\mathcal{K} = \mathrm{Mod}(\forall x Rxx \vee \exists x \forall y \neg Rxy)$. This is the class of all frames that are either reflexive or have a world with no access to any world. It is easy to see that $\mathcal{K}$ is closed under disjoint unions and surjective bounded morphisms, and reflects ultrafilter extensions. But, $\mathcal{K}$ is not closed under generated interframes, thus this condition cannot be omitted in Theorem 2.

To see this, let $\mathfrak{F}_1 = (\{w\}, \{(w,w)\})$, and $\mathfrak{F}_2 = (\{w,v,u\}, \{(w,w),(v,w)\})$. Obviously $\mathfrak{F}_1, \mathfrak{F}_2 \in \mathcal{K}$. Let $\mathfrak{F} = (\{w,v\}, \{(w,w),(v,w)\})$. Clearly, $\mathfrak{F}_1$ is a generated subframe of $\mathfrak{F}$, and $\mathfrak{F}$ is a generated subframe of $\mathfrak{F}_2$, but $\mathfrak{F} \notin \mathcal{K}$.

Example 7. Finally, the class $\mathcal{K} = \mathrm{Mod}(\forall x \exists y (Rxy \wedge Ryy))$, i.e. the property that every world has a reflexive R-successor, is closed under disjoint unions, generated subframes, generated interframes and surjective bounded morphisms, but does not reflect ultrafilter extensions. To prove the last claim, consider the frame $\mathfrak{F} = (\mathbb{N}, <)$, i.e. the set of natural numbers with the standard strict ordering. Obviously $\mathfrak{F} \in \mathcal{K}^c$. But, $\mathfrak{ue}\mathfrak{F} \in \mathcal{K}$. This follows from the fact that for each ultrafilter u over $\mathbb{N}$ and for each non-principal ultrafilter v over $\mathbb{N}$ we have $u <^{ue} v$ (see [1], p. 95).

The same frame shows in a similar way that the class $\mathrm{Mod}(\exists x Rxx)$ does not reflect ultrafilter extensions, and it is easy to see that it is closed under surjective bounded morphisms and reflects generated subframes. This shows that the reflection of ultrafilter extensions cannot be omitted in any of the characterizations.

6   Link to the Universal Modality

Although the approach of this paper is to define $\exists$-definability as a metalingual notion, it should be noted that it can be included in the language itself. That is, the satisfiability of a modal formula under any valuation on a frame can be expressed by a formula of the modal language enriched with the universal modality (BMLU). The syntax is an extension of the basic modal language by a new modal operator $A\varphi$, and we can also define its dual $E\varphi := \neg A \neg \varphi$. We call A the *universal modality*, and E the *existential modality*. The semantics of the new operators is standard modal semantics, with respect to the universal binary relation $W \times W$ on a frame $\mathfrak{F} = (W, R)$. This means that the standard translation of universal and existential operators is as follows (cf. [5] and [11]):

$$\mathrm{ST}_x(E\varphi) = \exists y\, \mathrm{ST}_y(\varphi),$$
$$\mathrm{ST}_x(A\varphi) = \forall y\, \mathrm{ST}_y(\varphi).$$

Now, let $\mathcal{K}$ be a class of Kripke frames. Clearly, $\mathcal{K}$ is modally $\exists$-definable if and only if it is definable by a set of formulas of the existential fragment of BMLU, i.e. by a set of formulas of the form $E\varphi$, where φ is a formula of BML. This immediately follows from the clear fact that for any frame $\mathfrak{F}$ and any φ we have $\mathfrak{F} \Vdash E\varphi$ if and only if $\mathfrak{F} \models \forall P_1 \ldots \forall P_n \exists y\, \mathrm{ST}_y(\varphi)$, where $P_1, \ldots, P_n$ correspond to propositional variables that occur in φ, and the latter holds if and only if φ is satisfiable under any valuation on $\mathfrak{F}$.

Goranko and Passy [5] gave a characterization that an elementary class is modally definable in BMLU if and only if it is closed under surjective bounded morphisms and reflects ultrafilter extension. So, from Theorem 1 we conclude that reflecting generated subframes, not surprisingly, is what distinguishes existential fragment within this language, at least with respect to elementary classes. Also, the usual notion of modal definability clearly coincides with the universal fragment of BMLU, hence the Goldblatt-Thomason Theorem tells us that closure under generated subframes and disjoint unions is essential for this fragment. Furthermore, from Theorem 2 it follows that closure under generated interframes and disjoint unions characterizes the union of universal and existential fragment of BMLU, i.e. definability by sets of formulas of the form $A\varphi$ or $E\varphi$, where φ is in BML.

On the other hand, a question is which modally $\exists$-definable classes are elementary, and whether there is an effective procedure analogous to the one for Sahlqvist formulas, to obtain a first-order formula equivalent to a second-order translation $\forall P_1 \ldots \forall P_n \exists x\, \mathrm{ST}_x(\varphi)$ for some sufficiently large and interesting class of modal formulas. Goranko and Vakarelov [6] answer this, and more: they provide a generalization of Sahlqvist formulas to languages with hybrid operators, including universal modal operator.

As for some further questions that might be worth exploring, we may be able to obtain general characterization theorems, without the assumption of the first-order definability. Furthermore, the results of this paper are easily generalized to the multi-modal framework, but more work is needed to obtain similar results for particular modal logics, for example temporal, with some restrictions on accessibility relations, e.g. transitivity.

References

1. Blackburn, P., de Rijke, M., Venema, Y.: Modal Logic. Cambridge University Press (2001)
2. Chang, C.C., Keisler, H.J.: Model Theory. Elsevier (1990)
3. de Rijke, M., Sturm, H.: Global Definability in Basic Modal Logic. In: Wansing, H. (ed.) Essays on Non-classical Logic. World Scientific Publishers (2001)
4. Goldblatt, R.I., Thomason, S.K.: Axiomatic classes in propositional modal logic. In: Crossley, J. (ed.) Algebra and Logic, pp. 163–173. Springer (1974)
5. Goranko, V., Passy, S.: Using the Universal Modality: Gains and Questions. Journal of Logic and Computation 2, 5–30 (1992)
6. Goranko, V., Vakarelov, D.: Sahlqvist Formulas in Hybrid Polyadic Modal Logics. Journal of Logic and Computation 11, 737–754 (2001)
7. Hollenberg, M.: Characterizations of Negative Definability in Modal Logic. Studia Logica 60, 357–386 (1998)
8. Perkov, T., Vuković, M.: Some characterization and preservation theorems in modal logic. Annals of Pure and Applied Logic 163, 1928–1939 (2012)
9. Perkov, T.: Towards a generalization of modal definability. In: Lassiter, D., Slavkovik, M. (eds.) New Directions in Logic, Language, and Computation. LNCS, vol. 7415, pp. 130–139. Springer, Heidelberg (2012)
10. Sahlqvist, H.: Completeness and correspondence in the first and second order semantics for modal logic. In: Kanger, S. (ed.) Proceedings of the Third Scandinavian Logic Symposium, Uppsala 1973, pp. 110–143. North-Holland, Amsterdam (1975)
11. van Benthem, J.: The range of modal logic. Journal of Applied Non-Classical Logics 9, 407–442 (1999)
12. Venema, Y.: Derivation rules as anti-axioms in modal logic. Journal of Symbolic Logic 58, 1003–1034 (1993)

Monoid Automata for Displacement Context-Free Languages[*]

Alexey Sorokin[1,2]

[1] Moscow State University, Faculty of Mathematics and Mechanics, Moscow, Russia
[2] Moscow Institute of Physics and Technology,
Faculty of Innovations and High Technologies, Moscow, Russia

Abstract. In 2007 Kambites presented an algebraic interpretation of Chomsky-Schützenberger theorem for context-free languages. We solve an analogous task for the class of displacement context-free languages which are equivalent to well-nested multiple context-free languages giving an interpretation of the corresponding theorem for that class in terms of monoid automata. We also show how such automata can be simulated on two stacks, introducing the simultaneous two-stack automaton. We compare different variants of its definition and show their equivalence basing on geometric interpretation of its memory operations.

1 Introduction

Through last decades the theory of monoid automata attracts a great interest both from the specialists in the theory of formal languages and algebra. The first are looking for a fine algebraic characterization of formal languages, which simplifies studying their properties and shows known theoretical facts in a wider scope. The algebraists are interested in the questions of effective computations in groups and semigroups where different variants of automata can be useful. Also the theory of monoid automata has some connections with the combinatorial group theory, e.g. with studying word problems for groups. For a more detailed survey and references see [6] or [23].

A monoid automaton (or valence automaton) is a finite automaton augmented with a register storing an element of a particular monoid. Each transition of the automaton multiplies the current element in the register by the monoid element associated with this transition. The automaton accepts a word iff it reaches a final state after reading the word with the monoid identity in the register. The usage of the memory register allows to recognize more complex languages then the usual automata do. Evidently, the recognizing power essentially depends from the monoid serving as the register.

The notion of monoid automaton is very useful since it offers a uniform treatment of different computational models. Assume that every successful computation in a particular model starts and terminates with an empty memory storage

---

[*] The work was partially supported by RFFI grants 11-01-00958a and NSh-1423.2014.1.

© Springer-Verlag Berlin Heidelberg 2014

and the model operates transforming the memory contents using some final set of states. This condition holds for most standard models (pushdown automata, embedded pushdown automata and many others). Then the set of memory operations obviously form a partial monoid and composition of the operations executed during the successful computation obviously equals the identity element. So the monoid automaton which is equivalent to a considered computational model uses the monoid of admissible memory operations.

For example, for the family of context-free languages every admissible operation is a composition of pushing and popping some symbols from the stack. The monoid of such operations is just the polycyclic monoid $\mathcal{B}_n = \{x_i, \overline{x}_i \mid 1 \leq i \leq n\}^*/\{x_i\overline{x}_i = 1 \mid 1 \leq i \leq n\}$ ([13]) where $x_1, \ldots, x_n$ are the elements of the stack alphabet and the equality $x_i\overline{x}_i = 1$ reflects the fact that popping x_i immediately after pushing it on the stack is the same as doing nothing. But this approach is useless in more complicated cases since we the structure of monoid of memory operations cannot be recovered so easily.

The alternative approach was developed by Kambites ([6]). He showed that in the case of context-free languages we may use the Chomsky-Schützenberger theorem, which states that every context-free language is the rational transduction of Dyck language,, which is the language of correct bracket sequences. By Kambites theorem it suffices to find a monoid with an identity language isomorphic to the set of correct bracket sequences and use its elements as memory contents. It is not very interesting in the case of context-free languages because such a monoid is unsurprisingly a polycyclic monoid but very useful in general since Chomsky-Schützenberger theorem is known for different families of languages.

In our work we consider the family of displacement context-free languages ([19]). There are many weakly equivalent grammar formalisms, such as well-nested multiple context-free grammars ([8]), non-duplicating macro grammars ([3], [17]). The class of displacement context-free languages also coincides with the class of string languages for simple context-free tree grammars ([11], [9]). Some computer scientists consider this class as a possible formalization for the notion of mildly context-sensitive language ([4], see [10] for discussion). Chomsky-Schützenberger theorem for this family of languages was first mentioned in [22], it also follows from a recent work of Kanazawa ([9], see Lemmas 39 and 42 of it). We use this theorem to give a characterization of displacement context-free languages in terms of monoid automata and then show how the elements of monoid constructed are interpreted as operations on two stacks.

The paper is organized as follows: first we recall the definition of monoid automaton and formulate the Kambites theorem. Then we define the family of displacement context-free languages and state the Chomsky-Schützenberger theorem for it. We construct a monoid, whose identity language is the multibracket language from this theorem, thus giving the characterization of displacement context-free languages in terms of monoid automata. Afterwards we show that this monoid is isomorphic to a particular submonoid of the cartesian product of two polycyclic monoids which allows us to interpret its elements as operations on the pair of stacks. Then we study some variants of the obtained computational model which do not affect its power: the recognized class of languages does not depend on the possibility to observe the top symbols of the stacks before executing the command and other minor modifications of the definitions.

2 Preliminaries

2.1 Monoid Automata

In this section we introduce the definitions and concepts which would be useful in the further. We expect the reader to be familiar with basic notions of formal languages theory, such as finite automata, rational transductions and context-free grammars, also some knowledge of semigroup theory is required. For necessary information refer to any textbook on formal languages theory, such as [16], see [1] and [12] for the introduction into the theory of rational transductions and theory of semigroups respectively. In this section we focus the attention on monoid automata and their relationship with other objects of formal languages theory.

Definition 1. *A monoid automaton (M-automaton) over the alphabet Σ is a tuple $\mathcal{A} = \langle Q, \Sigma, M, P, q_0, F \rangle$ where Q is a finite set of states, Σ is a finite alphabet, M is a partial monoid with the identity 1, $q_0 \in Q$ is an initial state, $F \subseteq Q$ is the set of final states and $P \in Q \times M \times (\Sigma \cup \epsilon) \times Q$ is a set of transitions.*

Just in the case of finite automata the notion of a label can be extended from edges to paths in the automaton. The only difference is that we replace mere concatenation by the multiplication operation of the monoid. According to this definition, the usual finite automata are 1-automata. Note that in all the cases we consider nondeterministic automata, which means we allow multiple moves with the same label in one state. Also note that monoid automata are blind in the sense that they do not take the current element in the memory into account before multiplying it by the element associated with an edge.

Definition 2. *A word $w \in \Sigma^*$ is accepted by the M-automaton $\mathcal{A} = \langle Q, \Sigma, M, P, q_0, F \rangle$ iff there is a state $q \in Q$ such that the pair $\langle 1, w \rangle$ labels some path from q_0 to q. The language recognized by the automaton $\mathcal{A}$ is denoted by $L(\mathcal{A})$.*

Example 1. Let S_1 be a monoid with the generators $\{\alpha, \overline{\alpha}\}$ and the defining relation $\alpha \circ_1 \overline{\alpha} = 1$ and S_2 be a monoid with generators $\{\beta, \overline{\beta}\}$ and defining relation $\beta \circ_2 \overline{\beta} = 1$. Then the language $\{a^n b^n c^n \mid n \in \mathbb{N}^+\}$ is recognized by the automaton $\mathcal{A} = \langle \{q_0, q_1, q_2\}, \{a, b, c\}, S_1 \times S_2, P, q_0, \{q_2\} \rangle$ where $P = \{\langle q_0, \langle \alpha, 1 \rangle, a, q_0 \rangle, \langle q_0, \langle \overline{\alpha}, \beta \rangle, b, q_1 \rangle, \langle q_1, \langle \overline{\alpha}, \beta \rangle, b, q_1 \rangle, \langle q_1, \langle 1, \overline{\beta} \rangle, c, q_2 \rangle, \langle q_2, \langle 1, \overline{\beta} \rangle, c, q_2 \rangle\}$.

Let M be a finitely generated monoid and X be its system of generators. The identity language of M consists of all the words in X^* that represent identity. The proposition below enlightens the connection between monoid automata and finite transducers. It entails that the class of languages recognized by M-automata is closed under rational transductions for any finitely generated monoid M.

Proposition 1 (Kambites, 2007). *The following conditions are equivalent:*

1. *L is accepted by an M-automaton.*
2. *L is a rational transduction of the identity language of M with respect to some finite generating set X.*
3. *L is a rational transduction of the identity language of M with respect to every finite generating set X.*

Suppose some language family is closed under rational transductions, then Kambites theorem offers a powerful method of characterizing it in terms of monoid automata. To prove that all languages recognized by M-automata are, for an instance, context-free, it suffices to construct a context-free grammar for the identity language of M. To prove the opposite inclusion one may either characterize the language family in terms of some computational model (pushdown automata provide such a characterization for context-free languages) and then translate it to the language of monoid automata or find some "typical" languages in the family and show that they are recognized by an M-automaton for the monoid M under consideration. The typicality of the languages means that other languages of the family are their images under rational transductions.

For context-free languages it is reasonable to use Chomsky-Schützenberger theorem. The Dyck language of rank n is a language containing all correct bracket sequences on n types of brackets $a_1, \bar{a}_1, \ldots, a_n, \bar{a}_n$. It is generated by a context-free grammar with the rules $S \to a_i S \bar{a}_i S$, $S \to \epsilon$, where i ranges from 1 to n. The next theorem shows that it is in some sense "typical" among the context-free languages:

Theorem 1 (Chomsky-Schützenberger, [2]). *A language L is context-free if and only if it is a rational transduction of the language D_n for some $n \in \mathbb{N}$.*

The proof of this theorem can be found, for example, in [12]. Informally, the statement of the theorem roughly corresponds to the fact that the subtrees of the derivation tree are either embedded one into another or do not intersect. Now we want to show that in fact this theorem is about monoid automata.

Let X be a finite set of generators, then for every element $x \in X$ we define two operators P_x and Q_x on the free monoid X^*. P_x transforms a string w to the string wx simulating the push operation. Q_x conversely transforms a string of the form wx to the string w and is a right inverse of P_x. The set of all P_x, Q_x is extended to the submonoid $\mathcal{P}_X$ of the monoid of partial functions from X^* to X^*. This monoid, which is called polyclique, was first studied in the work of [13] and plays a great role in the structural theory of semigroups.

Polycyclic monoid automata obviously are capable to perform the operations "push" and "pop" of usual pushdown automata, which suffices to simulate its computations. Note that if we refer to the elements of X as types of brackets, then P_x naturally corresponds to the opening bracket, as well as Q_x to the closing. With respect to this translation the identity language of $\mathcal{P}_X$ is exactly the set of correct bracket sequences. Summarizing, the following theorem holds:

Theorem 2 (Kambites, 2007). *The following conditions are equivalent:*

1. *The language L is context-free.*
2. *The language L is recognized by some polycyclic monoid automaton.*

Note that this theorem can also be proved directly without any references to Chomsky-Schützenberger theorem, just in the same way as the equivalence between context-free grammars and pushdown automata is established.

2.2 Displacement Context-Free Grammars

In this section we introduce basic definitions concerning the class of well-nested MCFLs. We find it more convenient to define them in terms of displacement context-free grammars, mostly following the article [19].

Let Σ be a finite alphabet and 1 a distinguished separator, $1 \notin \Sigma$. For every word $w \in (\Sigma \cup 1)^*$ we define its rank $\mathrm{rk}(w) = |w|_1$. We define the j-th intercalation operation $\odot_j$ which consists in replacing the j-th separator in its first argument by its second argument. For example, $a1b11d \odot_2 c1c = a1bc1c1d$.

Let k be a natural number and N be the set of nonterminals. The function $\mathrm{rk} \colon N \to \overline{0,k}$ assigns every element of N its rank. Let $Op_k = \{\cdot, \odot_1, \dots, \odot_k\}$ be the set of binary operation symbols, then the ranked set of k-correct terms $Tm_k(N, \Sigma)$ is defined in the following way (we write simply Tm_k in the further):

1. $N \subset Tm_k(N, \Sigma)$,
2. $\Sigma^* \subset Tm_k(N, \Sigma)$, $\forall w \in \Sigma^* \ \mathrm{rk}(w) = 0$,
3. $1 \in Tm_k$, $\mathrm{rk}(1) = 1$,
4. If $\alpha, \beta \in Tm_k$ and $\mathrm{rk}(\alpha) + \mathrm{rk}(\beta) \leq k$, then $(\alpha \cdot \beta) \in Tm_k$, $\mathrm{rk}(\alpha \cdot \beta) = \mathrm{rk}(\alpha) + \mathrm{rk}(\beta)$.
5. If $j \leq k$, $\alpha, \beta \in Tm_k$, $\mathrm{rk}(\alpha) + \mathrm{rk}(\beta) \leq k+1$, $\mathrm{rk}(\alpha) \geq j$, then $(\alpha \odot_j \beta) \in Tm_k$, $\mathrm{rk}(\alpha \cdot \beta) = \mathrm{rk}(\alpha) + \mathrm{rk}(\beta) - 1$.

We refer to the elements of the set $N \cup \Sigma^* \cup \{1\}$ as basic subterms. We will often omit the symbol of concatenation and assume that concatenation has greater priority than intercalation, so $Ab \odot_2 cD$ means $(A \cdot b) \odot_2 (c \cdot D)$. This simplification allows us to consider words in the alphabet Σ_1^* as terms either. The set of k-correct terms includes all the terms of sort k or less that also do not contain subterms of rank greater than k.

A term is ground if it contains no nonterminals. We associate with a ground term α its value $\nu(\alpha)$, mapping the elements of Σ_1^* to themselves and interpreting the connectives from Op_k as corresponding language operations. A context $C[]$ is a term with a distinguished placeholder $\#$ instead one of its leaves. If β is a term, then $C[\beta]$ denotes the result of replacing $\#$ by β (provided the created term is correct). For example, $C[] = b1 \odot_1 (a \cdot \#)$ is a context and $C[A \cdot c] = b1 \odot_1 aAc$.

Definition 3. *A k-displacement context-free grammar (k-DCFG) is a quadruple $G = \langle N, \Sigma, P, S \rangle$, where Σ is a finite alphabet, N is a finite ranked set of nonterminals and $\Sigma \cap N = \emptyset$, $S \in N$ is a start symbol such that $rk(S) = 0$ and P is a set of rules of the form $A \to \alpha$. Here A is a nonterminal, α is a term from $Tm_k(N, \Sigma)$, such that $rk(A) = rk(\alpha)$.*

Definition 4. *The derivability relation* $\vdash_G \in N \times Tm_k$ *associated with the gram-mar G is the smallest reflexive transitive relation such that the facts $(B \to \beta) \in P$ and $A \vdash C[B]$ imply that $A \vdash C[\beta]$ for any context C. If the set of words deriv-able from $A \in N$ is $L_G(A) = \{\nu(\alpha) \mid A \vdash_G \alpha, \alpha \in GrTm_k\}$, then $L(G) = L_G(S)$.*

Example 2. Let the i-DCFG G_i be the grammar $G_i = \langle \{S, T\}, \{a, b\}, P_i, S \rangle$. Here P_i is the following set of rules (notation $A \to \alpha|\beta$ means $A \to \alpha, A \to \beta$):

$$S \to \underbrace{(\ldots(}_{i-1 \text{ times}} aT \odot_1 a) + \ldots) \odot_1 a \mid \underbrace{(\ldots(}_{i-1 \text{ times}} bT \odot_1 b) + \ldots) \odot_1 b$$

$$T \to \underbrace{(\ldots(}_{i-1 \text{ times}} aT \odot_1 1a) + \ldots) \odot_i 1a \mid \underbrace{(\ldots(}_{i-1 \text{ times}} bT \odot_1 1b) + \ldots) \odot_i 1b \mid 1^i$$

G_i generates the language $\{w^{i+1} \mid w \in \{a, b\}^+\}$. For example, this is the deriva-tion of the word $(aba)^3$ in G_2: $S \to (aT \odot_1 a) \odot_1 a \to (a((bT \odot_1 1b) \odot_2 1b) \odot_1 a) \odot_1 a \to (a((b((aT \odot_1 1a) \odot_2 1a) \odot_1 1b) \odot_2 1b) \odot_1 a) \odot_1 a \to (a((b((a11 \odot_1 1a) \odot_2 1a) \odot_1 1b) \odot_2 1b) \odot_1 a) \odot_1 a = (a(b(a1a1a \odot_1 1b) +_2 1b) +_1 a) \odot_1 a = (aba1ba1ba \odot_1 a) \odot_1 a = abaabaaba.$

We have already noted that k-DCFGs are equivalent to well-nested $(k + 1)$-multiple context free grammars. In the case of $k = 1$ the intercalation operation is simply the wrapping operation of head grammars ([14], [15]), which are equiv-alent to tree adjoining grammars (TAGs), as proved in [20]. We will not recall the definitions of these classes due to the lack of space. The interested reader may consult [18] and [8] for the definitions of wMCFGs and [5] for the definition of TAGs.

Comparing the definition of DCFG with the definition of wMCFG it is nec-essary to mention that wMCFGs does not impose any condition on the rank of subterms which are well-nested substructures of the righthand side of the rule in terms of wMCFGs. However, this restriction can be also removed in the case of DCFGs: it is possible to show that for every term α which do not contain leaves of sort greater then k and is of sort k itself an equivalent term $\beta \in Tm_k(N, \Sigma)$ can be constructed. Equivalence in this case means that both terms have the same value under arbitrary assignment of values to nonterminals. We omit the details of the proof. So the condition on subterm ranks is redundant in general but we leave it for the sake of consistency.

2.3 Chomsky-Schützenberger Theorem and Correct Multibracket Sequences

To present the Chomsky-Schützenberger theorem we should replace brackets with multibrackets. Let X be a ranked alphabet with the arity function $\rho \colon X \to \overline{1, L}$, where L is a positive integer called the rank of X. We define the set of multibrackets $B(X) = \{x^j, \overline{x}^j \mid x \in X, j \in \overline{1, \rho(x)}\}$. Let $w[j]$ denotes j-th letter in a word $w \in B(X)^*$ (the numeration starts with zero) and $Pos(w) = \{0, 1, \ldots, |w| - 1\}$.

Definition 5. $w \in B(X)^*$ *is called a correct multibracket sequence if the set* $Pos(w)$ *can be partitioned into disjoint sets* $H_1, \ldots, H_m$ *such that:*

1. *Every* H_t *contains an even number of elements. If* $i_1 < j_1 < i_2 < \ldots < i_r < j_r$ *are its elements, then there is an element* $x \in X$ *such that* $r = \rho(x)$ *and for every* $l \leq r$ *it holds that* $H[i_l] = x^l, H[j_l] = \overline{x}^l.$
2. *If* H *and* H' *are two sets in the partition and* $i_1 < j_1 < \ldots < i_r < j_r$ *and* $i'_1 < j'_1 < \ldots < i'_s < j'_s$ *are their elements, then one of the following alternatives holds:*
   - $j_r < i'_1$ *or* $j'_s < i_1$,
   - *There exists* $l \in \overline{1, r-1}$ *such that* $j_l < i'_1 < j'_1 < \ldots < i'_s < j'_s < i_{l+1}$ *or there exists* $l' \in \overline{1, s-1}$ *such that* $j'_{l'} < i_1 < j_1 < \ldots < i_r < j_r < i'_{l'+1}$.
   - *For every* $l \in \overline{1, r}$ *there exists* $l' \in \overline{1, s}$ *such that* $i'_{l'} < i_l < j_l < j'_{l'}$ *or for every* $l' \in \overline{1, s}$ *there exists* $l \in \overline{1, r}$ *such that* $i_l < i'_{l'} < j'_{l'} < j_l.$

The generalized Dyck language $D(X)$ over the alphabet X is the language of all correct multibracket sequences $w \in B(X)^*$. Informally, let the set H in the partition consist of the positions $i_1 < j_1 < \ldots < i_s < j_s$. Let the elements of H define a closed curve on the plane as it is shown on the figure below ($s = 3$), we refer to the set of such curves as the induced curves of the partition:

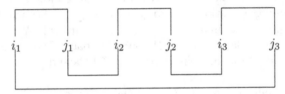

w is a correct multibracket sequence if it is possible to divide its set of positions into groups so, that the induced curves of these groups do not intersect. There is another geometrical intuition behind this definition: every set H in the partition of correct multibracket sequence divides the sequence into its "interior" and "exterior". For any other set H' in the partition there are four possibilities: H' is in the interior of H; H is in the interior of H'; H' lies entirely in one of the intervals of the exterior of H; H lies entirely in one of the intervals of the exterior of H'. Let H consist of the elements $i_1 < j_1 < \ldots < i_s < j_s$ and H' consist of $i'_1 < j'_1 < \ldots < i'_t < j'_t$. The picture below illustrates the possible variants of their mutual position ($s = 3$ and $t = 2$).

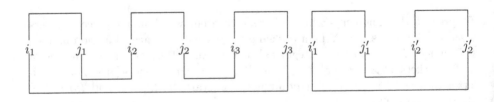

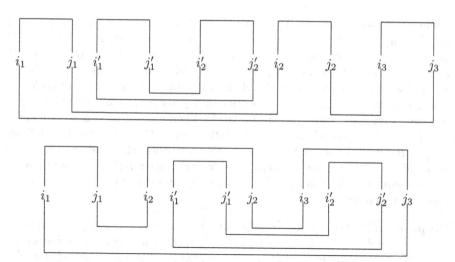

The next proposition offers $(rk(X) - 1)$-DCFG for $D(X)$, the proof follows from the definitions, so we omit it (we just reformulate the wMCFG-grammar from [22] in terms of DCFGs):

Proposition 2. *Let X be the ranked alphabet of rank L, then the language of correct multibracket sequences over X is generated the $(L-1)$-DCFG $G_X = \{\{S_i \mid i \in \overline{0, L-1}\}, B(X), P_X, S_0\}$ where P_X contains the following rules:*

- $S_{i+j} \to S_i S_j$, $i + j < L$,
- $S_{i+j-1} \to S_i +_l S_j$, $i + j \le L$, $l \le i < L$,
- $S_r \to x^1 \underbrace{(\ldots(S_r +_1 (\overline{x}^1 1 x^2)) +_2 \ldots) +_r (\overline{x}^r 1 x^{r+1}))\overline{x}^{r+1}}_{r \ times}$, $x \in X$, $r = \rho(x) - 1$,
- $S_0 \to \epsilon$, $S_1 \to 1$.

Below we formulate the Chomsky-Schützenberger theorem for the class of k-DCFGs. We omit the proof, since, as mentioned in [22], it can be recovered from the analogous theorem for the class of all MCFGs with natural modifications.

Theorem 3. *The language L is a k-displacement context-free language if and only if it is a rational transduction of generalized Dyck language $D(X)$ for some alphabet X of the rank $k+1$.*

3   Monoid Automata for Displacement Context-Free Grammars

In this section we characterize the class of k-displacement context-free languages in terms of monoid automata. For any set of X of multibrackets we construct a monoid whose identity language is exactly $D(X)$ and then use Chomsky-Schützenberger and Kambites theorems to prove the desired result.

Let X be a generating set, $rk(X) = L$ and $ar\colon X \to [1, L]$ be the arity function. Let A be the set $A = \{a_{x,i} \mid x \in X, 1 \le i \le rk(x)\}$. We define two homomorphisms[1] $\phi_1, \phi_2 : B(X) \to P(A)$, setting $\phi_1(x^i) = a_{x,i}$, $\phi_1(\overline{x}^i) = \overline{a}_{x,i}$, $\phi_2(x^1) = a_{x,1}$, $\phi_2(\overline{x}^{i-1}) = a_{x,i}, \phi_2(x^i) = \overline{a}_{x,i}$, $i \in \overline{2, \mathrm{ar}(x)}$, $\phi_2(\overline{x}^{\mathrm{ar}(x)}) = \overline{a}_{x,0}$. We introduce the factor-monoid $S_X = B(X)/\mathrm{Ker}\phi$ where $\phi(x) = \langle \phi_1(x), \phi_2(x) \rangle$: $B(X) \to P(A) \times P(A)$. Then $B(X)$ can be considered as the generating set for S_X and we want to prove that the identity language of S_X is exactly $D(X)$.

Let w be a word representing identity in S_X and w_1, w_2 be the words representing its images under ϕ_1, ϕ_2 respectively. Then w_1 and w_2 represent identity in $P(A)$. Let R_1, R_2 be the binary relations over $Pos(w)$ defined as follows: $(i, j) \in R_l$ iff $w_l[i]$ and $w_l[j]$ contract with each other when reducing the word w_l to identity. Since there is only one "contracting relation" for any correct bracket sequence, the relations R_1, R_2 are uniquely defined by the word w which represents identity.

Proposition 3. *Let $x \in X, r = \mathrm{ar}(x)$ and $i_1 < j_1 < \ldots < i_r < j_r$ be such that $w[i_1] = x^0$ and it holds that $(i_l, j_l) \in R_1$ for any $l \leq r$ and $(j_l, i_{l+1}) \in R_2$ for any $l < r$. Then $(i_1, j_r) \in R_2$ and for any $l < r$ it holds that $w[i_l] = x^l$, $w[j_l] = \overline{x}^l$.*

Proof. The second statement is established according to the definitions of ϕ_1, ϕ_2 and R_1, R_2. It remains to prove the first one. There is a cycle of numbers $p_1 = i_1, q_1 = j_1, \ldots, p_r = i_r, q_r = j_r, p_{r+1}, q_{r+1}, \ldots, p_{2r}, q_{2r}, \ldots, p_{dr}, q_{dr}, p_{dr+1} = p_1$ such that $(p_l, q_l) \in R_1$ and $(q_l, p_{l+1}) \in R_2$ for any $l \leq dr$. We prove that actually $r = 1$ which implies the theorem. Suppose the converse and let i_1 be the leftmost element i in this cycle such that $w[i] = x^0$, then $p_{r+1} > p_1$. It is easy to prove by induction on t using the planarity of R_1, R_2 that for every $t > r$ there exists some $l \leq r$ such that $i_l < p_t < q_t < j_l$ and $p_{r+1} \leq p_t < q_t < q_r$. This contradicts the equality $p_{dr+1} = p_1$. The proposition is proved.

Let us refer to the set $H = \{i_1, j_1, \ldots, i_r, j_r\}$ from the proposition as the chain cycle. The chain cycles form a partition of $Pos(w)$ since R_1, R_2 are total one-to-one relations. The proposition above and the planarity of relations R_1, R_2 imply that chain cycles can serve as sets H_l from the definition of multibracket sequence. So we have proved:

Lemma 1. *Any element of the identity language of S_X with respect to the set $B(X)$ is a correct multibracket sequence over the set X.*

Lemma 2. *Any correct multibracket sequence over the set X is an element of the identity language of S_X with respect to the set $B(X)$.*

Proof. Recall the grammar G_X from the previous section generating the set $D(X)$. To prevent confusion we denote the separator in the grammar by $\#$ instead of 1 We extend the mappings ϕ_l to the set $(B(X) \cup \{\#\})^*$ defining $\phi_1(\#) = \phi_2(\#) = 1$. We denote by $\mu_l(w)$ the value of the word $\phi_l(w)$ in $P(A)$, obviously μ_l is a homomorphism. We want to prove by induction that if $S_i \vdash w$, $w = w_0 \# w_1 \ldots \# w_i$, then $\mu_1(w_0) = \mu_1(w_1) = \ldots = \mu_1(w_i) = \mu_2(w) = 1$.

---

[1] Analogous construction was used in [21] to prove the Schützenberger theorem for the class of tree adjoining languages.

Consider the rule applied in the root of the derivation tree. The basis of induction if the obvious case of the rules $S_0 \to \epsilon$ or $S_1 \to \#$. In case of the rules $S_{i+j} \to S_i \cdot S_j$ and $S_{i+j-1} \to S_i +_k S_j$ the induction statement follows from the fact that the inverse homomorphic image of 1 is closed under concatenation and intercalation.

In the case of the rule $S_i \to x^1(\ldots (S_{i+1}(\overline{x}^1 \# x^2) +_2 \ldots) +_i (\overline{x}^i \# x^{i+1})\overline{x}^{i+1})$ we consider the components of the word w. There exists a word $u = u_0 \# \ldots \# u_i$, derivable from S_i, such that for any $j \leq i$ it holds that $w_i = x^{i+1} u_i \overline{x}^{i+1}$. So $\mu_1(w_l) = \mu_1(x^{l+1})\mu_1(u_l)\mu_1(\overline{x}^{l+1}) = a_{x,l+1}1\overline{a}_{x,l+1} = 1$. Let us prove $\mu_2(v) = 1$, indeed $\mu_2(v) = \mu_2(x^1)\mu_2(u^0)\mu_2(\overline{x}^1 \# x^2)\mu_2(u^1) \ldots \mu_2(\overline{x}^i \# x^{i+1})\mu_2(u_i)\mu_2(\overline{x}^{i+1}) = a_{x,1}\mu_2(u_0)a_{x,2}\overline{a}_{x,2}\mu_2(u_1) \ldots a_{x,i+1}\overline{a}_{x,i+1}\mu_2(u_i)\overline{a}_{x,1} = a_{x,1}\mu_2(u)\overline{a}_{x,1} = a_{x,1}\overline{a}_{x,1} = 1$. The last case is verified and the lemma is proved.

Theorem 4. *The class of languages recognized by k-DCFGs is exactly the class of languages recognized by S_X-automata for the generating sets X of rank $k+1$.*

Proof. By the lemmas above the language S_X coincides with the set of multi-bracket sequences $D(X)$, which is generated by a $(rk(X)-1)$-displacement grammar. Other languages recognized by S_X-automata are its images under rational transductions and, hence, displacement context-free languages since the latter are closed under rational transductions. From Theorem 3 it follows that all k-displacement context-free languages are rational transductions of $D(X)$ for some set X of rank $k+1$ and then by Theorem 2 they are all recognized by S_X-automata.

4   Simultaneous Two-Stack Automata

In the case of usual bracket sequences the opening and closing brackets naturally correspond to push and pop operations. In the case of multibracket sequences each bracket is in fact a pair of brackets, so every multibracket is an operation on the pair of stacks. The full power of two-stack machines allows to simulate every recursively enumerable language, but in our case there are some restrictions on possible operations. The most principal limitation is that our operations are synchronized: every move changes the length of each stack by 1. In general, there are only four possible types of operations: push the same symbol on both stacks, move the symbol from the first stack to the second, return a symbol back to the first stack from the second and remove the same symbol from both the stacks. Also the rank of a symbol determines the number of times it may be exchanged between the stacks.

Note that Proposition 3 in fact postulates that if a symbol a of arity k is pushed on the stack together with its copy a' then after transferring it $2(k-1)$ times between the stacks, then this symbol would be removed together with the same instance of the symbol a'. Therefore we should trace only the number of exchanges the symbol participated in, so we will keep in stacks not the symbols alone but the pairs consisting of a symbol and a counter of its number of exchanges. This counter is incremented every time the symbol is moved from one stack to another and equals 1 after the first push. Executing the remove operation, we verify that the top element of the first stack is $\langle a, 2k-1 \rangle$ and the top element of the second stack is $\langle a, 1 \rangle$ with the same a. We call this model of computation a simultaneous two-stack automata.

Definition 6. *A simultaneous two-stack automaton of rank k (k-STSA) is a tuple $\mathcal{A} = \langle Q, \Sigma, \Gamma, \mathrm{ar}, P, q_0, F \rangle$ where Q is a finite set of states, Σ is a finite alphabet, Γ is a finite stack alphabet, $\mathrm{ar} \colon \Gamma \to \overline{1, k}$ is the arity function, P is the set of transitions, $q_0 \in Q$ is an initial state and $F \subseteq Q$ is a set of final states. Transitions has the form $(\langle q_1, a \rangle \to \langle q_2, \tau, \alpha \rangle)$, where q_1, q_2 are states, $a \in \Sigma \cup \{\epsilon\}$ is an input symbol (or an empty word), $\tau \in \langle PUSH, MOVE, RETURN, POP \rangle$ is a command and $\alpha \in \Gamma$ is a stack symbol.*

As in the case of usual finite automata the formal definition of the acceptance relation is given through the notion of configuration, which is the instantaneous description of the automaton.

Definition 7. *A configuration of a simultaneous two-stack automaton $\mathcal{A} = \langle Q, \Sigma, \Gamma, \mathrm{ar}, P, q_0, F \rangle$ is a tuple $\langle q, u, \beta_1, \beta_2 \rangle$ where $q \in Q$ is the current state, u is the current suffix of input, which has not been processed yet, and β_1, β_2 are the words in the alphabet $\Sigma_{\mathbb{N}} = \Sigma \times \mathbb{N}$. A transition relation $\vdash_{\mathcal{A}}$ is the smallest transitive reflexive relation such that:*

- *If $(\langle q_1, a \rangle \to \langle q_2, PUSH, \alpha \rangle) \in P$ then $\langle q_1, au, \beta_1, \beta_2 \rangle \vdash \langle q_2, u, \beta_1(\alpha, 1), \beta_2(\alpha, 1) \rangle$ for any words $u \in \Sigma^*$ and $\beta_1, \beta_2 \in \Sigma_{\mathbb{N}}$.*
- *If $(\langle q_1, a \rangle \to \langle q_2, MOVE, \alpha \rangle) \in P$ then $\langle q_1, au, \beta_1(\alpha, 2i-1), \beta_2 \rangle \vdash \langle q_2, u, \beta_1, \beta_2(\alpha, 2i) \rangle$ for any words $u \in \Sigma^*$ and $\beta_1, \beta_2 \in \Sigma_{\mathbb{N}}$ and any counter value $i < \mathrm{ar}(\alpha)$.*
- *If $(\langle q_1, a \rangle \to \langle q_2, RETURN, \alpha \rangle) \in P$ then $\langle q_1, au, \beta_1, \beta_2(\alpha, 2i) \rangle \vdash \langle q_2, u, \beta_1(\alpha, 2i+1), \beta_2 \rangle$ for any words $u \in \Sigma^*$ and $\beta_1, \beta_2 \in \Sigma_{\mathbb{N}}$ and any counter value $i < \mathrm{ar}(\alpha)$.*
- *If $(\langle q_1, a \rangle \to \langle q_2, POP, \alpha \rangle) \in P$ then $\langle q_1, au, \beta_1(\alpha, 2\mathrm{ar}(\alpha) - 1), \beta_2(\alpha, 1) \rangle \vdash \langle q_2, u, \beta_1, \beta_2 \rangle$ for any words $u \in \Sigma^*$ and $\beta_1, \beta_2 \in \Sigma_{\mathbb{N}}$.*

The language $L(\mathcal{A})$ recognized by the automaton equals $L(\mathcal{A}) = \{w \in \Sigma^* \mid \exists q \in F(\langle q_0, w, \epsilon, \epsilon \rangle \vdash \langle q, \epsilon, \epsilon, \epsilon \rangle)\}$.

The condition on counter parity reflects the fact that only the symbols that were moved from the first stack to the second can be returned back. If a symbol was initially placed on the second stack then it would be removed from it only by the pop operation. That is done in order to keep the structure of multibracket chain which has one "embracing" link in the lower half plane and k small links in the upper half plane. So the stacks are not completely symmetric in their roles, in fact the first stack is the basic one and the second is just an additional memory register, which stores the placeholders of symbols pushed to the first stack to ensure the correct order of MOVE and RETURN operations.

Note that the set of possible memory operations can be extended by the KEEP command which does not change the contents of the stacks. In order to simulate an edge with KEEP between states q_1 and q_2 we add a dummy state q' in the middle and a new symbol Z of arity 1 to the stack alphabet and replace the edge under consideration with two transitions $\langle q_1, a \rangle \to \langle q', \mathrm{PUSH}, Z \rangle$ and $\langle q', \epsilon \rangle \to \langle q_2, \mathrm{POP}, Z \rangle$. Such procedure decreases the number of "keeping" edges so we proceed by induction. In the further the assume that rules of the form $\langle q_1, a \rangle \to \langle q_2, \mathrm{KEEP} \rangle$ are also allowed in the set of transitions.

Example 3. The rank 2 two-stack simultaneous automaton $\mathcal{A} = \langle \{q_i \mid 0 \le i \le 6\}, \{a, b\}, \{A, B\}, \mathrm{ar}, P, q_0, \{q_6\} \rangle$ where $\mathrm{ar}(A) = \mathrm{ar}(B) = 2$ with the set of transitions specified below recognizes the crossing copy language $\{a^m b^n a^m b^n \mid m, n \in \mathbb{N}\}$.

$\langle q_0, a \rangle \to \langle q_0, \mathrm{PUSH}, A \rangle$ $\qquad$ $\langle q_0, \epsilon \rangle \to \langle q_1, \mathrm{KEEP} \rangle$

$\langle q_1, b \rangle \to \langle q_1, \mathrm{PUSH}, B \rangle$ $\qquad$ $\langle q_1, \epsilon \rangle \to \langle q_2, \mathrm{KEEP} \rangle$

$\langle q_2, \epsilon \rangle \to \langle q_2, \mathrm{MOVE}, B \rangle$ $\qquad$ $\langle q_2, \epsilon \rangle \to \langle q_3, \mathrm{KEEP} \rangle$

$\langle q_3, a \rangle \to \langle q_3, \mathrm{MOVE}, A \rangle$ $\qquad$ $\langle q_3, \epsilon \rangle \to \langle q_4, \mathrm{KEEP} \rangle$

$\langle q_4, \epsilon \rangle \to \langle q_4, \mathrm{RETURN}, A \rangle$ $\qquad$ $\langle q_4, \epsilon \rangle \to \langle q_5, \mathrm{KEEP} \rangle$

$\langle q_5, b \rangle \to \langle q_5, \mathrm{RETURN}, B \rangle$ $\qquad$ $\langle q_5, \epsilon \rangle \to \langle q_6, \mathrm{KEEP} \rangle$

$\langle q_6, \epsilon \rangle \to \langle q_6, \mathrm{POP}, B \rangle$ $\qquad$ $\langle q_6, \epsilon \rangle \to \langle q_6, \mathrm{POP}, A \rangle$

For the sake of clarity we describe the computation process of this automaton in details. It is not difficult to see that if the automaton reaches the final state after reading the word, then this word is of the form $a^{m_1} b^{n_1} a^{m_2} b^{n_2}$, otherwise some of the reading operations would be impossible. Assume we have a word $a^{m_1} b^{n_1} a^{m_2} b^{n_2}$ that is accepted by the automaton, let us prove that $m_1 = m_2$ and $n_1 = n_2$. In the first part of its computation the automaton reads all the a's from the first segment of the word and both of its stacks contain the words $(A, 1)^{m_1}$. Afterwards the automaton passes the edge to q_1 and reads all the b's from the second segment, so both the stacks contain $(A, 1)^{m_1} (B, 1)^{n_1}$ when the automaton is entering the state q_2. Note that the state q_3 requires A on the top of the first stack, so we should move all the B's to the second stack in q_2 and the number of necessary moves is exactly n_1. Hence the first stack contains $(A, 1)^{m_1}$ and the second stack contains $(A, 1)^{m_1} (B, 1)^{n_1} (B, 2)^{n_1}$ before reading the second segment of a's in q_3. In q_3 the automaton should read all the remaining a's, so the stack contents are $(A, 1)^{m_1 - m_2}$ and $(A, 1)^{m_1} (B, 1)^{n_1} (B, 2)^{n_1} (A, 2)^{m_2}$ when the automaton is leaving the state q_3. In q_4 all the A's moved on the previous step should be returned, so the stacks contain $(A, 1)^{m_1 - m_2} (A, 3)^{m_2}$ and $(A, 1)^{m_1} (B, 1)^{n_1} (B, 2)^{n_1}$ when the automaton enters q_5. Note that in q_5 the automaton must read all the b's in the word in order to finish reading. So if this stage is successful, the stacks contain $(A, 1)^{m_1 - m_2} (A, 3)^{m_2} (B, 3)^{n_2}$ and $(A, 1)^{m_1} (B, 1)^{m_2} (B, 2)^{n_1 - n_2}$. Since in q_6 the automaton executes only POP operations there should be no $(A, 1)$'s on the first stack and no $(B, 2)$'s on the second stack implying that $m_1 = m_2$ and $n_1 = n_2$ which was required. The correctness of the automaton is proved.

Recall the definition of S_X-automata from Section 3. Since the notion of simultaneous two-stack automata is just a reformulation of S_X-automata and the rank of the automata equals the rank of the generating set, the following theorem holds:

Theorem 5. *Simultaneous two-stack automata of rank k recognize exactly the family of $(k-1)$-displacement context-free languages, which is the family of k-well-nested multiple context-free languages.*

It follows that simultaneous two-stack automata of rank 2 recognize exactly the family of tree-adjoining languages.

5   Generalized Simultaneous Two-Stack Automata

Though the introduced notion of simultaneous two-stack automata of rank k directly corresponds to the notion of $(k-1)$-displacement context-free language, the formulation itself seems to be not satisfactory. Its greatest disadvantage is the lack of flexibility: note that, for example, the recognizing power of pushdown automata remains the same, no matter whether the lookup of an arbitrary finite number of top stack symbols is allowed, the lookup of only the top symbol is possible or there is no lookup at all. We want to gain analogous flexibility in our case.

The first inconvenient restriction is that we are bound to push and pop the same symbols from both the stacks and it is not possible, for example, to push A to the first stack and B to the second. Analogously we cannot remove A from the first stack adding B to the second, the pushed symbol must be also A. If we weaken this restriction and allow to combine arbitrary symbols in such operations it is impossible to trace the rank of particular element of stack alphabet. However, we still want to distinguish, say, 2-DCFLs from 3-DCFLs so the notion of rank cannot be completely omitted. So we keep on associating a counter with every symbol on the stacks and incrementing this counter during every MOVE and RETURN operation. This counter is required to be less than $2K$ during the computation, where K is the rank of the automaton. The discussion above leads us to the following definition:

Definition 8. *A generalized simultaneous two-stack automaton of rank k (k-GSTSA) is a tuple $\mathcal{A} = \langle Q, \Sigma, \Gamma, P, q_0, F \rangle$ where Q is a finite set of states, Σ is a finite alphabet, Γ is a finite stack alphabet, P is the set of transitions, $q_0 \in Q$ is an initial state and $F \subseteq Q$ is a set of final states. Transitions has the form $(\langle q_1, a \rangle \to \langle q_2, \tau, \alpha_1, \alpha_2 \rangle)$, where q_1, q_2 are states, $a \in \Sigma \cup \{\epsilon\}$ is an input symbol (or an empty word) $\tau \in \langle PUSH, MOVE, RETURN, POP \rangle$ is a command and $\alpha_1, \alpha_2 \in \Gamma$ are stack symbols.*

The notion of configuration for k-GSTSAs is the same that for usual k-STSAs, the configuration includes the current state, the suffix of input to be read and the contents of the stacks. Since we have changed the format of automaton commands we should also modify the transition relation.

Definition 9. *A transition relation $\vdash_A$ is the smallest transitive reflexive relation such that*

- *If $(\langle q_1, a \rangle \to \langle q_2, PUSH, \alpha_1, \alpha_2 \rangle) \in P$ then $\langle q_1, au, \beta_1, \beta_2 \rangle \vdash \langle q_2, u, \beta_1(\alpha_1, 1), \beta_2(\alpha_2, 1) \rangle$ for any words $u \in \Sigma^*$ and $\beta_1, \beta_2 \in \Sigma_{\mathbb{N}}$.*

- If $(\langle q_1, a \rangle \to \langle q_2, MOVE, \alpha_1, \alpha_2 \rangle) \in P$ then $\langle q_1, au, \beta_1(\alpha_1, 2i-1), \beta_2 \rangle \vdash \langle q_2, u, \beta_1, \beta_2(\alpha_2, 2i) \rangle$ for any words $u \in \Sigma^*$ and $\beta_1, \beta_2 \in \Sigma_{\mathbb{N}}$ and any counter value $i < k$.

- If $(\langle q_1, a \rangle \to \langle q_2, RETURN, \alpha_1, \alpha_2 \rangle) \in P$ then $\langle q_1, au, \beta_1, \beta_2(\alpha_1, 2i) \rangle \vdash \langle q_2, u, \beta_1(\alpha_2, 2i+1), \beta_2 \rangle$ for any words $u \in \Sigma^*$ and $\beta_1, \beta_2 \in \Sigma_{\mathbb{N}}$ and any counter value $i < k$.

- If $(\langle q_1, a \rangle \to \langle q_2, POP, \alpha_1, \alpha_2 \rangle) \in P$ then $\langle q_1, au, \beta_1(\alpha_1, 2i-1), \beta_2(\alpha_2, 1) \rangle \vdash \langle q_2, u, \beta_1, \beta_2 \rangle$ for any words $u \in \Sigma^*$, $\beta_1, \beta_2 \in \Sigma_{\mathbb{N}}$ and any counter value $i < k$.

The language $L(\mathcal{A})$ recognized by the automaton equals $L(\mathcal{A}) = \{w \in \Sigma^* \mid \exists q \in F(\langle q_0, w, \epsilon, \epsilon \rangle \vdash \langle q, \epsilon, \epsilon, \epsilon \rangle)\}$.

Note that we can simulate keeping transitions in the automaton as well as earlier.

We use the values of counters not only to trace the number of MOVE and RETURN operations performed in a chain, but also use their parity for the same purpose as in the case of STSA-s. In fact, we want to keep untouched the multibracket geometric structure of the stack contents since this particular structure reflects the order and embedding of constituents.

Now we want to prove that k-GSTSAs have the same recognizing power as k-STSAs for any natural k. First note that the latter are just a particular case of the former since we can set $\alpha_1 = \alpha_2$ in all the transitions of the automaton. To prove the opposite inclusion we again refer to multibracket sequences. In this case we will not embed this approach into monoid framework to escape unnecessary technicalities.

Let $A = \{a_1, \overline{a}_1, \ldots, a_m, \overline{a}_m\}$ be the alphabet of brackets and $Y \subseteq A \times A$ be the set of admissible pairs. For any letter $a \in Y$ we denote by $\pi_i(a)$, $i = 1, 2$, its i-th coordinate. The mapping π_i is naturally extended to words in Y^*, we call $\pi_i(w)$ the i-th projection of the word w. The notion of k-garland introduced below is a generalization of the notion of multibracket sequence for the case of arbitrary set Y. Recall that if u is a correct multibracket sequence, then the contraction relation $R(u)$ consists of all such pairs $\langle i, j \rangle$ that the letters $u[i]$ and $u[j]$ contract with each other in u when reducing it to an empty word. Note that R is always a symmetric bijection and for every correct bracket sequence there is only one such relation. We define also an asymmetric contraction relation $R_<(u)$; a pair $\langle i, j \rangle$ belongs to $R_<(u)$ if it belongs to $R(u)$ and the inequality $i < j$ holds.

Definition 10. *The word $w \in Y^*$ is a k-garland over the alphabet Y if the following conditions hold:*

1. *$\pi_1(w), \pi_2(w)$ are correct bracket sequences.*
2. *For any indexes i_1, j_1, i_2, j_2, such that $j_1 < i_2$, $(i_1, j_1), (i_2, j_2) \in R(\pi_1(u))$ and $(j_1, i_2) \in R(\pi_2(u))$ holds one of the statements $i_1 < j_1 < i_2 < j_2$, $j_1 < i_1 < i_2 < j_2$ or $i_1 = j_2$ (in this case also $j_1 = i_2$).*
3. *The inequality $l \le k$ holds for any ascending chain $i_1 < j_1 < i_2 < j_2 < \ldots < i_l < j_l$ of indexes, such that $(i_t, j_t) \in R(\pi_1(w))$ for any $t \le l$ and $(j_t, i_{t+1}) \in R(\pi_2(w))$ for any $t < l$.*

Let $R_0(w)$ define the relation $(R_<(\pi_1(w)) \cup R(\pi_2(w)))^*$. Then the following lemma holds:

Lemma 3. *Any vertex in the set $Pos(w) = \overline{0, |w| - 1}$ belongs to some simple cycle in the graph $G_R = \langle Pos(w), R_0 \rangle$.*

Proof. Since the number of vertexes is finite, it suffices to proof that every edge in R_0 belongs to some infinite path with no edges traversed in both directions. Then it suffices to show that there is in infinite path in G_R with the edges from $R_<(\pi_1(w))$ (called the edges of the first type) and the edges from $R(\pi_2(w))$ (the edges of the second type) being alternated. Let us start from an arbitrary edge (i_1, j_1) of the first type and show we can always add two more edges. Indeed, there is some edge (j_1, i_2) of the second type because the $R(\pi_2(w))$ is a bijection. Then there is an edge $(i_2, j_2) \in R(\pi_1(w))$, we need to show that $i_2 < j_2$. In both the cases it follows from the second part of the definition of k-garland. Then we have added to more edges to the path and the lemma is proved.

Lemma 4. *If w is a k-garland, then every vertex $i \in Pos(w)$ belongs to some cycle in the graph $G_R = \langle Pos(w), R \rangle$ containing the indexes $i_1 < j_1 < \ldots < i_l < j_l$ such that for any $t \leq l$ it holds that $(i_t, j_t) \in R(\pi_1(w))$ and for any $t < l$ it holds that (j_t, i_{t+1}) belongs to $R(\pi_2(w))$. It also holds that $(j_t, i_1) \in R(\pi_2(w))$ and $l \leq k$.*

Proof. Consider the cycle which contains i, such a cycle exists due to Lemma 3. Take the leftmost vertex i_0 in this cycle and consider the longest ascending path containing i_0, according to the definition of $R_0(w)$ it starts and ends with en edge of the first type. Then the proof of the statement $(j_t, i_1) \in R(\pi_2(w))$ repeats the proof of the Proposition 3. The condition $l \leq k$ follows from the definition of k-garland.

Since the structure of states is the same for automata of all kinds, we should concentrate on the structure of their transitions. Let $\mathcal{T}$ be some transition of the generalized two-stack simultaneous automaton $\mathcal{A} = \langle Q, \Sigma, \Gamma, P, q_0, F \rangle$. Its stack image of $\psi(\mathcal{T})$ is a pair of symbols in the alphabet $\Gamma \cup \{\overline{A} \mid A \in \Gamma\}$ defined as follows:

1. If $\mathcal{T} = (\langle q_1, a \rangle \to \langle q_2, \text{PUSH}, \alpha_1, \alpha_2 \rangle)$ then $\psi(\mathcal{T}) = \langle \alpha_1, \alpha_2 \rangle$,
2. If $\mathcal{T} = (\langle q_1, a \rangle \to \langle q_2, \text{MOVE}, \alpha_1, \alpha_2 \rangle)$ then $\psi(\mathcal{T}) = \langle \overline{\alpha}_1, \alpha_2 \rangle$,
3. If $\mathcal{T} = (\langle q_1, a \rangle \to \langle q_2, \text{RETURN}, \alpha_1, \alpha_2 \rangle)$ then $\psi(\mathcal{T}) = \langle \alpha_1, \overline{\alpha}_2 \rangle$,
4. If $\mathcal{T} = (\langle q_1, a \rangle \to \langle q_2, \text{POP}, \alpha_1, \alpha_2 \rangle)$ then $\psi(\mathcal{T}) = \langle \overline{\alpha}_1, \overline{\alpha}_2 \rangle$.

We denote by $\psi(\mathcal{A}) = \{\psi(\mathcal{T}) \mid \mathcal{T} \in P\}$ the set of stack images of the transitions of the automaton $\mathcal{A} = \langle Q, \Sigma, \Gamma, P, q_0, F \rangle$. Two transitions of the GSTSA are called consecutive if the destination state of the first transition equals the source state of the second one. We call a computation a sequence of consecutive transitions. The computation is identity-preserving if there is nothing in the stacks after its termination provided the stacks are empty before it starts. Note that a word w is accepted by an automaton iff there is an identity-preserving computation of this automaton which starts in the initial state, terminates in some of the final states and reads exactly the word w.

Definition 11. *The stack image $\psi(\mathcal{C})$ of the computation $\mathcal{C} = \mathcal{T}_1 \ldots \mathcal{T}_r$ is the sequence $\psi(\mathcal{T}_1) \ldots \psi(\mathcal{T}_r)$.*

Proposition 4. *The identity-preserving computations of the k-GSTSA $\mathcal{A} = \langle Q, \Sigma, \Gamma, P, q_0, F \rangle$ are exactly all k-garlands over the set $\psi(\mathcal{A})$.*

Proof. Consider some sequence of "push" and "pop" operations executed on a single stack. The emptiness of the stack if preserved under this sequence of operations iff a natural encoding of operations maps this sequence to a correct bracket sequence. Since the projections of k-garlands are correct bracket sequences every k-garland is identity-preserving.

The opposite implication uses the specificity of k-GSTSA operations. Let a computation be identity-preserving then the first part of the k-garland definition is obviously valid. Let R_i, $i = 1, 2$ denote the contraction relation of the sequence of operations on the i-th stack. If $(i_1, j_1), (i_2, j_2) \in R_1$, $(j_1, i_2) \in R_2$ and $j_1 < i_2$; it means that in the i_2-th step of the computation we pop from the second stack the element pushed there on the j_1-th step. There are two possibilities: first, if this pop is a part of the RETURN operation then by the definition of GSTSA only the MOVE operation is possible in the j_1-th transition of the computation, also the symbol pushed on the first stack during the RETURN operation must be removed somewhen later. It means that $i_1 < j_1$ and $i_2 < j_2$. The second variant is that the POP operation is executed on the i_2-th step, it implies that the operation on the step i_1 is PUSH which implies $j_1 < i_1$ and $j_2 < i_2$. Both possibilities are allowed in the definition of k-garland so the second step is proved. To prove the third part of the definition note that all the intermediate elements of the ascending chains considered in that part are linked by MOVE and RETURN operations. Since every such operation increments the value of the same counter the number of intermediate operations is not greater then $2k - 2$ and the total number of vertexes in this chain is not greater then $2k$ which was required. The lemma is proved.

Corollary 1. *For any k-GSTSA $\mathcal{A}$ the language $L(\mathcal{A})$ is a rational transduction of the set of k-garlands over the alphabet $\psi(\mathcal{A})$.*

Proof. Evidently $L(\mathcal{A})$ is the rational transduction of the set of identity-preserving computations. Then we should apply the Proposition 4.

Lemma 5. *The set of k-garlands over the alphabet $\psi(\mathcal{A})$ is recognized by some k-STSA.*

Proof. Consider the finite set $\mathcal{D}$ of all possible closed chains in k-garlands and some chain $d \in \mathcal{D}$. Let $l(d)$ denote its number of vertexes in the chain and $d[i]$ denote its i-th leftmost vertex. Consider $\mathcal{D}$ as the ranked alphabet with the arity function l and define the set of multibrackets $B(X) = \{d[i] \mid d \in \mathcal{D}, 1 \leq i \leq l(D)\}$. It is easy to prove that the set of k-garlands is the homomorphic image of the generalized Dyck language $D(\mathcal{D})$ of correct multibracket sequences which is a $(k-1)$-DCFL. Then it is recognized by some k-STSA due to Theorem 5.

Theorem 6. *Any language recognized by some k-GSTSA is recognized by some k-STSA.*

Proof. The languages recognized by k-STSAs are closed under rational transductions. By Corollary 1 it suffices to show that the language of k-garlands is recognized by a k-STSA which was proved in Lemma 5.

We have proved that the permission for STSA commands to combine arbitrary pairs of symbols does not affect its recognizing power. It is worth mentioning that in fact k-garlands are a natural generalization of multibracket sequence under the same permission. Hence the method of Section 3 can also be used to find another version of Chomsky-Schützenberger theorem for the class of DCFLs.

6   Blind and Sighted Automata

There is another major disadvantage in our initial definition of STSA: the automaton is not able to observe top symbols of the stacks during the computation. Certainly, these symbols are significant in the case of POP operation since the automaton halts if the command to execute is, say, $\langle POP, A, B \rangle$ and current top symbols are C and D. In the same way the MOVE command takes into account the content of the first stack, as well as the RETURN operation — of the second. However, there is no possibility to refer to the top elements of the stack in the case of PUSH operation. This limitation seems to be unnatural and unpleasant, so we should develop some modification of the automaton to overcome this difficulty.

Let us first discuss the same problem in the case of usual pushdown automaton. Assume we have a command of the kind "in the state q_1 if A is the top symbol of the stack then read a from the input stream, push B to the stack and move to the state q_2" (we abbreviate this by $\langle q_1, a, A \rangle \to \langle q_2, PUSH, B \rangle$). The common way to simulate this instruction is to create two fresh states q' and q'' and add the following transitions: $(\langle q_1, \epsilon, \epsilon \rangle \to \langle q', POP, A \rangle), (\langle q', \epsilon, \epsilon \rangle \to \langle q'', PUSH, A \rangle)$ and $(\langle q'', a, \epsilon \rangle \to \langle q_2, PUSH, B \rangle)$. However, it is troublesome to adapt this approach to k-GSTSA since it is hard to ensure that the number of move/return operations would not exceed k. Therefore we choose another way to simulate top symbol observations.

Let k be the maximal number of stack symbols which are observed in the transitions of the pushdown automaton. Then it has the transition of the following two forms, where l is a natural number not greater than k:

$$\langle q_1, a, A_{l+1} \dots A_k \rangle \to \langle q_2, PUSH, B \rangle$$
$$\langle q_1, a, A_{l+1} \dots A_k \rangle \to \langle q_2, POP, A_k \rangle$$

Let Γ be the set of old stack symbols and Q be the set of states. First, we enrich the set of stack symbols with k new symbols $Z_1, \dots Z_k$ which serve as bottom markers and treat them as elements of Γ. Then the new set of states is $Q' = \{q_0, q_f\} \cup Q \times \Gamma^n$ and the new stack alphabet is $\Gamma \times \Gamma^n$. q_0 and q_f are distinguished initial and final states, respectively, and the second component of all other states contains the top k symbols of the stack. Analogously, the second component of the stack symbol always keeps the k symbols below it starting from the deepest. The symbols $Z_1, \ldots Z_k$ were added in order to ensure that there are always at least k symbols in the stack. Then it is straightforward to simulate the dependence from k top symbols by the means of the states only, the only difficulty is to maintain the invariant we announced.

The automaton always starts from the initial state q_0 and pushes the symbols $Z_1 \ldots Z_k$ on the stack, moving to the state $(q_0, Z_1 \ldots Z_k)$ to start the computation. Every transition of the form $\langle q_1, a, A_{l+1} \ldots A_k \rangle \to \langle q_2, \mathrm{PUSH}, B \rangle$ is simulated by a new transition

$$\langle (q_1, A_1 \ldots A_k), a \rangle \to \langle (q_2, A_2 \ldots A_k B), \mathrm{PUSH}, (B, A_1 \ldots A_k) \rangle.$$

Note that the deepest of the symbols observed on the previous stage in the first component of the automaton state is now observed as the deepest symbol of the second component of the stack top. That allows us to update the top k symbols when the POP operation is executed: every transition of the form $\langle q_1, a, A_{l+1} \ldots A_k \rangle \to \langle q_2, \mathrm{POP}, A_k \rangle$ is replaced by the transition

$$\langle (q_1, A_1 \ldots A_k), a \rangle \to \langle (q_2, A_0 A_1 \ldots A_{k-1}), \mathrm{POP}, (A_k, A_0 \ldots A_{k-1}) \rangle.$$

It is straightforward to prove that the desired invariant is maintained. In the end of the computation we should remove the bottom markers, so we add the transitions of the form $\langle (q, Z_1 \ldots Z_k), \epsilon \rangle \to \langle q_f, \mathrm{POP}, Z_1 \ldots Z_k \rangle$ (it is trivial to simulate immediate pop of k symbols by successively removing them one by one so we simplify the notation) for every former final state q. Then it is easy to prove that the new automaton without lookup recognizes exactly the same language as the old automaton did.

Then the same approach can be applied to k-GSTSAs. The only modification to be made is to trace the contents of both the stacks, not the single one. So we have proved the following theorem:

Theorem 7. *The generating power of k-GSTSAs is the same whether or not it is allowed to take into account the top k symbols.*

7   Conclusions and Future Work

We give the algebraic interpretation of Chomsky-Schützenberger theorem for the class of displacement-context free languages which are another realization of well-nested multiple context-free languages. We present their characterization in terms of monoid automata. Then we introduce the computational interpretation of the introduced monoid, showing how the multiplication operation of the monoid can be simulated on two stacks by specific combinations of PUSH and POP operations. The flexibility of the introduced notion of two-stack automata shows the vitality of our approach.

There are at least two directions of future work: the first is two develop fast analyzers on the base of GSTSAs for the class of DCFGs or for a significant subclass of them. For example, it is interesting to adopt the machinery of LR or Earley algorithms for DCFLs (see [7] for the variant of Earley analyzer for well-nested MCFGs). This question is especially important in the light of applying well-nested MCFGs for natural language processing. The other direction is the further investigation of underlying algebraic structure. The most straightforward question is to provide the same characterization in terms of monoids for the variants of generalized STSAs as it is done for simple STSAs. Also it is interesting to answer, whether the ϵ-moves are redundant, like it was done by Zetzsche for automata based on graph products of polycyclic monoids ([24]).

References

1. Berstel, J.: Transductions and context-free languages, vol. 4. Teubner Stuttgart (1979)
2. Chomsky, N., Schützenberger, M.P.: The algebraic theory of context-free languages. computer programming and formal languages, pp. 118–161. North-Holland (1963)
3. Fischer, M.J.: Grammars with macro-like productions. In: IEEE Conference Record of 9th Annual Symposium on Switching and Automata Theory, 1968, pp. 131–142. IEEE (1968)
4. Joshi, A.K.: Tree adjoining grammars: How much context-sensitivity is required to provide reasonable structural descriptions? University of Pennsylvania, Moore School of Electrical Engineering, Department of Computer and Information Science (1985)
5. Joshi, A.K., Schabes, Y.: Tree-adjoining grammars. In: Rozenberg, G., Salomaa, A. (eds.) Handbook of Formal Languages, pp. 69–123. Springer (1997)
6. Kambites, M.: Formal languages and groups as memory. Communications in Algebra 37(1), 193–208 (2009)
7. Kanazawa, M.: A prefix-correct Earley recognizer for multiple context-free grammars. In: Proceedings of the Ninth International Workshop on Tree Adjoining Grammars and Related Formalisms (TAG+ 9), pp. 49–56 (2008)
8. Kanazawa, M.: The pumping lemma for well-nested multiple context-free languages. In: Diekert, V., Nowotka, D. (eds.) DLT 2009. LNCS, vol. 5583, pp. 312–325. Springer, Heidelberg (2009)
9. Kanazawa, M.: Multi-dimensional trees and a Chomsky-Schützenberger-Weir representation theorem for simple context-free tree grammars. Technical report (2013)
10. Kanazawa, M., Salvati, S.: MIX is not a tree-adjoining language. In: Proceedings of the 50th Annual Meeting of the Association for Computational Linguistics: Long Papers, vol. 1, pp. 666–674. Association for Computational Linguistics (2012)
11. Kepser, S., Mönnich, U.: Closure properties of linear context-free tree languages with an application to optimality theory. Theoretical Computer Science 354(1), 82–97 (2006)
12. Lallement, G.: Semigroups and combinatorial applications. John Wiley & Sons, Inc. (1979)
13. Nivat, M., Perrot, J.F.: Une généralisation du monoıde bicyclique. CR Acad. Sci. Paris Sér. A 271, 824–827 (1970)
14. Pollard, C.: Generalized phrase structure grammars, head grammars, and natural languages. PhD thesis, Stanford University, Stanford (1984)

15. Roach, K.: Formal properties of head grammars. Mathematics of Language, 293–348 (1987)
16. Rozenberg, G., Salomaa, A. (eds.): Handbook of formal languages. Word, Language, Grammar, vol. 1. Springer, New York (1997)
17. Seki, H., Kato, Y.: On the generative power of multiple context-free grammars and macro grammars. IEICE Transactions on Information and Systems 91(2), 209–221 (2008)
18. Seki, H., Matsumura, T., Fujii, M., Kasami, T.: On multiple context-free grammars. Theoretical Computer Science 88(2), 191–229 (1991)
19. Sorokin, A.: Normal forms for multiple context-free languages and displacement Lambek grammars. In: Artemov, S., Nerode, A. (eds.) LFCS 2013. LNCS, vol. 7734, pp. 319–334. Springer, Heidelberg (2013)
20. Vijay-Shanker, K., Weir, D.J., Joshi, A.K.: Tree adjoining and head wrapping. In: Proceedings of the 11th coference on Computational linguistics, pp. 202–207. Association for Computational Linguistics (1986)
21. Weir, D.J.: Characterizing mildly context-sensitive grammar formalisms. PhD thesis, University of Pennsylvania (1988)
22. Yoshinaka, R., Kaji, Y., Seki, H.: Chomsky-schützenberger-type characterization of multiple context-free languages. In: Dediu, A.-H., Fernau, H., Martín-Vide, C. (eds.) LATA 2010. LNCS, vol. 6031, pp. 596–607. Springer, Heidelberg (2010)
23. Zetzsche, G.: On the capabilities of grammars, automata, and transducers controlled by monoids. In: Aceto, L., Henzinger, M., Sgall, J. (eds.) ICALP 2011, Part II. LNCS, vol. 6756, pp. 222–233. Springer, Heidelberg (2011)
24. Zetzsche, G.: Silent transitions in automata with storage. In: Fomin, F.V., Freivalds, R., Kwiatkowska, M., Peleg, D. (eds.) ICALP 2013, Part II. LNCS, vol. 7966, pp. 434–445. Springer, Heidelberg (2013)

True Precision Required

Adjectives of Veracity in Spanish as Imprecision Regulators*

Melania S. Masià

Spanish National Research Council (ILLA-CSIC)
melania.sanchez@cchs.csic.es

Abstract. The aim of this paper is to offer an analysis of adjectives of veracity in Spanish (*verdadero* 'true', *auténtico* 'authentic') that accounts for their modification of nouns in terms of imprecision regulation. Slack regulators are elements that signal the intended degree of precision in the use of an expression to describe a situation. In order to account for this fact, I will adopt [26]'s framework, which allows to directly compare and modify degrees along a scale of imprecision. Under this framework, expressions denote sets of alternatives whose size depends on the degree of precision of the context. *Verdadero* and *auténtico* are argued to be degree modifiers affecting this scale of imprecision by setting the degree of precision of the context to a high value, forcing the modified noun to be interpreted in a strict sense.

Keywords: adjectives of veracity, imprecision, degrees, alternatives.

1 Introduction

Language is normally used with varying degrees of (im)precision, and we employ expressions in circumstances in which they would be considered to be false, strictly speaking. Slack regulators are expressions that serve to fix the amount of slack that is afforded in judging an utterance 'close enough to true' in a concrete situation (in [22]'s terms) and, in this sense, they affect the truth conditions of the sentence in which they appear. They can be grouped according to whether they increase or reduce the degree of allowed imprecision: hedges such as *loosely speaking* or *sorta* expand the set of permitted referents of an expression to normally ignorable ones (see [21]; [3] for *sorta*); other regulators such as *exactly* or *perfectly* shrink that set to those referents in the strict denotation of the modified predicate. This paper focuses on adjectives of veracity (*verdadero* 'true',

---

* I would like to thank audiences at ESSLLI 2013 Student Session and UiL OTS seminar in Utrecht for their feedback. I am grateful to two anonymous reviewers for their valuable comments. I also thank Violeta Demonte, Carme Picallo, Elena Castroviejo, Camelia Constantinescu, Rick Nouwen, and Galit Sassoon for helpful discussion. All remaining errors are my own. This research has been partially supported by research project FFI2012-32886, funded by the Spanish Ministry of Economy and Competitiveness, and by grant FPU2010-6022 from the Spanish Ministry of Education.

M. Colinet et al. (Eds.): ESSLLI 2012/2013, LNCS 8607, pp. 174–193, 2014.
© Springer-Verlag Berlin Heidelberg 2014

auténtico 'authentic') in Spanish[1] as belonging to the latter class of slack regulators. Specifically, I will argue that they set the degree of precision of a context to a high value.

The aim of this paper is to offer an analysis of adjectives of veracity that accounts for their modification of nouns in terms of imprecision regulation. Slack regulators are interesting because, as they signal the intended degree of precision in the use of an expression to describe a situation, they can be understood as part of a pragmatic mechanism. However, at the same time, they have influence in truth conditions and they must be thus part of compositional semantics. In other words, slack regulation stands in the border between semantics and pragmatics. In order to account for these facts, I will adopt [26]'s framework, which reformulates the pragmatic-halos theory of imprecision of [22] in terms of a Hamblin alternative semantics ([11]). As a consequence, I will assume that expressions denote sets of alternatives[2] whose size depends on the degree of precision of the context. This framework allows to directly compare and modify degrees along a single scale of imprecision.[3]

This paper is organized as follows: Section 2 presents the data on adjectives of veracity in Spanish and argues for a slack regulation account. In Sect. 3, I provide the theoretical background, which is formalized in Sect. 4. Section 5 concludes.

2   Adjectives of Veracity

2.1   Interpretation and Distribution

Adjectives of veracity such as Spanish *verdadero* 'true' or *auténtico* 'authentic', when in prenominal position, have an intensifying effect on the modified noun. The natural interpretation of an example like (1) is that Paloma is an artist in a strict sense, this is to say that Paloma is not just someone who merely paints or works with her hands, but presents every quality the context associates with being an artist: creativity, originality, perspicacity, maybe success.

---

[1] Adjectives in Spanish and Romance languages can appear both prenominally and postnominally. The type of modification we are dealing with here is only present in prenominal position (see (2)-(3)). In any case, the analysis may be extended to equivalent modifiers in English and other languages.

[2] The alternatives in the denotation of an expression need not to be lexical items. In some cases, we use a slack regulator beca use we lack a lexical item to refer to a specific object, such as for *sorta kick the ground* ([3]). As a reviewer points out, however, sometimes the context does not require us to be precise, as happens in the use of round numerals (*The distance between Amsterdam and Vienna is 1,000 kilometres* vs. *The distance between Amsterdam and Vienna is 965 kilometres*) ([20]).

[3] We are considering here vagueness and imprecision to be two different phenomena. Both involve uncertainty about where cut-off points in the denotation are located, but a vague predicate shows contextual variability in truth conditions, borderline cases, and gives rise to the Sorites paradox, whereas an imprecise use of a predicate has the two former characteristics, but it is not easily associated with Sorites sequences and can be given natural precisifications (see [16], [31], a.o.).

(1)   Paloma es una verdadera / auténtica artista.[4]
      Paloma is a   true   / authentic artist

      'Paloma is a true / real artist.'

This type of modifiers appear only in prenominal position in Spanish. Their modification is different from that of *true* or *authentic* in their literal sense ('not false'), which is mainly restricted to postnominal or predicative position. For instance, according to (2a), the pain Paloma felt is a real one, not imaginary; whereas for (2b), the pain is a true pain, an intense one, not simple discomfort, or a twinge. The distribution of postnominal *verdadero*, on the contrary, is restricted to those entities that can be either true or false (3).

(2)   a. Paloma sintió dolor auténtico. / El   dolor era auténtico.
          Paloma felt   pain authentic / The pain was authentic

          'Paloma felt real pain.' / 'The pain was real.'

      b. Paloma sintió auténtico dolor.
          Paloma felt   authentic pain

          'Paloma felt real pain.'

(3)   a. ?? Un periodista verdadero.
          a   journalist true

          'A real journalist (not a fake one).'

      b. ?? Una tortura auténtica.
          a   torture authentic

          'A real torture (not a fake one).'

Modification by *verdadero* and *auténtico* has a scalar flavour, in the sense that Paloma seems to have a greater amount of 'artistness' (whatever that may consist of) than any other relevant artist, so she is in the upper part of a scale of artists ordered by this salient property. Adjectives of veracity appear with nouns that have been considered candidates of gradable nouns: nouns that categorize individuals based on a gradable property (4) and abstract mass and count nouns (5) ([7], [8]).

(4)   a. Lucía es una verdadera entusiasta de las tragedias clásicas.
          Lucía is a   true   enthusiast of the tragedies classic

          'Lucía is a true Greek tragedy enthusiast.'

      b. Juan es un auténtico idiota.
          Juan is a authentic idiot

          'Juan is a real idiot.'

---

[4] Note that the indefinite article slightly changes the sense of the sentences. *Paloma es artista* (lit. 'P. is artist') simply states Paloma's occupation, while *Paloma es una artista* 'P. is an artist' adds an affective nuance to the statement (see [23], and references therein).

(5) a. Tutoriza a   sus alumnos con verdadera dedicación.
         tutors   PREP her students with true   dedication
         'She tutors her students with true dedication.'

     b. Hay   una auténtica dificultad en distinguir   una explosión
         there.is a   authentic difficulty in distinguish-INF a   explosion
         nuclear y   un terremoto.[5]
         nuclear and a   earthquake
         'There is a real difficulty in telling apart a nuclear explosion from an
         earthquake.'

However, adjectives of veracity combine with other types of nouns as well,
such as concrete count nouns (6) or deverbal nouns (7). The sense of scalarity is
also present in these examples: the referent of the modified noun is understood as
close to the core notion denoted by the predicate. For example, a true revolution
(7a) fulfils every requirement to be considered so, i.e. is a revolution in a strict
sense.

(6) a. Quiero   que seas   un verdadero padre para mi hijo.[6]
         want-1S-PRES that be-2S-SUBJ a   true   father for my son
         'I want you to be a true father for my child.'

     b. La   casona es una auténtica casa   rural al   estilo del
         The villa   is a   authentic house rural PREP.the style of.the
         siglo   XIX.[7]
         century 19th
         'The villa is a real rural house with a 19th century style decoration'.

(7) a. Las compañías de bajo coste han   supuesto una
         the companies of low cost have-3PL-PRES supposed a
         verdadera revolución en el   transporte aéreo.[8]
         true   revolution in the transport aerial
         'Low-cost companies have brought about a true revolution to air
         transport.'

     b. Aquella victoria se   ha convertido en una auténtica derrota. [9]
         that   victory REFL has turned   in a   authentic defeat
         'That victory has become a real defeat.'

Finally, with a small set of nouns and the definite article, *verdadero* (*auténtico*
only to a lesser extend) receives a literal interpretation ('not fake') (8). For

---

[5] http://eldia.es/2012-03-24/AGENDA/3-D-decia-marzo.htm
[6] *The Angels' Share* (Ken Loach, 2012)
[7] http://www.toprural.com/Miguel/opini%C3%B3n-Mas-Masaller_278426_o.html
[8] http://www.iet.turismoencifras.es/transporte/item/
    89-la-revoluci%C3%B3n-de-las-low-cost.html
[9] http://www.tonibosch.com/la-lucidez-del-perdedor/

example, in (8a), the person Paloma wants to know is her biological father, not any other man who may have raised her.[10]

(8) a. Paloma quiso   conocer a   su verdadero padre. (cf (6a))
         Paloma wanted know-INF PREP her true   father
         'Paloma wanted to know her actual father.'

     b. Impuso   la condición de que   se   ocultara   al
         Imposed-3s the condition of COMP IMPRS hide-3S-SUBJ PREP.the
         niño su verdadera identidad.[11]
         child his true   identity
         'He imposed the condition that the child should never know his true identity.'

Adjectives of veracity combine with a wide range of nouns with an intensifying effect that involves some sense of ordering. As opposed to what happens in the adjectival domain, the issue of whether gradability is represented in the lexical semantics of nouns is a controversial issue.

[30] puts forward that all nouns are gradable at the conceptual level (entities in their denotation are ordered according to their typicality), but that this ordering is not accessible by linguistic means, except for a small class of adjective-like nouns, such as *idiot*. Some other authors ([7], [24], [25]) have also acknowledged the existence of a class of degree nouns based on tests such as modification by size adjectives ((9a), cf. (9b)) or combination with the degree operator *such* ((10a), cf. (10b)). An opposite view is that of [8], who argues that these environments are actually sensitive to factors other than the presence of a degree argument, such as expression of a value judgement.

(9) a. George is an enormous idiot. [25]

     b. # This is an enormous room.

(10) a. The calculation was no good at all, he made such a mistake! [7]

     b. * This man is such a person! [8]

In some of their uses, adjectives of veracity seem to be modifying the degree of the property denoted by the noun, like in *un verdadero idiota* 'a true idiot' and the examples in (4). We could then posit two *verdaderos*: a degree modifier of gradable nouns (see [25], [33]) and a slack regulator for non-gradable nouns.[12] However, this option is less economical than having a sole entry for *verdadero*

---

[10] I am assuming that the possessives in (8) are definite ([12]), as their impossibility to appear in existential constructions shows (i).

(i) * Hay   su padre en la cocina.
       Have-3S-PRES his father in the kitchen
       Lit. 'There is his father in the kitchen.'

[11] *Cien años de soledad*, Gabriel García Márquez (1967)

[12] I come back to this option in Sect. 4.2.

and, as I will argue in Sect. 2.2, the type of modification adjectives of veracity perform in the noun is better captured under a slack regulator analysis.

2.2 Adjectives of Veracity as Slack Regulators

Slack regulators ([21], [22]; see Sect. 3.1) are modifiers that control the imprecision that is required to interpret an utterance and can be ordered according to how precise they force the modified expression to be. The example in (11) shows an ordering from Paloma being an artist in a strict sense [maximal degree of precision] (11a) to being sort of an artist, but not really so [low degree] (11c).

(11)   a. Paloma es una artista en sentido estricto.
           Paloma is a   artist in sense   strict
           'Strictly speaking, Paloma is an artist.'

       b. En cierto modo, Paloma es una artista.
           In certain manner Paloma is a   artist
           'In a way, Paloma is an artist.'

       c. Paloma es algo   así   como una artista.
           Paloma is something like.that as   a   artist
           'Paloma is sort of an artist.'

In this scale, adjectives of veracity are close to slack regulators that restrict the afforded amount of imprecision, such as *strictly speaking* or *perfectly*. Combining any of these modifiers with *verdadero* results in redundancy (12a). Also, there is a contradiction in stating that someone is a true artist but not strictly speaking (12b).

(12)   a. ??Paloma es una verdadera artista en sentido estricto.
           'Paloma is a true artist in a strict sense.'

       b. Paloma es una verdadera artista, #pero no en sentido estricto.
           'Paloma is a true artist, but not in a strict sense.'

However, compared to *strictly speaking*, adjectives of veracity seem to require a high rather than a maximal precision in the interpretation of the modified predicate. In (13), other referents are allowed to be ranked higher in the precision scale than the referent of the noun: here, Lucía is said to be an artist in a stricter sense than Paloma is, although Paloma is already an artist in a strict sense.[13] Observe, though, that the sentences are felicitous when used with *but* but they would be odd when *and* is used instead. This may point to the fact that a maximal precision is expected from the use of *verdadero*, but this expectation is cancelled by means of the adversative connective (see [4], [34], a.o.).

(13)   a. Paloma es una verdadera artista, {pero/??y} Lucía lo es más.
           Paloma is a   true   artist but/and   Lucía it is more
           'Paloma is a true artist, but/and Lucía is more of an artist than her.'

---

[13] I thank two anonymous reviewers for pointing this out.

    b. Es un verdadero placer   tenerla   entre   las manos.
        is a   true   pleasure have.ACC between the hands.
        {Pero/??Y} todavía lo es más disfrutar de sus resultados.[14]
        But/And   still   it is more enjoy.INF of its results

        'It's a true pleasure having it in your hands. But it is even more to
        enjoy its results.'

More evidence pointing in this direction is shown in (14): whereas *auténtico* or
verdadero permit a figurative or metaphorical interpretation of the noun (14a),
maximal slack regulators force a literal reading (14b). Entities that have properties associated with a palace (big size, luxury, etc.) without being strictly one
are allowed in the denotation of *palace* and there is no contradiction in asserting
that the house is not an actual palace, although it resembles one (14a). However, maximal precision is required in the case of *strictly speaking*, and no entities
other than actual palaces can be in the denotation of the modified noun (14b).

(14)   a. Su   casa es un auténtico / verdadero palacio, pero no es un
          their house is a   authentic / true   palace, but not is a
          palacio de verdad.
          palace of truth

          'Their house is a real palace, but it's not an actual palace.'

    b. Su   casa es un palacio en sentido estricto, #pero no es un
          their house is a   palace in strict   sense, but   not is a
          palacio de verdad.
          palace of truth

          'Their house is a palace in a strict sense, but it's not an actual palace.'

Modification by adjectives of veracity in prenominal position in Spanish can
thus be analyzed in terms of slack regulation. My proposal is that *verdadero*
and *auténtico* fix the degree of precision of the context to a high value, and,
consequently, the modified expression, whose set of alternatives has been shrunk,
is interpreted in a stricter sense. To model this idea I will adopt [26]'s alternative
semantics for imprecision framework, which is presented in Sect. 3. But before
that, I will address a possible analysis based on modality, which I ultimately
reject.

2.3 An Epistemic Analysis

Expressions with similar properties such as English *real(ly)* have been analyzed
as epistemic operators ([28], [8]; see [6] for Washo *šemu*). Focusing on English
real, [8] argues that the contribution of this adnominal modifier is to emphasize
the speaker's commitment to the claim that the properties characteristically
associated with the predicate P undoubtedly apply to the individual x.

(15) $[\![real]\!] = \lambda P \lambda x \lambda w . P(x)$ in $w \wedge \forall w' \in Dox_{w,holder} : P(x)$ in w'

---

[14] http://www.finepix-x100.com/es/reviews/others/all?page=20

In particular, *x is a real P* is true only if x is in the positive extension of P in the speaker's belief worlds. A desirable consequence of this analysis is that it accounts for adjectives of veracity's wide distribution — not restricted to gradable nouns. Moreover, it allows to capture the epistemic commitment that their adverbial counterparts (*really, truly*) express in some of their sentential positions (see fn. 21; Sect. 4.3).

However, if adjectives of veracity were epistemic modals, it would be expected that they behave alike. The distribution of epistemic modals, especially regarding their embeddability, is restricted in some attitude contexts (see [27], [10], [2], a.o.). [2] show that epistemics are markedly degraded in the complement of desideratives and directives (16) in three Romance languages, including Spanish. By contrast, adjectives of veracity are licensed in these contexts.

(16)   ??Juan {quiere / ha exigido}   que sea   probable que María
       Juan   wants / has demanded that is-SUBJ probable that María
       haya   conocido a   su asesino.[15]
       has-SUBJ known   PREP her murderer
       'John {wants / demanded} that it is probable that Mary knew her killer.'

(17)   a. Kojima quiere que [su película] sea   una auténtica
          Kojima wants that [his film]   is-SUBJ a   authentic
          superproducción de Hollywood.[16]
          superproduction of Hollywood
          'Kojima wants his film to be a true blockbuster.'

       b. A un periodista se   le   exige   que sea   un verdadero
          To a journalist REFL DAT demands that is-SUBJ a   true
          detonante de puntos de vista.[17]
          trigger   of points of view
          'A journalist is required to really spark off new perpectives.'

Furthermore, modal quantification over doxastic worlds and quantification over contexts are not equivalent (see [22]) and there are reasons to believe that adjectives of veracity operate over contextual variables, such as the precision parameter. Under the epistemic view, *Paloma is a true artist* means that she is an artist in all believe worlds of the speaker, i.e. the speaker always considers Paloma to be in the positive denotation of *artist*. That sentence is felicitous in a situation where Paloma has prepared the perfect cappuccino (well-balanced, compact foam, with latte art). However, in a different context, with a different comparison class (for instance, piano players in an audition), Paloma would no longer be considered an artist, even though that world is consistent with the speaker's beliefs.

---

[15] [2] only provide the French examples, I have reconstructed the Spanish versions.
[16] http://www.otromas.com/otras/pelicula-de-metal-gear-solid-confirmada-por-el-propio-kojima/
[17] http://digitaliatec.blogspot.com/2008/11/las-nuevas-exigencias-para-el.html

In sum, an epistemic account of adjectives of veracity would have to explain why they can be embedded in contexts where epistemics are not generally licensed and is not appropriate to capture context shifts.

3 Alternative Semantics for Imprecision

3.1 Pragmatic Halos

[22] models imprecision in terms of pragmatic halos. The denotation of each expression is associated with a set of objects of the same logical type that differ from the denotation only in some 'pragmatically ignorable' respect. For instance, the halos of *3 o'clock* would include times that are close enough to 3 o'clock not to make a difference, such as 2:57 and 3:02 and, as a consequence, in usual contexts, it is acceptable to utter (18), even if Mary arrived shortly after 3:00.

(18)   Mary arrived at 3 o'clock.

The degree of deviation or imprecision allowed is determined by the context, but can be also manipulated by some specific regulators. A slack regulator such as *exactly* in (19) shrinks the halo to those times that are closest to *3 o'clock* and forces the expression to be interpreted precisely. In this way, (19) is infelicitous in a situation where Mary arrived at 2:57.

(19)   Mary arrived at exactly 3 o'clock.

Adjectives of veracity will be analysed as slack regulators, with a shrinking effect in the halos of the modified expression. But first, an implementation of Lasersohn's proposal is detailed in the next section.

3.2 Alternative Implementation

[26] recasts [22]'s pragmatic-halos theory of imprecision in terms of a Hamblin-style alternative semantics ([11]) to account for metalinguistic comparatives. For [26], the intuition behind metalinguistic comparatives is that they measure how precise a speaker is when using a particular word, i.e. they involve a comparison of degrees of precision. What (20) does then is to compare how precise is referring to George as dumb, rather than crazy.

(20)   George is more dumb than crazy. [26]

In this proposal, the cross-categorial 'approximates' relation $\approx$ holds between two objects in the model if they are sufficiently similar (21). To determine whether two objects are similar, a standard of similarity and a context that provides the scale of similarity are required, as different contexts impose different similarity orderings. The standard of similarity is construed as a degree d, a real number in the interval $[0,1]$.

(21) $\alpha \approx_{d,C} \beta$ iff, given the ordering imposed by the context C, α resembles β to (at least) the degree d and α and β are of the same type.

This similarity relation is the basis of denotations that reflect degrees of imprecision. The interpretation function is parameterized to a degree of precision and a context, $[\![.]\!]^{d,C}$, and denotations are partially ordered sets of alternatives ranging from the d-resembling alternative to the perfectly resembling one. An expression such as *dumb* thus denotes the set of alternatives that resemble *dumb* sufficiently (22a). When *dumb* is interpreted in the highest degree of precision, 1, it will denote the singleton set containing only *dumb* (22b); when it is interpreted in the lowest degree of precision, it will denote all the alternatives of the same semantic type (22c).

(22) a. $[\![dumb]\!]^{d,C} = \{f_{\langle e,t\rangle} : f \approx_{d,C} dumb\}$
      b. $[\![dumb]\!]^{1,C} = \{dumb\}$
      c. $[\![dumb]\!]^{0,C} = D_{\langle e,t\rangle}$

Accordingly, higher imprecision corresponds to a widening of a pragmatic halo, and higher precision to a narrowing of the denotation. To model pragmatic halos, [26] adopts [19]'s approach to Hamblin alternatives, according to which alternatives are part of the compositional semantics.[18] This sort of alternative framework requires some way of mapping a sentence denotation — a set of propositional alternatives — to a single proposition. [19] assume an existential closure operation (23) which can take place at intermediate points of the tree as well as at the top.

(23) $[\![\exists \alpha]\!]^d = \lambda w.\exists p[p \in [\![\alpha]\!]^d \wedge p(w)]$

As such, degrees of imprecision are not available for composition and do not play a role in the semantic derivation. In order to have access to this scale, [26] introduces a typeshift, called PREC (24) in his system.

(24) $[\![\text{PREC } \alpha]\!]^d = \lambda d'.[\![\alpha]\!]^{d'}$

PREC binds the degree of imprecision and makes it available as an argument. This typeshift applies as a last resort whenever there are certain type-theoretical or structural environments that require to make use of the imprecision scale, such as modification by *verdadero*, as I propose in the following section.

4 Proposal

4.1 Detour: Adverbs of Veracity

In order to determine what adjectives of veracity quantify over, I will first observe the behaviour of their adverbial counterparts *verdaderamente* 'truly',

---

[18] This idea connects metalinguistic comparatives with work on focus (e.g. [29]).

auténticamente 'authentically', and *realmente*[19] 'really' with adjectives. I will assume a degree approach to gradability ([9], [32], [13], [15], a.o.), according to which degrees are part of the ontology, and gradable predicates include a degree argument in their structure and are of type $\langle d, \langle e, t \rangle \rangle$.[20] The degree argument is to be bound by an overt degree operator (comparative morphology, degree modifiers) or by a null degree operator POS for the positive form. Syntactically, gradable adjectives project an extended functional structure headed by degree morphology (25) ([1], [15], a.o.).

(25)

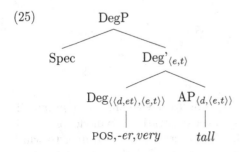

Verdaderamente, when combined with adjectives, is a degree modifier. As such, it occupies the degree head in the structure and, therefore, other degree morphology cannot appear in the same position, as the ungrammaticality of (26) shows. When the Degree head is occupied by another degree morpheme ((26b) and (26d)), *verdaderamente* is understood as affecting the whole proposition, with an epistemic reading.[21]

---

[19] The adjective *real* 'real' used as a slack regulator is restricted to a few nouns (i)-(ii), probably due to homonymy with *real* 'royal'. The adverb is nevertheless widely used as a degree/epistemic modifier.

  (i)   Luego de casarse, su vida se le convirtió en un real tormento. [www.mujertuvalesmucho.org/testimoniojuanitalovil.html]
       'After she got married, her life became a real torture.'

  (ii)   Él nunca deja de contestar nuestras plegarias, cuando son hechas con real intensidad. [books.google.es/books?isbn=9501701468]
       'He never stops attending our prayers when they are said with real passion.'

[20] Gradable adjectives have been alternatively analyzed as measure functions $\langle e, d \rangle$ ([5], [15]). Although I do not adopt the measure function analysis, nothing in my proposal hinges on this decision.

[21] *Verdaderamente*, as well as in English *really*, has at least two readings depending on its position and what it quantifies over: an epistemic one affecting propositions (first *verdaderamente* in (i)) and a degree one modifying properties (second *verdaderamente*) (see [17]). I will focus on the latter. The former is marked with # in the examples. I will return to this distinction in Sect. 4.3.

  (i)   Verdaderamente estaba verdaderamente satisfecha con el trabajo.
       'I really was really satisfied with the work.'

(26)   a. *Lucía es más verdaderamente alta que Paloma.
        Lucía is more truly   tall than Paloma

        'Lucía is more truly tall than Paloma.'

       b. # Lucía es realmente más alta que Paloma.
          Lucía is really   more tall than Paloma

          'Lucía is really taller than Paloma.'   [epistemic reading only]

       c. *Lucía es {completamente / muy} auténticamente alta.
          'Lucía is completely / very / authentically tall.'

       d. Lucía es verdaderamente {*completamente / #muy /} alta.
          'Lucía is truly completely / very / tall.'   [epistemic reading only]

The restriction adverbs of veracity impose on the degree argument of the adjective is similar to that of *very*, in that the relative standard is raised by some amount (27a)-(27b).[22] As was observed for adjectives of veracity in (13)-(14), the degree of the property denoted by the predicate is not fixed to its maximum in the scale. As (27b) and (27c) show, a fuller glass is conceivable when a closed scale adjective such as *full* is modified by *realmente* or *muy* 'very', but not when modified by a maximality modifier such as *completamente* 'completely'.

(27)   a. Lucía es {verdaderamente / muy} alta, pero podría serlo   más.
          Lucía is {truly   / very} tall, but could-3s be.PRN more

          'Lucía is truly / very tall, but she could be taller.'

       b. El vaso está {realmente / muy} lleno, pero podría   estarlo más.
          the glass is {really   / very} full, but could-3s be.PRN more

          'The glass is really / very full, but it can be fuller.'

       c. El vaso está completamente lleno, #pero podría   estarlo más.
          the glass is completely   full,   but could-3s be-PRN more

          'The glass is completely full, but it can be fuller.'

I will assume that the pairs of adverbs and adjectives of veracity such as *verdaderamente* and *verdadero* are instances of the same lexical root. As such, I will consider *verdadero* a degree modifier. The difference will lie in the type of degrees they quantify over: whereas the former is sensitive to the degree of a property that holds of an individual, the latter cares about degrees of imprecision in the use of a nominal expression.

---

[22] The regular standard for a predicate is a degree calculated on the basis of a contextually determined comparison class. *Very* calculates the new standard by restricting the comparison class to entities which already have the property G in the context of utterance (see [18], [17], a.o.).

(i) $[\![very]\!]^c = \lambda G \lambda x. \exists d[\mathbf{standard}(d)(G)(\lambda y.[\![pos(G)(y)]\!]^c) \wedge G(d)(x)]$ [17]

4.2 Modification by Adjectives of Veracity

Adverbs of veracity have been shown to be degree modifiers that raise the standard of the adjective by some amount. Their adjectival counterparts have a similar effect on nouns, in the sense that the denotation is also restricted to individuals closer to the maximal value of the predicate. As discussed at the end of Sect. 2.1, (most) nouns do not seem to be lexically associated with a scale onto which they map their arguments, as adjectives do. If this is so, there should be a type mismatch between *verdadero* and the noun it modifies.

However, how appropriate or precise it is to apply a certain noun to a referent is something that can be graded ((28a) see also (11)) and compared (28b). This points to some sort of ordering associated with nouns. This ordering can be modelled as a scale of imprecision [26] and I propose that adjectives of veracity operate on this scale.[23]

(28) a. Paloma es una artista {en sentido estricto / en cierto   modo}.
          Paloma is a   artist {in sense   strict   / in certain manner}
          '{Strictly speaking / in a way}, Paloma is an artist.'

      b. Paloma es más una artesana que una artista.
          Paloma is more a   artisan than a   artist
          'Paloma is more an artisan than an artist.'

In this framework, denotations consist of sets of alternatives (29) whose size depends on the standard of similarity of the particular context construed as a degree d ranging in the interval $[0,1]$ (30). As can be observed in (30), the higher the degree of precision, the narrower the denotation of the predicate.[24] By abstracting over d and applying the $\approx$ relation, an ordering over sets of

---

[23] As an anonymous reviewer points out, adjectives of veracity may have a degree use. If the modified noun had a degree argument, nothing in this analysis would prevent *verdadero* to target that degree, instead of forcing a typeshift that makes the precision parameter available for composition. In that case, the denotation in (31) would remain essentially the same, and the difference between *una verdadera casa* and *un verdadero artista* would be that, in the first case, it is the degree of the precision what is quantified over, whereas in the latter it is the degree of the property ('artistness' in this case) what is set to a high value. This would explain cases where stereotypical, rather than defining characteristics of the category denoted by the noun seem to be target by *verdadero*, such as with nationality nouns (i), an example pointed out by the reviewer. Another option is to understand this example as a metaphorical interpretation of the noun, as the one discussed in (14).

(i)   Arnold Schwarzenneger es un verdadero americano.
      'Arnold Schwarzenneger is a real American.'

Whether (some) nouns have a degree argument in their lexical representation is an issue beyond the purpose of this paper, so I will assume that adjectives of veracity are always slack regulators (see also Sect. 2.1; fn. 26).

[24] For the sake of illustration, I use lexical items in the representation of alternatives, but see fn. 2.

alternatives is generated. In the sense that these sets of alternatives can be understood as points in a general scale of imprecision, the denotation of any expression is gradable.

(29) $[\![artista]\!]^{d,C} = \{f_{\langle e,t\rangle} : f \approx_{d,C} artist\}$

(30)   a. $[\![artista]\!]^{0.9,C} = \{artist, creator, author\}$

   b. $[\![artista]\!]^{0.8,C} = \{artist, creator, author, artisan, designer\}$

   c. $[\![artista]\!]^{0,C} = D_{\langle e,t\rangle}$

The denotation of *artista* is thus build of partially ordered sets of alternatives of type $\langle e,t\rangle$ ranging from the d-resembling set of alternatives to the perfectly resembling one. This scale of imprecision is what provides a degree argument that can be targeted by slack regulators such as adjectives of veracity.

Adjectives of veracity can be analysed as modifiers fixing the degree of precision in a context to a very high value. This intuition can be formalized as follows: the standard of similarity is construed as a degree d, a real number in the interval $[0,1]$, so what *verdadero* does is to set the value of d to a value much higher (represented here by $>!$) than the standard of the context. As a degree modifier, *verdadero* takes an expression of type $\langle d, \langle e, st\rangle\rangle$ and returns a property $(\langle e, st\rangle)$, which applies to an individual in a particular world or context.

(31) $[\![verdadero]\!]^{d,C} = \lambda P_{\langle d,\langle e,st\rangle\rangle}\lambda x\lambda w.\exists d'[d' >! d \wedge P(d')(x)(w)]$

The degree of precision being a parameter of the interpretation function is not accessible by any modifier. As mentioned in Sect. 3, typeshift PREC (24) is required to make that degree available for composition. However, PREC cannot apply to a *set* of properties by pointwise functional application[25] because PREC does not denote itself any set. Before PREC can apply to the denotation of the predicate, existential closure turns the set of alternative properties into one property (32). Then PREC transforms this property into something of type $\langle d, \langle e, st\rangle\rangle$ (33).

(32) $[\![\exists\ artista]\!]^{d,C} = \lambda x\lambda w.\exists f \in [\![artista]\!]^{d,C}\wedge f(x)(w)$

(33) $[\![\text{PREC}\ \exists\ artista]\!]^{d,C} = \lambda d'.[\![\exists\ artista]\!]^{d',C} =$
   $= \lambda d'\lambda x\lambda w.\exists f \in [\![artista]\!]^{d',C} \wedge f(x)(w)$

Now *verdadero* can apply to the noun (34). The result is the property of being an artist in a very precise sense in the given context, as *verdadero* fixes the degree of precision of being an artist higher than the standard of precision of the context (the index d).

---

[25] In alternative semantics, compositionality makes use of a pointwise (or Hamblin) function application to generate alternative sets. The rule of composition adopted in this system is (i).

(i) HAMBLIN FUNCTIONAL APPLICATION:
   If α is a branching node with daughters β and γ, and $[\![\beta]\!]^{d,C} \subseteq D_\sigma$ and $[\![\gamma]\!]^{d,C} \subseteq D_{\langle\sigma,\tau\rangle}$, then $[\![\alpha]\!]^{d,C} = \{b(c) : b \in [\![\beta]\!]^{d,C} \wedge c \in [\![\gamma]\!]^{d,C}\}$

(34) $[\![verdadero\ \text{PREC}\ \exists\ artista]\!]^{d,C} =$
      $= \lambda x \lambda w. \exists d'[d' >! \ d \wedge \exists f \in [\![artista]\!]^{d',C} \wedge f(x)(w)]$

Consequences. As predicted by the analysis, using the noun modified by *verdadero* with a degree of precision lower than required produces infelicitous utterances. Imagine a context where you are in your Spanish class and the teacher asks you to write a composition. The example in (35a) would be felicitous, while (35b) would be considered inappropriate, as not having a pen is not a problem in a strict sense in that context where other students can lend you one.[26]

(35) a. Tengo   un problema: me he   dejado el boli en
          Have.1S.PRS a   problem: DAT have-1S-PRES left   the pen in
          casa.
          house
          'I have a problem: I forgot my pen at home.'

      b. # Tengo   un verdadero problema: me he   dejado
          Have.1S.PRS a true   problem: DAT have-1S-PRES left
          el   boli en casa.
          the pen in house
          'I have a real problem: I forgot my pen at home.'

As the degree of precision is high, but not maximal, other referents with a higher degree of precision are possible, even if not expected. This explains the felicitousness of the example in (13a), where Lucía is said to be more of an artist than Paloma, who already is a true artist, with *but* but not *and*.

The scale of imprecision is a general one and the same for all expressions. Because of this fact, *verdadero* and *auténtico* show no restriction in the type of noun they modify (4)-(7). This also accounts for the absence of incommensurability effects in metalinguistic comparisons (*Clarence is more tall than boring*), in contrast with the ill-formedness of regular comparatives constructed from adjectives that measure along distinct scales (**Clarence is taller than he is boring*) (see [26]).

As we mentioned, a small group of nouns including *father* and *identity* receive a literal interpretation when combined with prenominal *verdadero* and the definite article ((8), repeated here).

(8b) Impuso la condición de que se ocultara al niño su verdadera identidad.
      'He imposed the condition that the child should never know his true identity.'

---

[26] Here, as an anonymous reviewer notes, a slack regulation analysis, as opposed to a degree analysis of *verdadero* makes the prediction that small problems are not problems in a strict sense. My intuition is that that is right: A felicitous answer to the example in (35a) (without *verdadero*), would be (i).

(i) That's not a problem! I can lend you one.

One option is that *verdadero*, in combination with the definite article, turns into a maximal slack regulator, i.e. one that sets the degree of precision of the context to 1. In fact, if we compare the sentence with the adjective of veracity and the same sentence with *strictly speaking*, the meaning seems to be the same.

(36)   a.   Arcadio Buendía es la   verdadera identidad del   niño.
            Arcadio Buendía is the true   identity   of.the child
            'Arcadio Buendía is the child's true identity.'

       b.   En sentido estricto, Arcadio Buendía es la   identidad del   niño.
            in sense   strict,   Arcadio Buendía is the identity   of.the child
            'Strictly speaking, Arcadio Buendía is the identity of the child.'

However, to maintain compositionality, I will assume that the superlative reading is derived from the combination of *verdadera identidad* with the definite article (38). Following [14], I will take the definite article to be of type $\langle\langle e,t\rangle,e\rangle$ and a function that returns the unique individual in the denotation of the property (37).

(37) $[\![the]\!] = \lambda f : f \in D_{<e,t>} \wedge \exists!x[f(x) = 1] \; . \; \iota y[f(y) = 1]$ [14]

(38) $[\![la\ verdadera\ identidad]\!]^{d,C} =$
       $= \lambda w.\iota x \exists d'[d' >! d \wedge \exists f \in [\![identidad]\!]^{d',C} \wedge f(x)(w)]$

In an imprecise context, we refer to both fake and true identities by means of the noun *identity*. With the presence of *verdadero*, the degree of precision of the context increases, excluding from the denotation most fake identities (*identities* only in a loose sense). Then, the definite article introduces the requirement that the denotation of *verdadera identidad* has a sole individual. In this way, at degree of precision d' (which is high, but not necessarily maximal), only one identity remains in the denotation of the noun, the identity in the strictest sense in the context, which is equivalent to the actual identity.

To sum up, what adjectives of veracity do is to quantify over the degree of precision of the context and rise it to a very high value. As a consequence, the denotation (or halo) of the noun is shrunk to entities that resemble the predicate to at least this new degree, so this results in a stricter interpretation.

4.3   Back to Adverbs of Veracity

Now the analysis for adjectives of veracity has been developed, we can revisit adverbs of veracity and see whether the same denotation may apply to them. *Verdaderamente* and its kin appear with both open-scale and close-scale adjectives ((39a) and (39b) respectively) and their effect is similar to that of *very*, in that they raise the standard of the adjective to some amount.

(39)   a.   Lucía es verdaderamente alta.
            'Lucía is truly tall.'

b. El vaso está realmente lleno.
'The glass is really full.'

In this case, the predicate already includes a degree argument, so this will be the degree pointed by the modifier. The denotation for *verdaderamente* would be basically the same as for *verdadero* (40).

(40) $[\![verdaderamente]\!]^{d,C} = \lambda P \lambda x \lambda w. \exists d'[d' >! \mathbf{standard}(P) \wedge P(x)(d')(w)]$

I will assume also here that existential closure maps the predicate's denotation (a set of alternatives) to a single predicate (42).

(41) $[\![alta]\!]^{d,C} = \{f_{<d,<e,t>>} : f \approx_{d,C} alta\}$

(42) $[\![\exists\, alta]\!]^{d,C} = \exists f : f \in [\![alta]\!]^{d,C}$

Now *verdaderamente* can modify the gradable adjective (43). The original value for the degree argument of the adjective is given by the **standard** function, as in degree accounts for gradable adjectives (e.g. [17]). *Verdaderamente* sets this degree to a much higher value.

(43) $[\![verdaderamente\ \exists\ alta]\!]^{d,C} =$
$= \lambda x \lambda w. \exists d[d >! \mathbf{standard}(\text{tall}) \wedge \exists f \in [\![alta]\!]^{d,C} \wedge f(d)(x)(w)]$

Adverbs of veracity modifying non-gradable adjectives (44), as well as gradable ones already modified by a degree modifier ((26b), (26d)), results in an epistemic reading. In these cases, the predicates do not include a degree argument in their denotation. In contrast to other degree modifiers such as *very*, this modification does not result in coercion of the predicate into a degree one (45). The same epistemic reading is found in (46).

(44)   a. María está verdaderamente embarazada.
María is   truly   pregnant
'Truly, María is pregnant.' (never means 'she's very pregnant')

b.   ? Este es un asunto realmente geopolítico.
This is a   issue   really   geopolitical
'Really, this is a geopolitical issue.'

(45)   María está muy embarazada.
María is   very pregnant
'María is very pregnant.' (she's in her last months of pregnancy)

(46)   a. Verdaderamente, el   sacerdocio establecido no mostraba
truly   the priesthood established not showed-3s
afecto   alguno hacia   Santiago.[27]
affection any   towards Santiago
'Really, the official priesthood showed no affection at all for Santiago.'

---

[27] http://hemeroteca.abc.es/nav/Navigate.exe/hemeroteca/madrid/cultural/
1992/03/27/022.html

b. Realmente me   he   quedado sin   palabras.
   really   REFL have-1S-PRS remained without words
   'Really, it has left me speechless.'

The examples in (44) and (46) might be accounted for if *verdaderamente* forces a type shift that makes the imprecision parameter of the whole proposition available for composition. I would like to suggest that, in these examples, the modifier is again setting the degree of precision to a high value, so that the proposition must be interpreted in a stricter sense. Nevertheless, the assimilation of epistemic modification by adverbs of veracity to imprecision regulation is an issue that deserves further study.

5 Conclusion and Further Issues

Alternative semantics has been shown to be useful to formalize the imprecise use of language and the phenomenon of slack regulation. It also brings together two manifestations of uncertainty in language — vagueness and imprecision – by associating them to gradability along different scales — lexical and imprecision. The proposal made here assumes the basis of the analysis of metalinguistic comparatives [26] and *sorta* [3], and applies them to related modifiers, such as adjectives of veracity in Spanish. *Verdadero* has been argued to be an imprecision regulator setting the degree of precision of the context to a high value.

The analysis may be extended to other related degree modifiers, such as *completa(mente)* 'complete(ly)' and *perfecta(mente)* 'perfect(ly)'. When combined with expressions associated with a lexical scale, such as adjectives and some verbs, they behave as regular degree modifiers (*The glass is completely / perfectly full*; *The army completely destroyed the city*). But whenever no lexical scale is available, they target the imprecision scale of the modified expression. This is the case with nouns and some verbs (*The complete family came* (cf. *The family came*); *Mary was perfectly convinced*). This suggests that there may be two types of gradability in language [26], and developing this idea would contribute to a better understanding of scalarity across grammatical categories and the difference between vagueness and imprecision.

References

1. Abney, S.: The English noun phrase in its sentential aspect. PhD thesis. MIT (1987)
2. Anand, P., Hacquard, V.: Epistemics and attitudes. Semantics and Pragmatics 6, 1–59 (2013)
3. Anderson, C.: Hedging verbs and nouns using an alternative semantics. In: Proceedings of ConSOLE XXI (2013)
4. Anscombre, J.C., Ducrot, O.: Deux *mais* en français? Lingua 43(1), 23–40 (1977)
5. Bartsch, R., Vennemann, T.: Semantic structures: A study in the relation between semantics and syntax. Athenäum, Frankfurt (1973)

6. Bochnak, M.R.: The non-universal status of degrees: Evidence from Washo. In: Keine, S., Slogget, S. (eds.) Proceedings of NELS 42, pp. 79–92. GLSA, Amherst (2013)
7. Bolinger, D.: Degree words. Mouton, The Hague (1972)
8. Constantinescu, C.: Gradability in the nominal domain. PhD thesis, Universiteit Leiden (2011)
9. Cresswell, M.J.: The semantics of degree. In: Partee, B.H. (ed.) Montague Grammar, pp. 261–292. Academic Press, New York (1976)
10. Hacquard, V., Wellwood, A.: Embedding epistemic modals in english: A corpus-based study. Semantics and Pragmatics 5, 1–29 (2012)
11. Hamblin, C.: Questions in Montague English. Foundations of Language 10, 41–53 (1973)
12. Heim, I.: The semantics of definite and indefinite noun phrases. PhD thesis, University of Massachussetts, Amherst (1982)
13. Heim, I.: Notes on superlatives. Ms. MIT (1995)
14. Heim, I., Kratzer, A.: Semantics in generative grammar. Blackwell, Oxford (1998)
15. Kennedy, C.: Projecting the adjective: The syntax and semantics of gradability and comparison. Garland, New York (1999)
16. Kennedy, C.: Vagueness and grammar: the semantics of relative and absolute gradable adjectives. Linguistics and Philosophy 30(1), 1–45 (2007)
17. Kennedy, C., McNally, L.: Scale structure and the semantic typology of gradable predicates. Language 81(2), 345–381 (2005)
18. Klein, E.: A semantics for positive and comparative adjectives. Linguistics and Philosophy 4(1), 1–45 (1980)
19. Kratzer, A., Shimoyama, J.: Indeterminate pronouns: The view from Japanese. In: Otsu, Y. (ed.) Proceedings of the 3rd Tokyo Conference on Psycholinguistics, pp. 1–25. Hituzi Syobo, Tokio (2002)
20. Krifka, M.: Be brief and vague! and how bidirectional optimality theory allows for verbosity and precision. In: Restle, D., Zaefferer, D. (eds.) Sounds and Systems. Studies in Structure and Change. A Festschrift for Theo Vennemann, pp. 439–458. Mouton de Gruyter, Berlin (2002)
21. Lakoff, G.: Hedges: A study in meaning criteria and the logic of fuzzy concepts. Journal of Philosophical Logic 2(4), 458–508 (1973)
22. Lasersohn, P.: Pragmatic halos. Language 75(3), 522–551 (1999)
23. Leonetti, M.: El artículo. In: Bosque, I., Demonte, V. (eds.) Gramática Descriptiva de la Lengua Española, vol. 1, pp. 787–890. Espasa Calpe, Madrid (1999)
24. Matushansky, O.: Tipping the scales: The syntax of scalarity in the complement of *seem*. Syntax 5(3), 219–276 (2002)
25. Morzycki, M.: Degree modification of gradable nouns: Size adjectives and adnominal degree morphemes. Natural Language Semantics 17(2), 175–203 (2009)
26. Morzycki, M.: Metalinguistic comparison in an alternative semantics for imprecision. Natural Language Semantics 19(1), 39–86 (2011)
27. Papafragou, A.: Epistemic modality and truth conditions. Lingua 116(10), 1688–1702 (2006)
28. Paradis, C.: Between epistemic modality and degree: the case of *really*. In: Facchinetti, R., Krug, M., Palmer, F. (eds.) Modality in Contemporary English. Mouton de Gruyter, Berlin (2003)
29. Rooth, M.: A theory of focus interpretation. Natural Language Semantics 1, 75–116 (1992)
30. Sassoon, G.W.: Vagueness, Gradability and Typicality. The Interpretation of Adjectives and Nouns. Brill, Leiden (2013)

31. Sauerland, U., Stateva, P.: Two types of vagueness. In: Égré, P., Klinedinst, N. (eds.) Vagueness and Language Use, pp. 121–145. Palgrave Macmillan, Houndmills (2011)

32. von Stechow, A.: Comparing semantic theories of comparison. Journal of Semantics 3(1-2), 1–77 (1984)

33. de Vries, H.: Evaluative degree modification of adjectives and nouns. MA thesis, Universiteit Utrecht (2010)

34. Winterstein, G.: What *but*-sentences argue for: An argumentative analysis of *but*. Lingua 122(15), 1864–1885 (2012)

Language Change and the Force of Innovation

Roland Mühlenbernd[1] and Jonas David Nick[2]

[1] Eberhard Karls Universität Tübingen, Germany
[2] Eidgenössische Technische Hochschule Zürich, Switzerland

Abstract. Lewis [L1] invented *signaling games* to show that semantic meaning conventions can arise simply from regularities in communicative behavior. The behavioral implementation of such conventions are so-called *signaling systems*. Previous research addressed the emergence of signaling systems by combining signaling games with learning dynamics, and not uncommonly researchers examined the circumstances preventing the emergence of signaling systems. It has been shown that by increasing the number of states, messages and actions for a signaling game, the emergence of signaling becomes increasingly improbable. This paper contributes to the question of how the invention of new messages and extinction of unused messages would change these outcomes. Our results reveal that this innovation mechanism does in fact support the emergence of signaling systems. Furthermore, we analyze circumstances that lead to stable communication structure in large spatial population structures of interacting players.

1 Introduction

Signaling games are a leading model to analyze the evolution of semantic meaning. Researchers in this field use simulations to explore agents' behavior in repeated signaling games. Within this field of study two different research approaches are apparent: first, the simulation of a repeated 2-players signaling game combined with agent-based learning dynamics, in the majority of cases with *reinforcement learning* (e.g. [B1], [BZ1], [S1]); second, evolutionary models of population dynamics, wherein signaling games are usually combined with population-based *replicator dynamics* (e.g. [HH1], [HSRZ1]). To fill the gap between both methods, recent work deals with applying repeated signaling games combined with agent-based dynamics on multi-agent populations, e.g. on social network structures (c.f. [Z1], [W1], [M1], [MF1]). With this paper we want to make a contribution to this line of research.

Barrett [B1] was able to show that the simplest variant of a signaling game, called *Lewis game*, combined with a basic version of the learning dynamics *reinforcement learning*, with 2-players which play the game repeatedly, conventions about meaningful language always emerge. But by extending the domains[1] of

---

[1] With domains we refer to the number of states, messages and action of a signaling game. It will be introduced in the following section.

M. Colinet et al. (Eds.): ESSLLI 2012/2013, LNCS 8607, pp. 194–213, 2014.
© Springer-Verlag Berlin Heidelberg 2014

the signaling game, those conventions become more and more improbable. Furthermore, the number of possible different perfect signaling systems increases dramatically. This might be the reason why previous research work basically dealt with very simple variants of signaling games, especially in multi-agent setups, and avoided domain-extended games. If even two players fail to learn a signaling system for a given game, multiple players would not only have this problem, but could ultimately end up in a confusion of tongues, where a lot of different incompatible signaling systems evolve.

With this article we will show that by extending the learning dynamics to allow for innovation we can observe i) an improvement of the probability that signaling systems emerge for domain-extended signaling games and extended population sizes, ii) the emergence of different evolving perfect signaling systems in a spatial population structure with local interaction and iii) the formation of regions of the same signaling system that form a spatial continuum.

This article is divided in the following way: in Section 2 we'll introduce some basic notions of repeated signaling games, reinforcement learning dynamics and multi-agent approaches; in Section 3 we'll take a closer look at the variant of reinforcement dynamics we used - a further development of Bush-Mosteller reinforcement; in Section 4 we show how innovation of new and extinction of unused messages significantly improves the outcome in terms of the emergence of signaling systems; in Section 5 we simulate agents on a two-dimensional toroid lattice to show the emergence of a dialect continuum; we'll finish with a conclusion and some implications of our approach in Section 6.

2 Signaling Games and Learning

A signaling game $SG = \langle \{S, R\}, T, M, A, Pr, U \rangle$ is a game played between a sender S and a receiver R. Initially, nature selects a state $t \in T$ with prior probability[2] $\Pr(t) \in \Delta(T)$, which only the sender observes. Therefore the current state remains a secret to the receiver. S then selects a message $m \in M$, and R responds with a choice of action $a \in A$. For each round of play, players receive utilities depending on the actual state t and the response action a. Here we will be concerned with a common variant of this game, where the number of states is on par with the number of actions ($|T| = |A|$). For each state $t_i \in T$ there is exactly one action $a_j \in A$ that leads to successful communication. This is expressed by the utility function

$$U(t_i, a_j) = \begin{cases} \alpha, & \text{if } i = j \\ -\beta, & \text{otherwise} \end{cases}$$

where $\alpha > 0$ and $\beta \geq 0$. In standard signaling games α is 1 and β is 0. This utility function expresses the particular nature of a signaling game, namely that because successful communication does not depend on the used message, there is no predefined meaning of messages. A signaling game with n states and n messages is called an $n \times n$ game and n is called the *domain* of the game.

---

[2] $\Delta(X) : X \to \mathbb{R}$ denotes a probability distribution over random variable X.

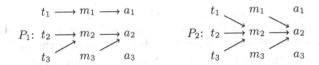

Fig. 1. Two perfect signaling systems of a 2 × 2 game, consisting of a pure sender and receiver strategy.

$$
\begin{array}{ccc}
t_1 \longrightarrow m_1 \longrightarrow a_1 \\
P_1: t_2 \longrightarrow m_2 \longrightarrow a_2 \\
t_3 \quad\quad m_3 \quad\quad a_3
\end{array}
\qquad
\begin{array}{ccc}
t_1 \quad\quad m_1 \quad\quad a_1 \\
P_2: t_2 \longrightarrow m_2 \longrightarrow a_2 \\
t_3 \quad\quad m_3 \quad\quad a_3
\end{array}
$$

Fig. 2. Two partial pooling systems. P_1 permits an information flow of 2/3, P_2 of 1/3.

2.1 Strategies and Signaling Systems

Although messages are initially meaningless in this game, meaning arises from regularities in behavior. Behavior is defined in terms of strategies. A *behavioral sender strategy* is a function $\sigma : T \rightarrow \Delta(M)$, and a *behavioral receiver strategy* is a function $\rho : M \rightarrow \Delta(A)$. A behavioral strategy can be interpreted as a single agent's probabilistic choice or as a population average. For a 2 × 2 game exactly two isomorphic strategy profiles constitute a perfect signaling system. In these, strategies are pure (i.e. action choices have probabilities 1 or 0) and messages associate states and actions uniquely, as depicted in Figure 1.

It is easy to see that for an $n \times n$ game the number of perfect signaling systems is $n!$. This means that while for a 2 × 2 game we get the 2 signaling systems as mentioned above, for a 3 × 3 game we get 6, for a 4 × 4 game 24, and for a 8 × 8 game more than 40,000 perfect signaling systems. Moreover, for $n \times n$ games with $n > 2$ there is a possibility of partial *pooling equilibria*, which transmit information in a fraction of all possible cases. Figure 2 shows different possibilities of partial pooling systems for a 3 × 3 game.

2.2 Models of Reinforcement Learning

The simplest model of reinforcement learning is *Roth-Erev reinforcement* [RE1] and can be captured by a simple model based on urns, known as *Pólya urns*, which works in the following way: an urn contains balls of different types, each type corresponding to an action choice. Now, drawing a ball means to perform the appropriate action. An action choice can be successful or unsuccessful and in the former case, the number of balls of the appropriate act will be increased by one, such that the probability for this action choice is increased for subsequent draws. All in all, this model ensures that the probability of making a particular decision depends on the number of balls in the urn and therefore on the success of past action choices. This leads to the effect that the more successful an action choice is, the more probable it becomes in following draws.

But Roth-Erev reinforcement has the property that after some time the learning effect[3] slows down: while the number of additional balls for a successful action is a static number α, in the general case $\alpha = 1$, as mentioned above, the overall number of balls in the urn is increasing over time. E.g. if the number of ball in the urn at time τ is n, the number at a later time $\tau + \epsilon$ must be $m \geq n$. Thus the learning effect is changing from α/n to α/m and therefore can only decrease over time.

Bush-Mosteller reinforcement (see [BM1]) is similar to Roth-Erev reinforcement, but without slowing the learning effect down. After a reinforcement step the overall number of balls in an urn is adjusted to a fixed value c, while preserving the ratio of the different balls. Thus the number of balls in the urn at time τ is c and the number at a later time $\tau + \epsilon$ is c and consequently the learning effect stays stable over time at α/c.

A simple yet powerful modification is the adoption of *negative reinforcement*: while in the standard procedures unsuccessful actions have no effect on the urn value, with negative reinforcement an unsuccessful action is punished by decreasing the number of balls that lead to that action.

2.3 Reinforcement Learning and Signaling Games

To apply reinforcement learning to signaling games, sender and receiver both have urns for different states and messages and make their decision by drawing a ball from the appropriate urn. In detail: the sender has an urn $\mho_t$ for each state $t \in T$, which contains balls for different messages $m \in M$. Let $m(\mho_t)$ denote the number of balls of type m in urn $\mho_t$ and $|\mho_t|$ denote the overall number of balls in urn $\mho_t$. If the sender is faced with a state t she draws a ball from urn $\mho_t$ and sends message m, if the ball is of type m. Accordingly, the receiver owns urn $\mho_m$ for each message $m \in M$, which contains balls for different actions $a \in A$. The number of balls of type a in urn $\mho_m$ is denoted as $a(\mho_m)$, the overall number of balls in urn $\mho_m$ as $|\mho_m|$. Upon perceiving message m the receiver draws a ball from urn $\mho_m$ and plays the action a, if the ball is of type a. Thus the sender's behavioral strategy σ and receiver's behavioral strategy ρ can be defined in the following way:

$$\sigma(m|t) = \frac{m(\mho_t)}{|\mho_t|} \qquad (1) \qquad \rho(a|m) = \frac{a(\mho_m)}{|\mho_m|} \qquad (2)$$

Recently, Franke and Jäger [FJ1] introduced the concept of *lateral inhibition* for reinforcement learning in signaling games in order to lead the system more speedily towards pure strategies. In the next section we will show that lateral inhibition also generally increases the probability that repeated signaling games lead to the emergence of signaling systems (as e.g. depicted in Figure 3).

The concept of lateral inhibition applied on reinforcement learning can basically describes as follows: drawing a successful action not only increases the

---

[3] The learning effect is the ratio of additional balls for a successful action choice to the overall number of balls.

number of corresponding balls, but also decreases the number of each other type of ball. Likewise, an unsuccessful action decreases its probability, while the probability of competing actions increases. E.g. using Roth-Erev reinforcement with lateral inhibition value $\gamma \in \mathbb{N} \geq 0$ the following update process is executed after each round of play: if communication via t, m and a is successful, the number of balls in the sender's urn $\mho_t$ is increased by $U(t,a) = \alpha \in \mathbb{N} > 0$ balls of type m and reduced by γ balls for each type $m' \neq m$. Similarly, the number of balls in the receiver's urn $\mho_m$ is increased by α balls of type a and reduced by γ balls for each type $a' \neq a$. Furthermore, negative reinforcement also changes urn contents in the case of unsuccessful communication in the following way: if communication via t, m and a is unsuccessful, the number of balls in the sender's urn $\mho_t$ is decreased by $U(t,a) = \beta \in \mathbb{N} \geq 0$ balls of type m and increased by γ balls for each type $m' \neq m$; the number of balls in the receiver's urn $\mho_m$ is decreased by β balls of type a and increased by γ balls for each type $a' \neq a$.

Some further remarks: the lateral inhibition value γ ensures that the probability of an action can become zero and therefore speeds up the learning process. Note that the number of balls can never become a negative value, what is ensured by a lower boundary of 0. Finally, note that in the same way lateral inhibition can be applied on Bush-Mosteller reinforcement.

2.4 Multi-agent Accounts

It is interesting not only to examine the classical 2-players sender-receiver game, but the behavior of agents in a society (e.g. [Z1], [W1], [M1], [MF1]), where more than 2 agents interact with each other and switch between sender and receiver role. In this way an agent can learn both a sender and a receiver strategy. If such a combination forms a signaling system, it is called a *signaling language*. Thus, the number of different possible signaling languages is defined by the number of possible signaling systems and for an $n \times n$ game an agent can learn one of $n!$ different signaling languages. Furthermore, if an agent's combination of sender and receiver strategy forms a pooling system, it is called a *pooling language*. It is easy to see that the number of possible pooling languages exceeds the number of possible signaling languages for any kind of $n \times n$ game.

3   Simulating Bush-Mosteller

Barrett [B1] simulated repeated signaling games with Roth-Erev reinforcement in the classical sender-receiver variant and calculated the *run failure rate* (RFR). The RFR is the proportion of runs not ending with communication via a perfect signaling system. Barrett started 10^5 runs for $n \times n$ games with $n \in \{2, 3, 4, 8\}$. His results show that 100% (RFR = 0) of 2×2 games were successful. But for $n \times n$ games with $n > 2$, the RFR increases rapidly (Figure 3, left).

To compare different dynamics, we started two series of simulation runs for Bush-Mosteller reinforcement in the sender-receiver variant with urn content parameter $c = 20$ and reinforcement value $\alpha = 1$. In the second series we additionally used lateral inhibition with value $\gamma = 1/|T|$. We tested the same games

Game	RFR
2 × 2	0%
3 × 3	9.6%
4 × 4	21.9%
8 × 8	59.4%

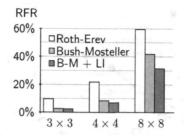

Fig. 3. *Left:* Barrett's results for different $n \times n$ games. *Right:* Comparison of different learning dynamics: Barrett's results of Roth-Erev reinforcement, results for Bush-Mosteller reinforcement without and with lateral inhibition.

as Barrett and correspondingly 10^5 runs per game. In comparison to Barrett's findings, our simulation outcomes i) also resulted in a RFR of 0 for the 2×2 game, but ii) revealed an improvement with Bush-Mosteller reinforcement for the other games, especially in combination with lateral inhibition (see Figure 3, right). Nevertheless, the RFR is never 0 for $n \times n$ games with $n > 2$ and gets worse for increasing n-values, independent of the dynamics.

To analyze the behavior of agents in a multi-agent society, we started experiments with the smallest group of agents in our simulations: three agents arranged in a complete network. In contrast to our first simulations, all agents communicate as both sender and receiver and can learn not only a perfect signaling system, but a signaling language. Furthermore, it was not only recorded if the agents learned a language, but how many agents learned one. With this approach we started between 500 and 1000 simulation runs using Bush-Mosteller reinforcement ($\alpha = 1$, $c = 20$) for $n \times n$ games with $n = 2 \ldots 8$. We stopped a simulation run when each agent in the network learned a signaling or pooling language. We measured the percentage of simulation runs ending with no, one, two or three signaling language learners.

We obtained the following results (Figure 4, left): in 2×2 games, all three agents learned the same signaling language in more than 80% of all simulation runs. But in 3×3 games in less than a third of all runs agents learned a signaling language; in more than 40% of all runs exactly two agents learned a signaling language. And it gets even worse for games with bigger n. E.g. for an 8×8 game in almost 80% of all runs no agents learned a signaling language and it never happened that all agents learned a signaling language.

In addition, we were interested in whether and how the results would change by extending the number of agents. Thus, in another series of experiments we tested the behavior of a complete network of 5 agents in comparison with the results of the 3 agent population. Figure 4 (right) shows the average number of agents who learned a signaling language per run for different $n \times n$ games. As one can see, the percentage of language learners declines rapidly with larger domains and is by and large the same for 3- and 5-agents populations.

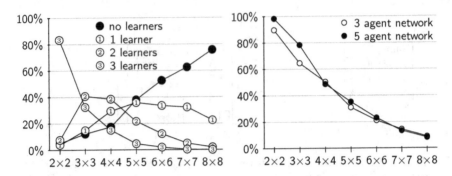

Fig. 4. Left: Percentage of simulation runs ending with a specific number of learners of signaling languages in a network with three agents for different $n \times n$ games with $n = 1 \ldots 8$. Right: Average percentage of agents learning a signaling language over all runs for different $n \times n$ games with $n = 1 \ldots 8$. Comparison of the results of a complete network of 3 agents (white circles) and 5 agents (black circles).

In a nutshell, the results for the classical sender-receiver game reveal that by extending learning dynamics, the probability of the emergence of perfect signaling systems can be improved but it is never one for an $n \times n$ game with $n > 2$. Moreover, the results of the multi-agent network with three agents show that even for the 2×2 game there are cases where not all agents learn a language. And for games with larger domains the results are worse. Furthermore, they don't get better or worse by changing the number of agents, as shown in a multi-agent population with 5 agents. A learning dynamics should be capable of dealing with environments with many states and a lot of interlocutors, because otherwise it does not yield a sufficient explanation for the emergence of many of the signaling systems we find in nature. We show in the next section that by allowing extinction of unused messages and emergence of new messages, perfect signaling systems will emerge with certainty in games with multiple agents and more states.

4   Innovation

The idea of reinforcement learning with innovation is basically as follows: messages can become extinct and new messages can emerge; thus the number of messages during a repeated play can vary, whereas the number of states is fixed. Pioneer work on innovation and extinction for reinforcement learning applied on signaling games stems from Skyrms [S1], further basic experiments with Roth-Erev reinforcement were made by Alexander et al. [ASZ1]. The main contribution of this paper is i) to combine it with Bush-Mosteller reinforcement plus negative reinforcement and ii) to use it for multi-agent accounts.

The process of the emergence of new messages works like this: in addition to the balls for each message type, each sender urn has an amount of *innovative balls* (according to Skyrms we call them *black balls*). If drawing a black ball the sender sends a completely new message. Because the receiver does not have a receiver urn of the new message, he chooses a random action. If action and state matches, the new message is adopted in the *set of known messages* of both interlocutors in the following way: i) both agents get a receiver urn for the new message, wherein the balls for all actions are distributed equiprobably, ii) both agents' sender urns are filled with a predefined amount of balls of the new message and iii) the sender and receiver urn involved in this round are updated according to the learning dynamics. If the newly invented message does not lead to successful communication, the message will be discarded and there will be no change in the agents' strategies.

As mentioned before, messages can go extinct, and that is realized in the following way: because of lateral inhibition, infrequently used or unused messages' value of balls in the sender urns will get lower and lower. At a point when the number of balls of a message is 0 in all sender urns of a particular agent, the message isn't existent in the active use of that agent (i.o.w. she cannot send the message anymore), and will also be removed from the agent's passive use by deleting the appropriate receiver urn. At this point the message isn't in this agent's set of known messages. Some further notes on this model are as follows:

- it is possible that an agent can receive a message that is not in her set of known messages. In this case she adopts the new message like described for the case of innovation. Note that in a multi-agent setup this allows for a spread of new messages
- the black balls are also affected by lateral inhibition. That means that the number of black balls can decrease and increase during runtime; it can especially be zero
- a game with innovation has a dynamic number of messages during a repeated play, but generally ends with $|M| = |T|$. Thus we call an innovation game with n states and n ultimate messages an $n \times n^*$ game

4.1   The Force of Innovation

Since an agent invents a new message if she draws a black ball, the proportion of black balls of an agent's sender urns represents the probability to invent a new message. We call this probability the *force of innovation*, defined as follows:

Definition 1: Given an agent's set of sender urns $\mho = \{\mho_t | t \in T\}$ for a set of states T. An agent's *force of innovation* FOI describes her proportion of black balls over her set of all sender urns:

$$FOI(\mho) = \frac{\sum_{\mho_t \in \mho} \frac{b(\mho_t)}{|\mho_t|}}{|\mho|} \tag{3}$$

where $b(\mho_t)$ is the number of black balls in urn $\mho_t$.

In the following study we investigated the way the force of innovation changes over time in a simulation run. Furthermore we wanted to find out if it correlates

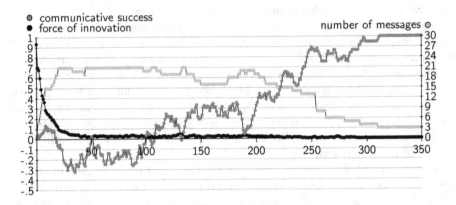

Fig. 5. Simulation run of a $3 \times 3^*$ game with innovation in a 3-agents population: communicative success, force of innovation (averaged over all agents) and the actual number of used messages in the population - alteration over time

with the agents' *communicative success*[4], since we expected a highly negative correlation between it and the force of innovation. We started 100 simulation runs with the following settings:

- network type: complete network with 3 agents
- game type: $3 \times 3^*$ game
- learning dynamics: Bush-Mosteller reinforcement with negative reinforcement, lateral inhibition value ($\alpha = 1$, $\beta = 1$, $\gamma = 1/|T|$) and innovation
- initial state: every urn of the sender is filled with black balls and the receiver does not have any a priori urn.
- break condition: simulation stops if all agents learned a signaling language

Note that the settings of the learning dynamics implicate that the communicative success value can be between -1 and 1[5], and because of the initial state of the sender urns, the force of innovation of all agents is 1 at the beginning.

The simulation results revealed first of all: all agents learn the same signaling language; and that really quickly: they need maximally 500 simulation steps. Now let's take a closer look at how a $3 \times 3^*$ game played in a 3-agents population develops during a simulation run by analyzing the *communicative success* and the average *force of innovation* of the population, plus the number of messages, used in the population.[6] Figure 5 shows an exemplary course of the resulting values' alteration over time for one of the simulation runs. It shows that in the beginning the agents are very innovative and create a lot of messages, which

---

[4] The communicative success is measured as the average utility value of all agents' utility value at a given simulation step

[5] Note that the range of an utility value is between $-\beta$ and α.

[6] This is the number of all messages that were i) once invented and ii) of which at least one agent has a non-zero probability to draw at the given simulation step.

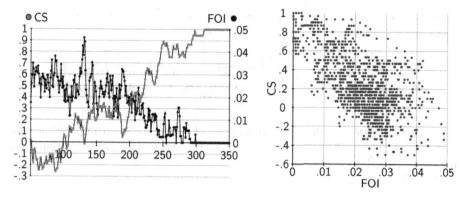

Fig. 6. Left: Simulation run of a $3 \times 3^*$ game with innovation in a 3-agents population, starting with simulation step 50. Comparison of communicative success (CS) and force of innovation (FOI) over time. Right: Data points of 10 simulation runs for FOI values $\leq .05$. CS and FOI reveal a very high negative correlation of $-.6$ for 40,000 data points.

reduces the number of black balls in the urns, because balls for the new messages are added and then the urn content is normalized. Note that for the first communication steps the force of innovation drops rapidly, while the number of messages rises until it reaches 21 messages. Furthermore, the communicative success is below zero at the beginning, since agents use a diversity of different messages and successful communication is less probable than chance. But once there evolved an agreement on which messages might be useful in terms of successful communication, further messages died out, so the number of known messages decreased. Finally, the communicative success reaches a perfect 1 on average, while the number of messages equals the number of states (3) and the force of innovation drops to zero.

By taking a closer look on the data, an interesting interplay between communicative success and force of innovation becomes evident: successful communication lowers the force of innovation, whereas unsuccessful communication raises it. That is not a surprise, since black balls can only change by lateral inhibition: increase in the case of unsuccessful communication and decrease in the case of successful communication. The relationship of both values is better seen in Figure 6 (left) that shows the force of innovation and the communication success of the same simulation runs as already depicted in Figure 5, but this time i) without the initial phase of the first 50 simulation steps and ii) the value of the force of innovation is displayed 20 times more fine-grained. The relationship between both values is clearly recognizable in this figure: one measure's peak is simultaneously the other measure's valley. Admittedly, the mirroring is not perfect, but is clearly reveals a plausible social dynamics: the higher the communicative success, the lover the force of innovation, and vice versa.

To get a more quantitative picture of this relationship, we analyzed the data points' correlation of all 100 simulation runs (about 40,000 data points). It turned out that force of innovation and communicative success reveal a very strong

Table 1. Runtime Table for $n \times n^*$ games with $n = 2 \ldots 8$; for a complete network of 3 agents and 5 agents

Game	$2 \times 2^*$	$3 \times 3^*$	$4 \times 4^*$	$5 \times 5^*$	$6 \times 6^*$	$7 \times 7^*$	$8 \times 8^*$
3 agents	1,052	2,120	4,064	9,640	21,712	136,110	> 500,000
5 agents	2,093	5,080	18,053	192,840	> 500,000	> 500,000	> 500,000

negative correlation: a *Pearson-Correlation* of $-.6$. To get an impression how the data correlate, Figure 6 (right) depicts the data points of ten simulation runs for FOI-values $\leq .05$.[7]

4.2   Learning Languages by Innovation: A Question of Time

In Section 3 we were able to show that the percentage of agents learning a signaling language in a multi-agent context decreases by increasing the domain size of the game. To find out whether innovation can improve these results we started simulation runs for games with different domains. We used the following settings:

- network types: complete network with 3 agents and with 5 agents
- learning dynamics: Bush-Mosteller reinforcement with negative reinforcement and lateral inhibition value ($\alpha = 1$, $\beta = 1$, $\gamma = 1/|T|$) and innovation
- initial state: every urn of the sender is filled with black balls and the receiver does not have any a priori urn.
- experiments: 100 simulation runs per $n \times n^*$ game with $n = 2 \ldots 8$
- break condition: simulation stops if the communicative success of every agent exceeds 99% or the runtime passes the runtime limit of 500,000 communication steps (= runtime)

These simulation runs gave the following results: i) for the 3-agents account in combination with $n \times n^*$ games for $n = 2 \ldots 7$ and the 5-agents in combination with $n \times n^*$ games for $n = 2 \ldots 5$ all agents learned a signaling language in each simulation run and ii) for the remaining account-game combinations all simulation runs exceeded the runtime limit (see Table 1). We expect that for the remaining combination all agents will learn a signaling language as well, but it takes extremely long.

All in all, we were able to show that the integration of innovation and extinction of messages leads to a final situation where all agents learned the same signaling language, if the runtime does not exceed the limit. Nevertheless, we expect the same result for account-game combinations where simulations steps of these runs exceeded our limit for a manageable runtime.

---

[7] The reason to illustrate only data points with a FOI value $\leq .05$ was to get a better depiction of the data. Note that more that 99% of all data points have a FOI value $\leq .05$ and therefore are depicted here.

4.3   Games with a Limited Message Set

As our previous experiments have shown, increasing the number of agents of the population and/or states of the game has a disastrous impact on the runtime. Especially the dependency of the runtime on the number of agents makes the game inapplicable to experiments with larger populations and network structures. The problem of the current account is as follows: whenever communication does not work well, agents' force of innovation increases and they invent new messages. And the more agents are interacting with each other, the more new messages might arise. Thus, the probability of all agents agreeing on a specific set of messages is virtually zero for a larger population. Of course, the probability is close to zero but non-zero and therefore you just have to wait long enough for an population-wide agreement to happen. But the larger the population, the closer is the probability to zero and the longer is the expected runtime.

A reasonable compromise that allows for innovation while keeping the computational complexity feasible is to limit the maximum number of messages.[8] Thus, we introduce a new signaling game: an $n \times n^m$ game has n states and actions and *maximally* m different messages. In such a game, agents that draw a black ball choose randomly a message from the limited message set, without the restriction that this message must be completely new to the population.

The new game has an intuitive analogy to actual signaling beings. In principle, nature might allow an infinite message set, but living beings are only capable of distinguishing a finite number of messages due to sensory, cognitive or motor imperfection. In this sense, each message represents a particular category of non-distinguishable messages.

By adopting this new feature in our game we made experiments to check the runtime improvement for larger population sizes. In particular, we analyzed a 3×3^m game with a set of 30 messages (3×3^{30} game) in comparison with a $3 \times 3^*$ game by using the following settings:

- network types: complete network with different sizes from 2 up to 9 agents
- learning dynamics: Bush-Mosteller reinforcement with negative reinforcement and lateral inhibition value ($\alpha = 1$, $\beta = 1$, $\gamma = 1/|T|$) and innovation
- initial state: every urn of the sender is filled with black balls and the receiver does not have any a priori urn.
- experiments: 100 simulation runs per network size and for a $3 \times 3^*$ game and 3×3^{30} game as well
- break condition: simulation stops if the communicative success of every agents exceeds 99% or the runtime passes the runtime limit of 500,000 communication steps (= runtime)

The result is depicted in Figure 7: the comparison of runtime behavior of the game with an unlimited message set and a limited message set of 30 messages. As already seen in the experiments of Section 4.2, for a $3 \times 3^*$ game the runtime is

---

[8] Note that since a new invented message extends the history of all messages ever used by one, the set of possible messages is virtually unlimited.

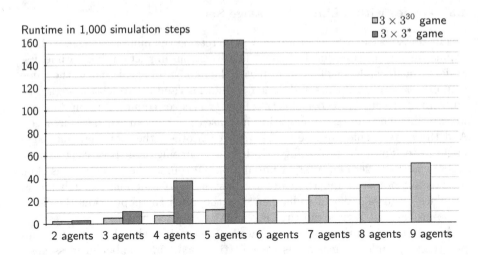

Fig. 7. Runtime comparison of games with limited (3×3^{30}) and unlimited ($3 \times 3^*$) message sets for different population sizes

only manageable for a population up to 5 agents and increases with the number of agents in a strong slope, whereas for a 3×3^{30} game the runtime increases slowly and is for 9 agents still manageable. All in all, the new feature improves the runtime behavior quite well by keeping the innovational nature of the game. This makes it also applicable for larger network structures, as we will show in the next section.

5 Spatial Dynamics

So far we have dealt with fully connected networks of a few agents. However, when modeling natural scenarios encompassing participants of whole populations, it is rather unreasonable that i) the number of population members is that small and ii) all members are connected to each other. To target a more realistic framework, we arranged experiments on a large population with local interaction structure: a *toroid lattice* of 100×100 agents; here each agent can only communicate with her eight direct neighbors (Moore neighborhood).

There are a number of previous studies that addressed signaling games on spatial structures: one of the first studies analyzed a simple 2×2 signaling game on a toroid lattice structure, whereby agents use *imitation* to guide their decisions [Z1]. Some consecutive studies take this analysis up by either changing the dynamics to reinforcement learning [M1] or by changing the interaction structure to small-world networks and by extending the game domains to a 3×3 signaling game [W1]. Another study entails experiments on social network structures as interaction structure plus incorporating reinforcement learning as update dynamics [MF1]. All these studies analyzed a simple variant: a 2×2 or 3×3 game.

The basic result of all these studies was the emergence of regional meaning: the lattice or network structure was split into local language regions.[9]

Note that in all these studies the number of possible signaling systems is quite small. The 2×2 game has only two signaling systems (as depicted in Figure 1) and the 3 × 3 game has 6 signaling systems. In the upcoming experiments we applied 3×3^{30} games and we also expect regional meaning to emerge. But as opposed to the before-mentioned studies, here not 2 or 6, but 6840 different signaling systems are possible! These prerequisites bring a number of questions about: do stable language regions emerge? And if so, how many different language regions emerge? And how are these regions arranged? Do they depict a specific pattern in terms of arrangements with other language regions? Are they stable? This section addresses these questions.

5.1   Spatial Structure: Dialect Regions

To find answers to the before mentioned questions, we started experiments with the following settings:

- network type: 10,000 agents placed on a 100 × 100 toroid lattice
- game type: 3×3^{30} game
- learning dynamics: Bush-Mosteller reinforcement with negative reinforcement and lateral inhibition value ($\alpha = 1$, $\beta = 1$, $\gamma = 1/|T|$) and innovation
- break condition: simulation stops after 50,000 simulation steps or every agent has learned a signaling language

Like in the previous experiments, we measured the average communicative success and the force of innovation of the whole population over time. Furthermore, since we were interested in the number of signaling languages that might emerge, we also measured the population-wide number of signaling languages over time. The resulting course of a simulation run for the first 10,000 simulation steps is depicted in Figure 8.

Like for the previous experiments of Section 4.1 (c.f. Figure 5) for a small population of 3 agents, the force of innovation decreases really fast down to (almost) zero, while the communicative success first decreases to a negative value and then increases again. Thus, the initial phase is quite similar. But while for the experiments of Section 4.1, the 3 agent-population quickly agrees on one signaling language and the communicative success reaches a perfect value of 1, here the population of 10,000 agents 'agrees' on more than 600 signaling languages and the communicative success reaches an average value of almost .8 after around 500 simulation steps. From this point on the number of signaling languages slowly decreases, while the value of communicative success slowly increases. Note that Figure 8 only shows the first 10,000 simulation steps. The whole simulation run showed that after 50,000 simulation steps the number of signaling languages has

---

[9] A language region is defined as a connected sub-network, of which each member has learned the same signaling language L, but each other agent connected to this regions hasn't learned L.

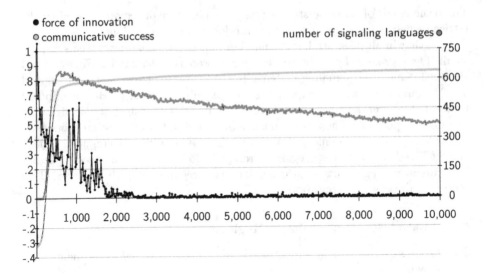

Fig. 8. Simulation run of a 3×3^{30} game in a population of 10,000 agents placed on a 100×100 toroid lattice: average communicative success, force of innovation and the number of society-wide signaling languages over the first 10,000 simulation steps.

decreased to around 170, while the communicative success reached a value of more than .9.

Communication cannot be perfectly successful because the population does not learn one unique, but multiple signaling languages. But how are these different signaling languages spatially arranged? If they would be arbitrarily spread over the whole lattice, we would expect a much lower communicative success value, basically lower than zero. But since the value is around .8, communication works quite well even with such a huge number of different signaling languages. The reason becomes visible if we take a look at the spatial arrangement of the different signaling languages. It turns out that they form what we call *language regions*. A language region is a connected subgraph for which each agent uses the same signaling language. Figure 9 shows the resulting pattern on a 100x100 toroid lattice, the left figure for the pattern after 2,000 simulation steps (more than 500 different language regions), the right figure for the pattern after 50,000 simulation steps (around 170 language regions).

Note that Figure 8 shows that the number of signaling languages slowly decreases over time. Furthermore, Figure 9 shows that each signaling language that evolved forms at least one language regions. Consequently, the number of language regions decreases over time: after 50,000 simulation steps there is only a third of the number of language regions than after 2,000 simulation steps. We haven't analyzed the concrete dynamics that lead to this decline of language regions, but we expect mechanisms like unification, melting, displacement and extinction at the borders of neighboring language regions. The exact dynamics behind this process remains to be analyzed in subsequent studies.

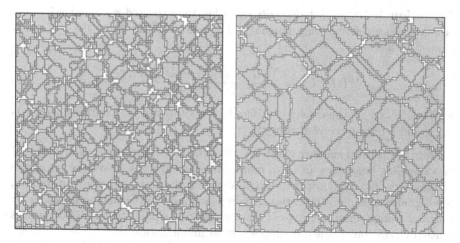

Fig. 9. The allocation of language regions on a 100x100 toroid lattice. A gray cell represents an agent that learned a signaling language. The borders between language regions are marked by darkgray lines. While after 2,000 simulation steps the map is segmented in over 500 language regions (left figure), after 50,000 simulation steps it is only one third of it, around 170 language regions (right figure).

In the upcoming section we present results of the analyses of the spatial relationship between language regions at one point in time to examine if their placement is randomly or follows particular patterns.

5.2 Spatial Relationships

In this section we want to analyze how the different language regions actually relate to each other. We hypothesize that there is an interaction between spatial distance of two language regions and the similarity of their signaling languages $L = \langle s, r \rangle$.[10] For that purpose we define two similarity measures, *lexical similarity* and *mutual intelligibility*, as follows:

Definition 2: *Lexical Similarity* describes the proportionally common items of lexical entries.[11] Thus between two given signaling languages $L_1 = \langle s_1, r_1 \rangle$ and $L_2 = \langle s_2, r_2 \rangle$ the lexical similarity is defined as follows:

$$LS(L_1, L_2) = \frac{|\{m \in M | \exists t \in T : m = s_1(t)\} \cap \{m \in M | \exists t \in T : m = s_2(t)\}|}{|T|}$$

$$(4)$$

---

[10] Note that a signaling language is defined as a strategy pair of a *pure* sender and receiver strategy, defined as $s : T \to M$ and $r : M \to A$, respectively. Note: while agents play according to behavioral strategies, once they have learned a signaling language, their behavioral strategy profile represents a pair of pure strategies.

[11] In the case of signaling languages, lexical entries are entailed messages.

Definition 3: *Mutual Intelligibility* describes the expected communicative success for two given signaling languages $L_1 = \langle s_1, r_1 \rangle$ and $L_2 = \langle s_2, r_2 \rangle$ and is defined as follows:

$$MI(L_1, L_2) = \frac{\sum_t (U^x(t, s_1, r_2) + U^x(t, s_2, r_1))}{2 \times |T|} \tag{5}$$

where $U^x(t, s, r)$ is the expected utility[12] for a given state t, a pure sender strategy s and a pure receiver strategy r.

Note that lexical similarity just describes the number of common messages of two signaling languages. In turn, mutual intelligibility also takes the semantics of messages into account: if messages describe the same state/action, mutual intelligibility is higher. But if two signaling languages have common messages for different states/actions, it gives advantage to lexical similarity, but disadvantages mutual intelligibility, since it supports miscommunication.[13]

To give an example of these similarity measures, lets take a look at the lattice distribution after 50,000 simulation steps as depicted in the left picture of Figure 10. There are three language regions that are marked by its signaling languages L_{55}, L_{72} and L_{139}. The concrete signaling languages are depicted in the right picture. It turns out that the close language regions 55 and 72 have a quite high lexical similarity value (.67) and an even higher mutual intelligibility value (.78). The distant language regions 139 and 55 have a low lexical similarity value (.33) and also a low mutual intelligibility value (.22). Similarly, language region 139 and 72 have no lexical similarity and a low mutual intelligibility value (.33).

To compare these similarity measures to spatial distances of language regions in a more systematic way, we introduce the measure *regional distance*, a value that describes the distance between two language regions. In detail, it describes the average distance of all members of one language region to all members of the other language region. It is defined as follows:

Definition 4: *Regional Distance* describes the distance between two connected subgraphs of a connected graph as the average distance over all members $n \in N_1$ of subgraph G_1 and $n \in N_2$ of subgraph G_2, defined as follows:

$$RD(G_1, G_2) = \frac{\sum_{n_i \in N_1} \sum_{n_j \in N_2} SP(n_i, n_j)}{|N_1| \times |N_2|} \tag{6}$$

where $SP(n_i, n_j)$ is the *shortest path length*[14] between node n_i and node n_j.

---

[12] Expected utility $U^x(t, s, r)$ is defined as follows:

$$U^x(t, s, r) = \begin{cases} U(t, r(s(t))) & \text{if } s(t) \in \{m \in M | \exists a \in A : m = r^{-1}(a)\} \\ \frac{\alpha}{|A|} + \frac{\beta \times (|A| - 1)}{|A|} & \text{else} \end{cases}$$

[13] It can be shown that two signaling languages can have a high value of lexical similarity as well as a low value of mutual intelligibility, and vice versa.

[14] The shortest path length of two nodes describes the length of a path (= number of edges) between them that has a minimal number of edges.

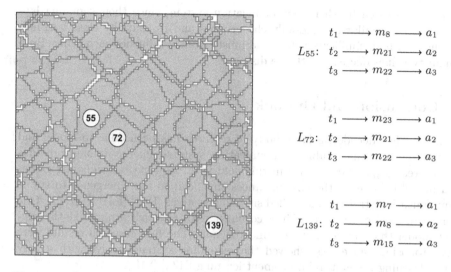

Fig. 10. Left: The language regions 55, 72 are next to each other, while 139 is far off. Right: strategy profiles of signaling languages L_{55}, L_{72} and L_{139}.

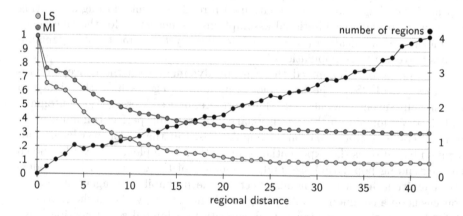

Fig. 11. The grey dots depict the average values of lexical similarity (LS) and mutual intelligibility (MI) between two language regions in dependence of the distance between them. The black dots depict a language region's average number of other language regions in a particular distance to it.

Following this approach we analyzed lexical similarity and mutual intelligibility depending on the distance of each pair of two language regions. The result is depicted in Figure 11: while the number of language regions increases with its distance to a specific language region, the similarity values decrease. Furthermore, both similarity measures follow a curve with falling slope to an expected random value. This result reveals that i) distant language regions seem to have no influence to each other's communication system, since their signaling languages are as similar as randomly chosen ones and ii) spatially close language regions have high similarity values and must strongly influence each other. Both similarity values decrease with the distance.

6   Conclusion and Outlook

In the last few decades, a large body of research has been done to model and analyze the way that stable communication systems emerge among individuals, whereas a popular account in this field is to use repeated signaling games as a model to analyze the circumstances that lead to the emergence to stable communication strategies, so-called signaling systems [BZ1] [FJ1] [HSRZ1]. One premise of most of the work is in accordance with *Occam's razor*: take the simplest model that can explain the phenomenon. The first research results were very promising, since they showed that signaling systems evolve with a very simple learning account: reinforcement learning [HZ1] [S1].

But further studies showed that this result holds basically for simple 2×2 games between two players, but not for more complex games [B1] or larger populations [M1]. Thus we proposed the question: by taking the model of a repeated signaling game in combination with reinforcement learning as starting point, what reasonable additional assumptions are necessary for the emergence of efficient communication (in terms of signaling systems) in complex signaling games played in large populations?

In a first step we extended the learning dynamics with the concept of *innovation* [ASZ1] [S1]. The basic plot is as follows: agents have the ability to occasionally invent new messages. Furthermore, unused messages get automatically lost. We found that these additional concepts enable perfect communication in more complex games and larger populations. But the major drawback of allowing innovation is the exponential computational complexity. It can be shown that extending population size and/or domains of the game even in a moderate magnitude has a tremendous effect on the probability of agents to find a consensus on a common signaling language, what strongly affects the runtime.

By limiting the game's innovation capacity to a limited set of possible messages, we created a new game that keeps the innovative character of the previous game, but solved the problem of computational complexity. With this new account, we varied the network structure, such that agents were arranged on a two-dimensional toroid lattice. This lead to the emergence of language regions, arranged in a particular pattern: the closer the language regions, the more similar their signaling languages.

All in all, we were able to show that, by starting with an account of repeated signaling games and reinforcement learning, one simple extensions is sufficient to realize the emergence of signaling systems for complex signaling games and large populations: innovation of new messages within a limited message set. Furthermore, our experiments showed that a local communication structure leads to local language regions arranged in a continuous way.

Further research should go in multiple directions. First of all, it might be worth to take a closer look at the way language regions change over time, especially at the border regions. Additionally, what remains to be shown is that our results in fact hold for higher numbers of domains of the game. Is our result general, or only true for specific values? It would also be interesting to see what kind of influence more realistic network-types (c.f. small-world networks) would have on the outcome. These are but a few extensions, as a multitude of further experiments addressing factors that might influence the way language regions emerge and interact readily suggest themselves.

References

[ASZ1]   Alexander, J.M., Skyrms, B., Zabell, S.L.: Inventing new signals. Dynamic Games and Applications 2(1), 129–145 (2012)

[B1]   Barret, J.A.: The Evolution of Coding in Signaling Games. Theory and Decision 67, 223–237 (2009)

[BZ1]   Barret, J.A., Zollman, K.J.S.: The Role of Forgetting in the Evolution and Learning of Language. Journal of Experimental and Theoretical Artificial Intelligence 21(4), 293–309 (2009)

[B2]   Bloomfield, L.: Language. George Allen & Unwin, London (1935)

[BM1]   Bush, R., Mosteller, F.: Stochastic Models of Learning. John Wiley & Sons, New York (1955)

[FJ1]   Franke, M., Jäger, G.: Bidirectional optimization from reasoning and learning in games. Journal of Logic, Language and Information 21(1), 117–139 (2012)

[HH1]   Hofbauer, J., Huttegger, S.M.: Feasibility of communication in binary signaling games. Journal of Theoretical Biology 254(4), 843–849 (2008)

[HSRZ1]   Huttegger, S.M., Skyrms, B., Smead, R., Zollman, K.J.: Evolutionary dynamics of Lewis signaling games: signaling systems vs. partial pooling. Synthese 172(1), 177–191 (2010)

[HZ1]   Huttegger, S.M., Zollman, K.J.S.: Signaling Games: Dynamics of Evolution and Learning. In: Benz, A., Ebert, C., Jäger, G., van Rooij, R. (eds.) Language, Games, and Evolution. LNCS (LNAI), vol. 6207, pp. 160–176. Springer, Heidelberg (2011)

[L1]   Lewis, D.: Convention. Harvard University Press, Cambridge (1969)

[M1]   Mühlenbernd, R.: Learning with Neighbours. Synthese 183(S1), 87–109 (2011)

[MF1]   Mühlenbernd, R., Franke, M.: Signaling Conventions: Who Learns What Where and When in a Social Network. In: Proceedings of EvoLang IX, pp. 242–249 (2011)

[RE1]   Roth, A., Erev, I.: Learning in extensive-form games: experimental data and simple dynamic models in the intermediate term. Games and Economic Behaviour 8, 164–212 (1995)

[S1]   Skyrms, B.: Signals: Evolution, Learning & Information. Oxford University Press, Oxford (2010)

[W1]   Wagner, E.: Communication and Structured Correlation. Erkenntnis 71(3), 377–393 (2009)

[Z1]   Zollman, K.J.S.: Talking to neighbors: The evolution of regional meaning. Philosophy of Science 72(1), 69–85 (2005)

(The) Most in Flemish Dutch: Definiteness and Specificity

Koen Roelandt*

KU Leuven HUBrussel / CRISSP
Warmoesberg 26, 1000 Brussel, Belgium
koen.roelandt@krowland.net

Abstract. This paper is concerned with *de meeste* '(the) most' and *het meeste* 'the most' in Flemish Dutch. I first give an overview of their distribution and the different readings they produce. I then submit them to definiteness and specificity tests. The ensuing analysis of *de/het meeste* builds on the theory set out in Hackl [2009], but proposes a more complex syntactic structure to account for the Flemish data.

Keywords: proportional quantification, specificity, definiteness.

1 Introduction

The denotation of the English proportional quantifier *most* has been a much-debated subject in the semantics and pragmatics literature (Ariel [2003, 2004], Horn [1996], Hackl [2009]). In this paper, I present data concerning two structures with *most* in Flemish Dutch: *de meeste* '(the) most' and *het meeste* 'the most'. The structures both contain definite determiners (*de* and *het*, respectively), but I demonstrate that DPs with *de meeste* are always definite and DPs with *het meeste* can be definite or indefinite depending on the head noun. A second point concerns the readings *de/het meeste* can receive: I claim that comparative readings are limited to the positions of non-specific indefinite DPs. The two claims lead to a new analysis of the structures underlying *de meeste* and *het meeste*.

This paper is organized as follows. Section 2 contains a short overview of the account for *(the) most* in Hackl [2009]. In Sect. 3, I present the Flemish data concerning the distribution of the definite determiners *de* and *het* and the different readings *de/het meeste* can receive. I submit *de/het meeste* to tests for definiteness and specificity in Sect. 4. The analysis in Sect. 5 accounts for the Flemish data, but some areas have to be dealt with in future research (Sect. 6).

* I am indebted to my supervisors Dany Jaspers and Hans Smessaert for their invaluable input and feedback. I also wish to thank the members of the CRISSP research group for their constructive comments and ideas: Marijke De Belder, Adrienn Jánosi, Tanja Temmerman, Jeroen van Craenenbroeck and Guido Vanden Wyngaerd. I would also like to thank the anonymous reviewers for their insightful remarks. All errors are mine.

M. Colinet et al. (Eds.): ESSLLI 2012/2013, LNCS 8607, pp. 214–232, 2014.
© Springer-Verlag Berlin Heidelberg 2014

2   English *(The) Most*: Hackl [2009]

Hackl [2009] distinguishes two interpretations for English *most*: proportional and comparative.[1] In order to explain the difference between them, I will introduce two sets. The first one, K, contains three persons: Allison, John and Bill (1a) and the second one, M, contains mountains (1b).

(1)   a. $K = \{a, j, b\}$
      b. $M = \{m_1, m_2, m_3, m_4, m_5\}$

In a situation where Allison, John and Bill go climbing, one could utter the following expressions:

(2)   a.   John climbed most mountains.
           'John climbed more mountains than he didn't climb.'

                                                        (proportional)
      b.   John climbed the most mountains.
           'John climbed more mountains than the other climbers did.'

                                                        (comparative)

As noted by Hackl [2009, p. 75], English sentences containing *most* have a proportional reading (2a), where John climbed more mountains than he didn't climb. If John climbed three mountains out of five (3), the expression in (2a) is true because three is greater than two. Since this reading refers to a proportion of mountains, I will call it "the proportional reading".

(3) $[\![\text{climbed}(j)]\!] = \{m_1, m_2, m_3\}$

Sentences with *the most*, on the other hand, can only receive a comparative reading (2b): John climbed more mountains than anyone else did. The comparative reading does not refer to a proportion of mountains. Instead, different climbers are compared relative to the number of mountains they climbed.

(4) $[\![\text{climbed}(j)]\!] = \{m_1, m_2, m_3\}$
      $[\![\text{climbed}(a)]\!] = \{m_4, m_5\}$
      $[\![\text{climbed}(b)]\!] = \{m_3\}$

In (4), John climbed more mountains than Allison or Bill. Since this reading compares different numbers of mountains climbed by individuals, I will call it "the comparative reading".

Superlative forms of gradable adjectives can have similar interpretations (Szabolcsi [1986], Heim [1999], Farkas and Kiss [2000]). In the absolute reading (5a), the height of the mountain is compared to the height of other mountains: the Mount Everest is the highest of all mountains. In the comparative reading (5b), climbers are compared relative to the height of the mountain(s) they climbed.

---

[1] Hackl uses "absolute/proportional" and "relative" for these interpretations. I refer to "proportional/comparative interpretations" of sentences containing *most* for expository purposes.

(5) John climbed the highest mountain.

    a. "Mount Everest"   (absolute)

    b. "a higher mountain than the other climbers"   (comparative)

Following Heim [1999], Hackl assumes that the superlative morpheme *-est* is a degree quantifier restricted by a comparison class C. The absolute reading of the superlative in example (5a) is derived by comparing the height of the mountains in the comparison class. Mountain x is the highest mountain if its maximal degree of height is greater than the maximal degree of height of any other mountain in the comparison class (6a, Hackl [2009, p. 80]). The comparative interpretation, on the other hand, compares people in the comparison class relative to the height of the mountains they climbed. Climber x climbed the highest mountain if he climbed a mountain with a maximal degree of height that is greater than the maximal degree of height of mountains climbed by any other climber in the comparison class (6b, Hackl [2009, p. 80]).

(6) a. $[\![$ [-est C]$_i$[d$_i$-high mountain]$]\!]$ = λx.$\forall$y $\in$ C [y $\neq$ x $\rightarrow$
    max {d: x is a d-high mountain} > max {d: y is a d-high
    mountain}]

    b. $[\![$ [-est C]$_i$climbed [d$_i$-high mountain]$]\!]$ = λx.$\forall$y $\in$ C [y $\neq$ x $\rightarrow$
    max {d: x climbed a d-high mountain} > max {d: y climbed a
    d-high mountain}]

I will follow Hackl's analysis of *most* as the superlative form of *many*, containing a superlative morpheme restricted by a comparison class C. In this analysis, *most* does not compare degrees, but the cardinality of pluralities consisting of atomic mountains. The semantics for the proportional reading in (7a) states that the maximal degree of cardinality of the plurality x (e.g. $m_1 \oplus m_2 \oplus m_3$) is greater than the maximal degree of cardinality of any other non-overlapping plurality y in the comparison class (i.e. $m_4 \oplus m_5$, m_4, m_5).

(7) a. $[\![$ [-est C]$_i$[d$_i$-many mountains]$]\!]$ = λx.$\forall$y $\in$ C [y $\neq$ x $\rightarrow$
    max {d:mountains(x)=1 & |x| $\geq$ d} >
    max {d:mountains(y)=1 & |y| $\geq$ d}]

    b. $[\![$ [-est C]$_i$ [climbed [$\varnothing$ d$_i$-many mountains]]$]\!]$ = λx.$\forall$y $\in$ C [y $\neq$ x $\rightarrow$
    max { d:$\exists$z [mountains(z) = 1 & |z| $\geq$ d & x climbed z]} >
    max { d:$\exists$z [mountains(z) = 1 & |z| $\geq$ d & y climbed z]}]

The comparative reading does not compare mountains, but climbers relative to the number of mountains they climb. The semantics for the comparative reading in (7b) states that the maximal degree of cardinality of the plurality climbed by x is greater than the maximal degree of cardinality of the plurality climbed by any other climber y in the comparison class. Applied to the situation in (4), John climbed the most mountains if the cardinality of the mountains he climbed ($m_1 \oplus m_2 \oplus m_3$) is greater than the cardinality of the mountains climbed by Allison or Bill ($m_4 \oplus m_5$ and m_3, respectively). In Hackl's analysis, the difference between proportional and comparative readings depends on the position of [-est C] at LF. It stays inside the DP in proportional readings, but moves to [SPEC, VP] in comparative ones.

Hackl claims that the presence or absence of *the* triggers the two different readings of *most*. In the case of proportional readings (2a), the definite article is undefined and cannot occur because it clashes with Link's maximality presupposition of *the* for plurals (Link [1983]). The maximality presupposition demands reference to the full set of objects ('the books' = 'all books'), whereas the proportional reading of *most* is incompatible with a full set on Hackl's view. In sentences with comparative readings (2b), *the* is indefinite. Following Heim [1999], Hackl assumes that indefinite DPs do not act as islands, allowing [-est C] to move out of the DP to [SPEC, VP].

In Hackl's analysis, the definite determiner is either undefined in proportional readings or indefinite in comparative readings. An analysis of *most* in Dutch challenges this view: the definite determiner is always definite in proportional readings and the comparative reading is associated with the position of non-specific DPs. In the next section, I introduce the two Flemish structures with *most*: *de meeste* and *het meeste*.

3 Dutch Data: *de/het meeste*

In Flemish, there are two structures containing *most*, namely *de meeste* '(the) most' and *het meeste* 'the most'. These constituents show behavior that is of interest to Hackl's analysis. Firstly, the determiner *de* 'the' of *de meeste* always agrees with the head noun it appears with, while this is not the case for the determiner *'het'* in *het meeste* (Sect. 3.1). Secondly, *de meeste* is ambiguous between proportional and comparative readings, whereas *het meeste* only has a comparative reading in combination with plural nouns (Sect. 3.2).

3.1 The Distribution of the Definite Determiners

In this section, I discuss the distribution of the definite determiner in DPs with *het meeste* and *de meeste*. Before turning to *de/het meeste*, however, I will give a brief overview of definite determiners in Dutch.

The Dutch grammatical gender system has three classes: feminine, masculine and neuter. There are two definite determiners: *de* and *het*. The former appears in combination with plural count nouns and singular feminine/masculine count nouns. The latter only occurs with singular neuter nouns. Both *de* and *het* appear with mass nouns. Table 1 gives an overview of the distribution of definite articles in Dutch.

Table 1. The distribution of definite articles in Dutch

	feminine	masculine	neuter
singular	de	de	het
mass	de	de	het
plural	de	de	de

In many contexts, *de meeste* and *het meeste* follow the distribution of the determiners they contain. *De meeste* combines with plural count nouns (8a) and masculine/feminine mass nouns (8b), while *het meeste* precedes neuter mass nouns (8c).

(8) a. de (meeste) vrouwen   - de (meeste) mannen   - de (meeste)
         the most   women$_{\text{pl.fem.}}$ - the most   men$_{\text{pl.masc.}}$ - the most
         huizen
         houses$_{\text{pl.neut.}}$

     b. de (meeste) muziek   - de (meeste) chocolade
         the most   music$_{\text{sing.fem.}}$ - the most   chocolate$_{\text{sing.masc.}}$

     c. het (meeste) geld
         the most   money$_{\text{sing.neut.}}$

A puzzling observation in Flemish Dutch, however, is that *het meeste* also combines with plural and with non-neuter mass nouns ((9a) and (9b)), despite the fact that *het* cannot appear in front of them ((9c) and (9d)).

(9) a. het meeste vrouwen   - het meeste mannen   - het meeste
         the most   women$_{\text{pl.fem.}}$ - the most   men$_{\text{pl.masc.}}$ - the most
         huizen
         houses$_{\text{pl.neut.}}$

     b. het meeste muziek   - het meeste chocolade
         the most   music$_{\text{sing.fem.}}$ - the most   chocolate$_{\text{sing.masc.}}$

     c. *het vrouwen   - *het mannen   - *het huizen
         the women$_{\text{pl.fem.}}$ - the men$_{\text{pl.masc.}}$ - the houses$_{\text{pl.neut.}}$

     d. *het muziek   - *het chocolade
         the music$_{\text{sing.fem.}}$ - the chocolate$_{\text{sing.masc.}}$

De meeste is not as liberal because it cannot appear with neuter nouns (10).

(10) *de meeste huis   - *de meeste geld
      the most   house$_{\text{sing.neut.}}$ - the most   money$_{\text{sing.neut.}}$

It is important to note that Flemish speakers consider *het meeste N* as one constituent, since it can be topicalized (11).[2]

(11) a.   Het meeste vrouwen/mannen/huizen heeft Jan gezien.
            the most   women/men/houses   has John seen
            'John saw the most women/men/houses.'

---

[2] Speakers of Dutch from the Netherlands, on the other hand, have different judgements. For them, *het meeste N* can never be a constituent and topicalizing it is ungrammatical. This means that both sentences in (11) are ungrammatical and the comparative reading of *het meeste* is not available to them. In the Netherlands, *het meeste* can only quantify over events, with some variation concerning the inflection -*e*.

    b.   Het meeste muziek/chocolade heeft Jan gemaakt.
           the most   music/chocolate has   John made
           'John made the most music/chocolate.'

In another reading available to Flemish speakers, *het meest* (without the *-e* inflection) quantifies over events, which yields a different reading altogether (12):

(12)   a. Jan heeft [het meest] vrouwen/mannen/huizen gezien.
          John has   the most women/men/houses   seen
          'John mostly saw women/men/houses (and not something else such as trees).'
       b. Jan heeft [het meest] muziek/chocolade gemaakt.
          John has   the most music/chocolate   made
          'John mostly made music/chocolate (and not something else such as paintings).'

The sentence in (12a) states that the event of John seeing women/men/houses took place more often than the event of seeing trees.[3] The cardinality of the different entities (women, men, houses etc.) is not taken into account. *Het meest* can also be topicalized:

(13)   a.   Het meest heeft Jan vrouwen/mannen/huizen gezien.
           the most has   John women/men/houses   seen
           'Mostly, John saw women/men/houses.'
       b.   Het meest heeft Jan muziek/chocolade gemaakt.
           the most has   John music/chocolate   made
           'Mostly, John made music/chocolate.'

In summary, *de meeste* must agree with the head noun in the DP. In Flemish Dutch, *het meeste N* is one constituent but *het meeste* does not have to agree with the head noun: it can appear with feminine/masculine mass nouns and plural count nouns.

3.2 Proportional and Comparative Readings

Flemish sentences with *de meeste* are ambiguous between proportional[4] and comparative readings (14).

(14)   a. Jan heeft de meeste vrouwen/mannen/huizen gezien.
          John has   the most women/men/houses   seen
          'John saw (the) most women/men/houses.'

---

[3] *Voornamelijk* 'mainly' has the same interpretation. It would be interesting to explore the differences between *het meest* and *voornamelijk* regarding their distribution and interpretation, but this a point for future research.

[4] Hackl [2009] points out that the correct paraphrase of the proportional reading is *more N than he didn't V* (cf. example (2a)). However, I will use *more than half* in the glosses for ease of reading.

    i. 'John saw more than half of the women/men/houses.'

          (proportional)

    ii. 'John saw more women/men/houses than anybody else did.'

          (comparative)

  b. Jan   heeft de   meeste muziek/chocolade gemaakt.

    John has   the most   music/chocolate   made

    'John made (the) most music/chocolate.'

    i. 'John made more than half of the music/chocolate.'

          (proportional)

    ii. 'John made more music/chocolate than anybody else did.'

          (comparative)

When *de meeste* is replaced with *het meeste*, the sentence yields different readings. The proportional reading is no longer available and only the comparative reading remains (15).

(15) a. Jan   heeft het meeste vrouwen/mannen/huizen gezien.

    John has   the most   women/men/houses   seen

    'John saw the most women/men/houses.'

    i. -   (proportional)

    ii. 'John saw more women/men/houses than anybody else did.'

          (comparative)

  b. Jan   heeft het meeste muziek/chocolade gemaakt.

    John has   the most   music/chocolate   made

    'John made the most music/chocolate.'

    i. -   (proportional)

    ii. 'John made more music/chocolate than anybody else did.'

          (comparative)

When *het meeste* combines with a neuter mass noun such as *geld* 'money', both readings are available (16).

(16) Jan   heeft het meeste geld   verloren.

    John has   the most   money lost

    'John lost (the) most money.'

    a. 'John lost more than half of the money.'   (proportional)

    b. 'John lost more money than anybody else did.'   (comparative)

To sum up, *de/het meeste* can produce different readings, depending on the definite determiner the speaker chooses. Sentences containing *de meeste* have both proportional and comparative readings. *Het meeste N* has the comparative reading in Flemish Dutch if *het* doesn't agree with the noun in number and gender. Finally, *het meeste* combined with a neuter mass noun can have both proportional and comparative readings.

4   Tests: Definiteness and Specificity

In the previous section, I identified syntactic and semantic differences between *de meeste* and *het meeste*. In this section, I dig a bit deeper and examine the quantifiers' relation to definiteness and specificity. I will concentrate on Flemish Dutch.

4.1   Definiteness

De meeste and *het meeste* will be submitted to two tests for definiteness: existential *there* clauses and expressions with DP-internal focus, i.e. focus on a PP postmodifier inside the DP.[5]

The definiteness effect context is a classic diagnostic for indefiniteness (Milsark [1974], Szabolcsi [1986]). Sentences that start with *there is/are* may contain indefinites (17a), bare plurals (17b) and bare mass nouns (17c-17d) in the associate position, but no definite DPs (18).

(17)   a.   Er   is een berg   in Canada.
            there is a   mountain in Canada
            'There is a mountain in Canada.'

       b.   Er   zijn bergen   in Canada.
            there are mountains   in Canada
            'There are mountains in Canada.'

       c.   Er   is chocolade in België.
            there is chocolate in Belgium
            'There is chocolate in Belgium.'

       d.   Er   is geld   in België.
            there is money in Belgium
            'There is money in Belgium.'

(18)   a.   *Er   is de berg   in Canada.
            there is the mountain in Canada

       b.   *Er   zijn de bergen   in Canada.
            there are the mountains in Canada

       c.   *Er   is de chocolade in België.
            there is the chocolate in Belgium

       d.   *Er   is het geld   in België.
            there is the money in Belgium

When this test is applied to *de/het meeste*, we see that *het meeste* follows the pattern of indefinite DPs, both in combination with a masculine noun in

---

[5] I will only use the plural count nouns *bergen* 'mountains' (masculine) and *platen* 'records' (feminine) in the examples, but the judgements can be extended to all plural nouns, regardless of their gender. I will also present data with the neuter mass noun *geld* 'money' because it yields different judgements.

examples (19a) and (19b) and a neuter mass noun in example (19c). Even though *het* is a definite article, *het meeste* is perfectly acceptable in these sentences. *De meeste*, however, is ungrammatical in combination with existential *there*, which suggests that it heads a definite DP.

(19) a.   Er   zijn het meeste bergen   in Canada.
           there are the most   mountains in Canada
           'There are the most mountains in Canada.'   (comparative)

      b.   Er   is het meeste chocolade in België.
           there is the most   chocolate in Belgium
           'There is the most chocolate in Belgium.'   (comparative)

      c.   Er   is het meeste geld   in België.
           there is the most   money in Belgium
           'There is the most money in Belgium.'   (comparative)

(20) a.   * Er   zijn de meeste bergen   in Canada.
           there are the most   mountains in Canada

      b.   * Er   is de meeste chocolade in België.
           there is the most   chocolate in Belgium

Milsark [1974] makes a distinction between determiners based on this test. Determiners that are acceptable in sentences with existential *there* fall in the group with a cardinality reading (*a, three, ten* etc.). Determiners that are ungrammatical have quantificational readings (*the, each, every, most* etc.). Following this line of reasoning, *het meeste* yields cardinality readings, while *de meeste* only produces quantificational readings.

The second test uses DP-internal focus. Example (2b) shows a comparative reading where the comparison class is determined by the verb *klimmen* 'to climb' and the comparison class contains climbers. In Slavic languages, however, the comparison class can also be provided by a focused PP inside the DP containing *most*, but only if this DP is indefinite [Pancheva and Tomaszewicz, 2012].[6] Dutch sentences with *het meeste* can also have DP-internal focus. Example (21b) shows a comparative reading with a PP providing the comparison class.

(21) ...dat Jan [DPhet meeste platen [PPvan   Zappa]] beluisterd heeft.
      ...that John   the most   records   of/by Zappa   listened to has
      '...that John listened to the most records of/by Zappa.'

      a.   John listened to more records of/by Zappa than anybody else did.
                                          (comparative - focus on *Jan*)

      b.   John listened to more records of/by Zappa than he listened to records
           of/by any other band.   (comparative - focus on *Zappa*)

Following the Slavic pattern, this suggests that the DP *het meeste N* is indefinite. Moreover, Pancheva and Tomaszewicz [2012] show that a DP with *most*

---

[6] I would like to thank the anonymous reviewer for pointing this out to me.

can be headed by a definite determiner in Macedonian and Bulgarian, but the comparative reading with a comparison class delivered by the DP-internal PP is not available then. If *de meeste N* is indeed definite, it should follow the pattern of definite DPs in Bulgarian and Macedonian. This prediction is borne out. The sentences with *de meeste platen* can only have a proportional reading (22a) or a comparative reading with the comparison class determined by the verb (22b). The DP-internal comparative reading is not available.

(22)   ...dat Jan [$_{DP}$de meeste platen [$_{PP}$van Zappa]] beluisterd heeft.
       ...that John   the most   records   of/by Zappa   listened to has

'...that John listened to (the) most records of/by Zappa.'

  a. John listened to more than half of the records of/by Zappa.

                                  (proportional)

  b. John listened to more records of/by Zappa than anybody else did.

                           (comparative - subject)

Taking the Macedonian and Bulgarian pattern into account, this shows that *het meeste N* is indefinite and *de meeste N* definite. Sentences containing *het meeste* and a neuter mass noun can have all the readings associated with both definite and indefinite DPs (23), which suggests that it is ambiguous and can be both definite and indefinite.

(23)   ...dat Jan [$_{DP}$het meeste geld   [$_{PP}$uit   zijn portefeuille]] verloren
       ...that John   the most   money   from his wallet   lost
       heeft.
       has

'...that John lost (the) most money from his wallet.'

  a. John lost more than half of the money from his wallet.

                                  (proportional)

  b. John lost more money from his wallet than anybody else did.

                   (comparative - focus on *Jan*)

  c. John lost more money from his wallet than from his account.

               (comparative - focus on *portefeuille*)

Tests with existential *there* and DP-internal focus show that *de meeste* is definite: it cannot appear in sentences with existential *there* and the comparative reading with DP-internal focus is not available. *Het meeste N$_{pl.}$* is indefinite: it can appear in sentences with existential *there* and can have a comparative reading with DP-internal focus. *Het meeste* in combination with a neuter mass noun can appear in sentences with existential *there*, and it can have a comparative reading with DP-internal reading. This shows that it is indefinite. On the other hand, it can also have a proportional reading, which suggests that it can also be definite.

In short, diagnostics for definiteness offer a first insight into the behavior of *de/het meeste*, but the mechanism behind the comparative and proportional readings has not been clarified yet. Specificity tests will turn out to be a more fine-grained tool.

4.2 Specificity

Specificity makes a further distinction between the various uses or interpretations of indefinite noun phrases. The concept has a long history and covers a wide range of data. von Heusinger [2011] distinguishes referential, scopal and epistemic specificity, specificity associated with familiarity and topicality, and specificity as noteworthiness and as discourse. In this paper, I limit myself to one type, namely epistemic specificity, where a specific indefinite NP refers to a particular referent, the referent "the speaker has in mind" (von Heusinger [2011]).

Scrambling is an excellent test for specificity in Dutch (de Hoop [1996], Broekhuis et al. [2012]). The direct object is base-generated in a position adjacent to that of the verb. Different types of DPs can appear there (24a), namely indefinite DPs with *een* 'a' or definite DPs with *de* 'the'. In (24a), the DP *een auto* 'a car' is non-specific: it can only refer to some car or other and not to a specific car. However, indefinite DPs are always specific when they are scrambled (24b). A sentence with *een auto* 'a car' in a position to the left of the adverb is degraded. A sentence with a definite DP is still correct when the DP is scrambled.

(24) a. ...dat Jan gisteren de/een auto gekregen heeft.
          ...that John yesterday the/a car gotten   has
          '...that John got the/a car yesterday.'

      b. ...dat Jan [de/?een auto] gisteren   t gekregen heeft.
          ...that John the/a   car   yesterday t gotten   has
          '...that John got the/a car yesterday.'

The non-specific indefinite pronoun *wat* 'something'[7] and the ambiguous (non-)specific indefinite pronoun *iets* 'something' also demonstrate this effect.

(25) a. ...dat Jan gisteren iets/wat gekregen heeft.
          ...that John yesterday something gotten   has
          '...that John got something yesterday.'

      b. ...dat Jan iets/*wat gisteren   t gekregen heeft.
          ...that John something yesterday t gotten   has
          '...that John got something yesterday.'

The pronouns can both appear in the base-generated position next to the verb, but sentences with non-specific *wat* become ungrammatical when *wat* is situated left of the adverb. However, *iets* is still felicitous in these cases because it can be specific. This makes scrambling a good diagnostic for the specificity of DPs.

Let us now turn to *de meeste* and *het meeste*. The former can have both proportional and comparative readings in the base-generated position (26a). When

---

[7] *Wat* can be both an indefinite and an interrogative pronoun, depending on its position in the sentence (Postma [1994]).

the DP appears to the left of the adverb, only the proportional reading remains (26b). This demonstrates that the comparative reading is limited to the position of non-specific DPs (compare with *wat* in (25b)).

(26) a. ...dat Jan gisteren de meeste bergen   beklommen heeft.
  ...that John yesterday the most   mountains climbed   has

  '...that John climbed (the) most mountains yesterday.'
                        (proportional/comparative)

  b. ...dat Jan [de meeste bergen]   gisteren   t beklommen heeft.

  ...that John the most   mountains yesterday t climbed   has

  '...that John climbed most mountains yesterday.' (proportional)

In case *het meeste* combines with plural nouns, it is restricted to the comparative reading in the base-generated position (27a). If the DP is scrambled, the sentence is ungrammatical (27b).

(27) a.   ...dat Jan gisteren het meeste bergen   beklommen heeft.
    ...that John yesterday the most   mountains climbed   has

    '...that John climbed the most mountains yesterday.'
                        (comparative)

  b. *...dat Jan [het meeste bergen]   gisteren   t beklommen

  ...that John the most   mountains yesterday t climbed

    heeft.

    has

Sentences with DP-internal focus are infelicitous when the DP is scrambled, no matter whether focus is on the subject (28b) or the PP (29b).

(28) a.   ...dat JAN   gisteren [het meeste platen [van   Zappa]]
    ...that JOHN yesterday the most   records of/by Zappa
    beluisterd heeft.
    listened to has

    '...that John listened to more records of/by Zappa than anyone
    else did.'   (comparative - subject)

  b. *...dat JAN   [het meeste platen [van   Zappa]] gisteren   t

  ...that JOHN the most   records of/by Zappa   yesterday t
    beluisterd heeft.

    listened to has

(29) a. ...dat Jan gisteren [het meeste platen [van ZAPPA]]
          ...that John yesterday the most   records of/by ZAPPA
          beluisterd heeft.
          listened to has

          '...that John listened to more records of/by Zappa than to records
          of any other band.'   (comparative - PP)

      b. *...dat Jan [het meeste platen [van ZAPPA]] gisteren t

          ...that John the most records of/by ZAPPA yesterday t
          beluisterd heeft.

          listened to has

Het meeste can receive both interpretations if it combines with neuter mass
nouns. It has both a proportional and a comparative reading in the base-generated position (30a), but only a proportional one when scrambled (30b).[8]

(30) a. ...dat Jan gisteren het meeste geld   verloren heeft.
          ...that John yesterday the most   money lost   has
          '...that John lost (the) most money yesterday.'
                                          (proportional/comparative)

      b. ...dat Jan [het meeste geld]   gisteren   t verloren heeft.

          ...that John the most   money yesterday t lost   has
          '...that John lost most money yesterday.'   (proportional)

---

[8] A reviewer pointed out that the effect of scrambling can also be caused by the
freezing principle [Wexler and Culicover, 1981, p. 542], represented in (a).

(a) A node is frozen if (i) its immediate structure is non-base, or (ii) it has been
     raised.

Assuming that the DP with *de/het meeste* actually moves when it is scrambled,
the node is frozen and nothing can move from it. The freezing principle thus prevents the superlative morpheme from moving to [SPEC, VP] at LF, which blocks
the comparative reading: only the proportional reading is available. However, there
is a counterexample: the freezing principle predicts that a comparative reading is impossible in case of topicalization. Example (1) however, shows that the comparative
reading is still present.

(1) Het meeste BERGEN heeft JAN beklommen.
     the most   mountains has   John climbed
     It is John who climbed more mountains (and not skyscrapers) than anybody
     else.   (comparative)

The example discards the account of the freezing principle as an explanation, but it
also presents challenges for the definiteness/specificity account. Firstly, non-specific
indefinites normally cannot be topicalized. Secondly, the sentence is only felicitous
with multiple stress on *bergen* 'mountains' and *Jan* 'John'. Future research will have
to clarify the effects and consequences of topicalization and focus.

The specificity tests show that comparative readings of *de/het meeste* are restricted to the positions of non-specific indefinite DPs and that they disappear when the DP is scrambled. When only the comparative reading is available in the first place, scrambling the DP yields ungrammatical sentences. The tests lead to the generalisation that DPs with comparative readings follow the distribution of non-specific indefinite DPs. The results of the definiteness and specificity tests will form the basis for my analysis in the next section.

5 Analysis

The Flemish data confront us with two intriguing questions. Why can *het meeste* occur in combination with plural nouns even though *het* is a singular neuter determiner? And why are sentences containing *de meeste* and a plural noun ambiguous between proportional and comparative readings, while sentences with *het meeste* always have the latter readings?

Based on the definiteness tests, I propose that there are two distinct structures underlying *de/het meeste*. The first one contains a definite determiner that agrees in gender and number with the noun it c-commands (31). Therefore I will label it the "AGR structure".

(31) a. de   meeste vrouwen   - de   meeste mannen   -
          the$_{pl.fem.}$ most   women$_{pl.fem.}$ - the$_{pl.masc.}$ most   men$_{pl.masc.}$ -
          de   meeste huizen
          the$_{pl.neut.}$ most   houses$_{pl.neut.}$
      b. het   meeste geld
          the$_{sing.neut.}$ most   money$_{sing.neut.}$

These AGR structures have comparative and proportional readings in the base-generated object position (26a/30a) and retain the proportional one in the scrambled position (26b/30b).

In the second structure, the determiner does not agree with the head noun. All nouns in (32a) would normally require *de* as a determiner because they are plural, but the singular neuter article *het* appears instead. I will term this the "*AGR structure" because of the lack of agreement. Note that constituents with neuter mass nouns are ambiguous between AGR (31b) and *AGR (32b) structures.

(32) a. het   meeste vrouwen   - het   meeste mannen   -
          the$_{sing.neut.}$ most   women$_{pl.fem.}$ - the$_{sing.neut.}$ most   men$_{pl.masc.}$ -
          het   meeste huizen
          the$_{sing.neut.}$ most   houses$_{pl.neut.}$
      b. het   meeste geld
          the$_{sing.neut.}$ most   money$_{sing.neut.}$

5.1 AGR Structures

For the analysis of AGR structures, I follow Hackl and analyze them as the superlative of *veel* 'many'. The superlative morpheme in Dutch is *-st*. The AGR structure thus consists of a definite determiner (*de/het*) agreeing with the head noun, an AP with the modifier *veel*, the superlative morpheme and the comparison class *C*, and the NP *bergen* (33).

(33)

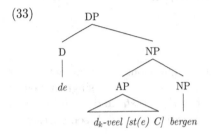

Following Hackl [2009], I assume that proportional readings require the superlative morpheme to stay inside the DP at LF. Example (34) shows how [-st(e) C] moves and has scope over the NP. The determiner of the DP is definite, which explains why *de meeste* is ungrammatical in sentences with existential *there*.

(34)

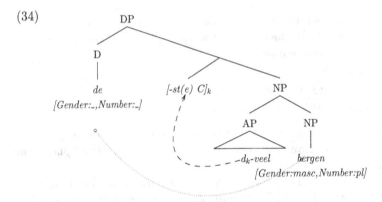

Although Hackl's analysis would predict that *de* is undefined in combination with *meeste* (very much like *most*), this is not borne out by the Flemish data: *de meeste* is definite. The definiteness also follows from the semantics of *de/het meeste*, since the proportional reading refers to a unique fraction in the comparison class, and uniqueness is a feature of definite DPs (Farkas and Kiss [2000]). In other words, the proportional reading is true if John climbed a unique plurality of mountains that is greater than any other non-overlapping plurality in the comparison class.

For the comparative readings, Hackl assumes that the DP is indefinite. This prevents it from acting as an island, which allows the superlative morpheme to move to [SPEC, VP] (35) (cf. Szabolcsi [1986], Heim [1999]). In Flemish, comparative readings are indeed associated with DPs in base-generated positions, where non-specific indefinite DPs may appear.

(35)

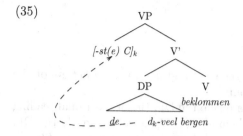

The indefiniteness of comparative readings may be connected to their semantics. In comparative readings, different pluralities are compared to each other (e.g. those climbed by John, Allison and Bill) and is not possible to pick out one unique plurality. The exact relation between definiteness and the semantics of proportional/comparative readings remains a matter for future research.

5.2   *AGR Structures

*AGR structures behave like indefinite DPs, but start with the definite determiner *het*. In order to resolve this problem, I propose that *het* is not the head of the DP, but that *het meeste* is nevertheless part of it. When we use topicalization as a diagnostic, the complete *AGR constituent can be fronted in Flemish Dutch (11). This indicates that *het meeste* must be located inside the DP.

I propose the structure in (36) for *AGR constructions. I follow Matushansky [2008] and assume that superlative phrases without an overt noun modify a null head noun.

(36)

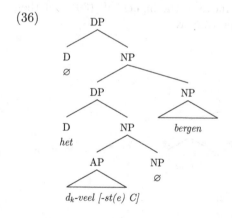

The *AGR structure explains the behavior of *het meeste*. Firstly, the head of the DP containing *bergen* has a null head determiner, which makes the whole constituent a bare plural. Bare plurals can indeed occur in existential *there-*constructions (37b).

(37)   a.   *Er   zijn de bergen.
           there are the mountains

    b.   Er   zijn ∅ bergen.
                there are ∅ mountains

                'There are mountains.'

Secondly, there is no agreement between *het* and *bergen* since *het* is not the head of the DP containing the head noun.

Because of the position of *het meeste* inside the DP, the proportional reading is not available. The superlative morpheme [-st(e) C] moves up inside the DP, but it can only have scope over the null head noun and not over *bergen* (38).

(38)

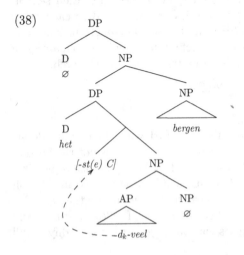

The superlative morpheme could also move to the higher DP (39), but then it is no longer dominated by a definite determiner.

(39)

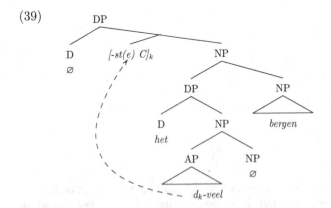

The proportional reading is thus blocked and the only option left for [-st(e) C] is to move out to [SPEC, VP], which produces the comparative reading. This explains why *AGR structures only have comparative interpretations in Flemish Dutch (15).

6   Future Research

In Hackl's analysis, the definite determiner was undefined in the case of *most* and indefinite in the case of *the most*. My discussion of Flemish *de/het meeste* has shown that the analysis should be more fine-grained to account for the Flemish data. Firstly, *de meeste*, the Dutch equivalent of *most*, is always definite. *Het meeste*, on the other hand, is indefinite in *AGR structures. Secondly, the proportional reading is associated with definite DPs, whereas the comparative reading is associated with non-specific indefinite DPs. In the previous section, I presented an analysis for the Dutch data. However, some issues remain.

Firstly, the structures proposed in examples (34) and (38) explain the (in)definiteness of *de meeste* and *het meeste* but not their (non-)specificity. The implementation of specificity in the grammar is no easy matter and remains the subject of debate (cf. von Heusinger [2011]). The Flemish data, however, add quantifiers such as *de/het meeste* to this debate and may contribute to the discussion on the internal structure of (non-)specific DPs.

Secondly, the analysis is based on the account for *most* in Hackl [2009], which in turn builds on the theory for superlatives in Heim [1999]. In these theories, the determiner is definite in proportional readings, which creates an island and prevents the superlative morpheme from moving. In comparative readings, the determiner is indefinite and can move to [SPEC,VP]. It remains unclear how and why the definite determiner *the* can be indefinite. My analysis of Dutch has the same issue. Moreover, the problem has become more pressing: if *de meeste* is indeed non-specific indefinite in comparative readings, then we would expect *de meeste* to appear in sentences with existential *there*. Example (18b) shows that this is not the case. The status of the definite determiner in superlative constructions and *de/het meeste* thus remains a matter for future research.

Thirdly, I did not show how feminine/masculine mass nouns behave in sentences with *het meeste* and existential *there*, DP-internal focus or scrambling. It is possible that these structures influence the analysis presented in Sect. 5, but they will be dealt with in future research.

7   Conclusion

The analysis of *most* in Hackl [2009] claims that the definite determiner is undefined in *most* and indefinite in *the most*. Moreover, the presence or absence of the definite determiner triggers the proportional and comparative readings: sentences with *most* always have a proportional reading and sentences with *the most* always have a comparative reading.

The Flemish data presented in this paper show that *de meeste* is always definite, but that it can have both proportional and comparative readings. *Het meeste* is always indefinite when combined with plural count nouns, but the comparative reading is only available in positions associated with non-specific indefinite DPs.

The analysis of the Flemish data shows that Hackl's analysis is compatible with *de meeste*. However, I propose a more complex syntactic structure for *het meeste* to account for its indefiniteness and the comparative reading it produces.

Some issues have to be dealt with in future research. Firstly, specificity has not been included in the analysis. Secondly, it remains unclear how definite *de meeste* can have a comparative reading. Thirdly, feminine and masculine mass nouns have to be included in the data. Finally, the relation between definiteness, specificity and the semantics of proportional/comparative readings remains an interesting area for future research.

References

Ariel, M.: Does most mean 'more than half'? In: Proceedings of the Annual Meeting, pp. 17–30. Berkeley Linguistics Society (2003)

Ariel, M.: Most. Language 80(4), 658–706 (2004)

Broekhuis, H., Keizer, E., Den Dikken, M.: Syntax of Dutch: Nouns and Noun Phrases. Comprehensive Grammar Resources. Amsterdam University Press (2012) ISBN 9789089644602

de Hoop, H.: Case Configuration and Noun Phrase Interpretation, Garland. Linguistics Series (1996) ISBN 9780815325604

Farkas, D.F., Kiss, K.É.: On the comparative and absolute readings of superlatives. Natural Language and Linguistic Theory 18, 417–455 (2000)

Hackl, M.: On the grammar and processing of proportional quantifiers: Most versus more than half. Natural Language Semantics 17(1), 63–98 (2009)

Heim, I.: Notes on superlatives. MIT lecture notes (1999), http://semanticsarchive.net/Archive/TI1MTlhZ/Superlative.pdf

Horn, L.R.: The border wars: A neo-Gricean perspective. In: Where Semantics Meets Pragmatics. Current research in the semantics/pragmatics interface, pp. 21–48. Elsevier (2006)

Link, G.: The logical analysis of plurals and mass terms: a lattice-theoretical approach. In: Meaning, Use and Interpretation of Language, pp. 302–323. de Gruyter (1983)

Matushansky, O.: On the attributive nature of superlatives. Syntax 11(1), 26–90 (2008) ISSN 13680005

Milsark, G.: Existential Sentences in English. PhD thesis, Massachusetts Institute of Technology, Cambridge, Massachusetts (1974)

Pancheva, R., Tomaszewicz, B.: Cross-linguistic differences in superlative movement out of nominal phrases. In: Arnett, N., Bennett, R. (eds.) Proceedings of the 30th West Coast Conference on Formal Linguistics, Cascadilla Proceedings Project (2012)

Postma, G.: The indefinite reading of wh. In: Bok-Bennema, R., Cremers, C. (eds.) Linguistics in the Netherlands 1994, pp. 187–198. John Benjamins (1994)

Szabolcsi, A.: Comparative superlatives. In: Sagey, E., Rapoport, T.R., Fukui, N. (eds.) Papers in Theoretical Linguistics. MIT WPL 8. MIT (1986)

von Heusinger, K.: Specificity. In: Semantics: An International Handbook of Natural Language Meaning, vol. 2, pp. 1024–1057. de Gruyter (2011)

Wexler, K., Culicover, P.: Formal principles of language acquisition. MIT Press, Cambridge (1981)

Toward a Discourse Structure Account of Speech and Attitude Reports

Antoine Venant[1,2]

[1] Universite Toulouse 3, France
[2] Institut de Recherche en Informatique de Toulouse, France

Abstract. This paper addresses the question of propositional attitude reports within Segmented Discourse Representation Theory (SDRT). In line with most SDRT discussions on attitudes reports, we argue that reported speech should be segmented as the rest of the discourse is, but we identify several issues raised by such a segmentation: first, the nature of some relations crossing the boundaries between main and embedded speech remains unclear. Moreover, such constructions are introducing a conflict between SDRT's Right Frontier Constraint (RFC) and well established facts about accessibility from factual to modal contexts. We propose two solutions for adapting discourse structure to overcome these conflicts. The first one introduces a new ingredient in the theory while the second one is more conservative and relies on continuation-style semantics for SDRT.

1 Introduction

From a semantic perspective, attitudes reports require solving several notorious puzzles. Among these are problems triggered by definites: substitution of directly co-referential expressions is generally not allowed under the scope of an attitude verb and neither is existential generalization (see the shortest spy problem raised by [1]). Closely related to those are effects of attitudes verbs on the availability of discourse referents. For instance, factive epistemic verbs like *to know* allow referents introduced under their scope to be later referred to from an external context, laying outside the scope of the modal operator. On the contrary, non-factive like *to believe* do not. These two issues are context-related which has naturally led to several accounts involving dynamic semantics such as [2,3].

In order to model discourse coherence, we need to understand how reporting someone's propositional attitude interacts with the overall discourse structure. The dynamic framework of Segmented Discourse Representation Theory (SDRT) [4] allows us to address both perspectives simultaneously by looking at the interaction between discourse structure and anaphoric phenomena. However there is in SDRT no semantic contribution for attitude reports that is as precise as the ones cited above and formulated within Discourse Representation Theory (DRT) [5]. SDRT however builds over a lower-level logical formalism, most often set as DRT, and enriches it with rhetorical relations. Elementary discourse units are given a semantics in the lower-level logical formalism, and

M. Colinet et al. (Eds.): ESSLLI 2012/2013, LNCS 8607, pp. 233–246, 2014.
© Springer-Verlag Berlin Heidelberg 2014

rhetorical relations holding between discourse units provide logical consequences that add to the meaning represention of the discourse and pragmatic constraints on accessible referents for anaphoric expressions and attachment of new information. One of SDRT's aims is to refine the predictions of the lower-lovel formalism with additional constraint, and one may wonder whether DRT-style accounts can be straightforwardly embedded into SDRT, and wether SDRT's ability to handle discourse relation might yield a more accurate interface between semantics and pragmatics than simpler DRT-style accounts. We want to address the question of how SDRT's treatment of embedded speech acts fares with respect to such considerations.

We attach particular attention to examples in the spirit of example (1) below:

(1)   The criminal parked his car somewhere near the airport. Therefore, detectives think that afterwards he tried to get into a plane.

We claim that such examples involve two things: an attitude report and a rhetorical relation (here triggered by the cue word *afterwards*) holding between a first unit of the factive context, *i.e.* that does not belong to the scope of the attitude verb, and a second unit, part of the reported speech. In SDRT's vision, speech acts are relational and bear two components: the performance of a new utterance and the picking of a previous discourse unit in the context together with a coherence relation holding between the latter unit and the newly uttered content. Hence, a speech act in SDRT gives rise to a speech-act level anaphora, which the theory models with the same tool as for "classical" correference: namely the right-frontier constraint (RFC). This treatment is supported by successful predictions of available referents for both kinds of anaphora in numerous examples (see [4]). This is true in particular concerning accessibility prediction in modal contexts, for instance both kind of anaphora are forbidden from a nonfactive modal context to a factive context (*A witch might leaves nearby. #She has a pointy hat — I might go to the movies. # Afterwards, I will have dinner.*), and both are subject to modal subordination of the kind found in [6] (*A wolf might enters. He would growl*, which involves a temporal relation between the two clauses) . We think however that the current analyses of attitude reports in SDRT are not fully satisfactory for examples of the like of example (1). More specifically, while DRT based approaches would very likely allow event correference from an embedded DRS to the main DRS, these examples seem to clash with the right frontier constraint. Distinguishing between Intentional/Evidential uses of reportative verbs as it is done in [7] brings some light but does not solve the problem.

After briefly introducing SDRT in section 2, we argue in section 3 for the segmentation of reported constructions. Section 4 deals with relations that links a reported speech act to a factual one. It shows that the discursive structure of intensional reports is closed to incoming relations, but still licences anaphoric links to the context. On this basis it exhibits a family of relations for which RFC makes bad predictions. Section 5 presents two ways of restoring the right accessibility conditions while still benefitting from SDRT more specific constraints.

2   Segmented Discourse Representation Structures

SDRT derives a structure for a given discourse following two steps: first the discourse is segmented into *elementary discourse units* (edus). these units are then linked to each other by means of coherence relations. Several units linked to each other can form a compound and recursively serve as the argument of other relations. Such a compound is called a *complex discourse units* (CDU). The level of segmentation for *elementary discourse units* merely corresponds to the clause level, *i.e.* utterances involving a single event or a single state[1].

Each discourse unit is assigned a label $(\pi_i,...\pi_n)$ in a countable vocabulary Π and a corresponding formula in a given language for the representation of atomic clauses $(K_{\pi_1},...,K_{\pi_n})$. This is the *lower-level* language and associated representations. Discourse Representation Structures (DRSs) from DRT are a classical choice for the lower-level representations. Labels in Π are used as arguments of rhetorical relations, like $Narration(\pi_i, \pi_j)$ or $Explanation(\pi_i, \pi_j)$. CDUs made of rhetorical relations and other subordinated labels are assigned a label in Π as well. An SDRS is a triple $\langle A, \mathcal{F}, Last \rangle$ where $A \subseteq \Pi$ is the set of labels of the SDRS, $\mathcal{F}$ a function mapping labels to contents. If $\pi \in A$ labels an elementary unit, then $\mathcal{F}(\pi) = K_\pi$ is a lower-level formula such as a DRS. If π labels a CDU, then $F(\pi)$ is a conjunction of discourse relation predicates holding between subordinated labels, for instance $Elaboration(\pi_1, \pi_2) \wedge Explanation(\pi_2, \pi_3)$. Finally, $Last \in A$ is the label of the last segment introduced. (See [4]:p.138 for the precise definition). A complex constituent is said to *immediately outscopes* every label that appears in its content. The transitive closure of this relation on A has a maximal element in A that we will often denote by π_{top}. π_{top} is the top-level complex consituent who contains every other labels. We will often abuse notations and write an SDRS as the content of its top element $\mathcal{F}(\pi_{top})$. For instance, we will write $Elaboration(\pi_1, \pi_2)$ to denote the SDRS $\langle \{\pi_1, \pi_2, \pi_{top}\}, \mathcal{F}, \pi_2 \rangle$ with $\mathcal{F}(\pi_{top}) = Elaboration(\pi_1, \pi_2)$.

SDRT makes a structural distinction between coordinating and subordinating relations. The former, like *Narration*, confer an equal status to their two arguments. The latter introduce a hierarchy between the related constituents. Such a distinction allows to define the so-called Right Frontier constraint. The Right Frontier is the set of labels $RF = \{\pi \mid \pi \prec^* Last\}$ where $\prec^*$ is the transitive closure of the dominance relation $\prec$ defined by $\pi \prec \alpha$ iff α is a complex consitutent which immediately outscopes π or there is subordinating edge $R(\alpha, \pi)$ in some constituent γ. The Right Frontier Constraint stipulates that labels accessible for discourse continuation are those of the Right Frontier, while the ones accessible for correference have to be DRS-accessible on the right frontier.

Finally, an SDRS has a truth-functional semantics which, as in DRT, is expressed in terms of context-change potential (*i.e* relation between world-assignments pairs). This informational content is recursively computed from the semantic consequences associated with each rhetorical relation applied to the

---

[1] How fine-grained segmentation should be is still a matter of discussion. The present work also contributes at this level since we argue in favor of segmenting attitudes.

content of their arguments, eventually relying on the lower-level logical forms to express the content of edus.

Consider the following example:

(2)   a.   John visited his friend.
       b.   Then he went to the cinema.
       c.   He watched *Pirates of the Caribbean*
       d.   #They talked for a long time

The structure of example (2)-abc is $Narration(\pi_a, \pi_1)$, with π_1 a complex constituent label whose content is $Elaboration(\pi_b, \pi_c)$ (labelling the utterances (2)-a , (2)-b and (2)-c respectively with π_a, π_b and π_c). Hence example (2) exhibits a temporal relation of *Narration* between the event mentionned in π_a (going to the cinema), and the complex structure $[elaboration(\pi_b, \pi_c)]$ yield by (2)-bc. Hence, this semantic representation implies that the semantic consequences of the *Narration* relation hold between the content of π_b and **both** the content of π_c and π_d. Generally, there is a major semantic difference between attaching to a complex constituent and to one of its constituent. However, in the very particular case of example (2), the same temporal consequences could be retrieved from the alternative "flat" structure $Narration(\pi_a, \pi_1) \wedge Elaboration(\pi_b, \pi_c)$ because of the specific semantics of *Elaboration* which makes π_c and π_d contributing to the same event. Moreover, both structure have the same consequences on accessibility. We detail the right frontier of the first structure: since *Elaboration* is a subordinating relation and *Narration* a coordinating one, the right frontier is $\{\pi_b, \pi_c, \pi_1\}$ and the discourse could not be felicitiously continued by (2)-d which intends to attach to π_a.

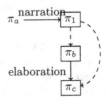

Fig. 1. Graphical representation of example (2)

Figure 1 gives a graphical representation of (2), with the convention that coordinating relation are drawn horizontally, subordinating one vertically, complex constituents are linked with dashed edges to their subconstituents, and nodes of the right frontier are red.

To conclude this section, consider the following example:

(3)   a.   John thinks that
       b.   Mary missed school
       c.   Because she is ill.

This example involves a reported speech. In SDRT the matrix clause and embedded speech form distinct segments, a choice that we motivate in the next section. The reported speech and the matrix clause are linked by an *Attribution* relation whose semantics places the informational content of the reported speech under the right attitude modality. *Attribution* relations and the example (3) above illustrate the crucial semantic difference provided by CDUs in a SDRS: the structure of example (3) is $S = \langle\{\pi_{top}, \pi_a, \pi_b, \pi_c, \pi_1\}, \mathcal{F}, \pi_c\rangle$ with $\mathcal{F}(\pi_{top}) = Attribution(\pi_a, \pi_1), \mathcal{F}(\pi_1) = Explanation(\pi_b, \pi_c)$. The modality introduced by the attribution scopes over both π_b and π_c, and neither the content π_b nor the one of π_c is implied by the the content of S.

In this framework, we now move to the discourse structure of attitude and speech reports.

3 Segmentation and Treatment of the Matrix Clause

There are at least two reasons for capturing the interaction between attitudes or speech reports and discourse structure. First, we need to account for discourse phenomena both inside the reports and across their boundaries. Then the treatment of intentional and evidential uses of attitude reports in the way of [7] also requires segmentation.

Regarding the first point, example (4) below is not felicitous, because the pronoun *it* cannot easily refer to the salmon in the given context. Such a behaviour is predicted by RFC. Therefore, even if the semantics of attitudes generally involves quantification over intensions or contents, and thus erases to some extent the structure of the logical form of the original speech act, the discourse structure of the report is needed anyway to build the logical form of the speech report.

(4) a. John told me that Marry had a wonderful evening last night.
     b. He said
     c. she ate salmon
     d. and then won a dancing competition
     e. #and that it was beautiful pink.

Consider now

(5) John says that he left after Mary did, but he left because she did.

In example (5) the reported speech introduces a *Narration* between two events while the non-reported discourse asserts a causal relation (*Result*) between the two same events. The contrast introduced by the cue word *but* is however coherent, partially because it is supported by the isomorphic structures of the reported speech and the non-reported one. SDRT treatment of contrast as a scalar relation, following [4,8] provides such an analysis, assuming that the structure of the embedded speech is accessible.

Regarding the segmentation of the matrix clause, we may consider it as nothing more than a kind of logical operator[2]. However, that would be inaccurate since the matrix clause can be fairly sophisticated. It generally includes a communication event or a mental state that can be modified by adverbs or prepositional phrases and therefore would be difficult to model as simple logical operator. Since removing the matrix-clause from the discourse representation is not an option neither, we have no option but to deal with a segment for the matrix-clause.

[7] addresses several issues raised by such a treatment of reported speech. The approach consists in segmenting apart matrix clause and reported speech and in identifying the relation between these elements themselves but also their relations with the surrounding context. It distinguishes between two uses of reportative verbs, namely *evidential* where the embedded content is asserted by the main speaker and *intensional* where the content of the report is not asserted by the main speaker. In evidential uses, the matrix clause is subordinated to the embedded content by a veridical *Evidence* relation.[3] In intensional uses, the embedded content is subordinated to the matrix via a relation of attribution which is non-veridical. Such a distinction makes very profitable the separation of the matrix clause and the reported speech, accounting for cases like example (6).

(6)   a.   The neighbours are gone.
      b.   John told me that
      c.   they went on vacation in an expensive hotel.
      d.   (i)   I have called it this morning.
           (ii) But he lied.

As [7] argues, we can see in the example (6-a)–(6-d-i) above that (6-c) is asserted by the speaker since (6-d-i) is carrying an anaphoric link to *the hotel* even though it has first been introduced under the scope of the attitude[4]. On the contrary, in example (6-a)–(6-d-ii), the author disagrees with what is reported, and the existence of *the hotel* is not ensured anywhere outside the scope of the attitude. Therefore *the hotel* should not be referred to later in the discourse. [7] also argues that the compositional semantics of both the reported speech and the matrix clause do not change from an intensional to an evidential report. Neither can it be deleted without loss of compositional content in the one nor the other case. But the way the two parts of speech are related can change. Furthermore, since the two first sentences are the same in both examples, the decision of choosing one or the other might only be a matter of context, as such it is essentially information packaging, and in SDRT, this level is kept aside from the logic of information content.

---

[2] This would still requires to modify the SDRT framework since all logical operators are delegated either to the lower-level logical forms or to the semantic effects of discourse relations.

[3] To be satisfied, veridical relations require their arguments to be true in the model. Non veridical relations do not have this requirement. [4]

[4] At least if we assume that *d* is not part of what John said here, but in that case that would be a very odd reading.

Following this analysis, the structure of example (6-a)–(6-d-i), links (6-d-i) to (6-b) with a veridical relation *Narration*, forcing the evidential reading. (6-c) is linked to (6-b) with the veridical relation *Evidence* and to (6-a) with the veridical relation *Explanation*. The structure of example (6-a)–(6-d-ii) is different. The continuation (6-d-ii) is attached to the whole report with a *Contrast* relation and yields an intensional reading (attaching (6-d-ii) to the embedded clause would entail that John said something incoherent, which unlikely is the intended meaning) and (6-b) is related to (6-c) using the non-veridical relation *Attribution(b, c)*. The two different type of structures are sketched below. (Left column is evidential, right one is intensional. We also give some of the semantics conditions associated with the two relations involved).

$$\mathcal{F}(\pi_b) = \boxed{\begin{array}{c} \phi \\ \hline A(x, \phi) \end{array}} \qquad\qquad \mathcal{F}'(\pi_b) = \boxed{\begin{array}{c} \phi \\ \hline A(x, \phi) \end{array}}$$

$$R_e(\pi_a, \pi_c) \wedge Evidence(\pi_c, \pi_b) \qquad\qquad R_i(\pi_a, \pi_b) \wedge attribution(\pi_b, \pi_c)$$

$$\Phi_{Evid(\pi_c,\pi_b)} \Rightarrow K_{\pi_c} \wedge K_{\pi_b} \qquad\qquad \Phi_{Attr(\pi_b,\pi_c)} \Rightarrow K_{\pi_b} \wedge \phi \sim \pi_c$$
$$\wedge\, \phi \sim \pi_c$$

$\sim$ may be understood as an equivalence relation between SDRS contents, where *content* means the context change potential. (Blocking substitution of logically equivalent expressions under the scope of an attitude verb may however require some amount of structure being kept in the notion of *content* [2]).

4 Relations Across Boundaries

We have introduced and motivated SDRT's current treatment of attitude reports. We now move to the main problem. In this section we focus on examples inspired by [9]. These examples involve speech reports where a discourse relation is attributed to the author of the embedded speech, with the particularity that this relation *crosses boundaries*: it holds between a discourse unit in the embedded speech and a unit introduced prior to the report, in the factive context. We first discuss the discourse structure of these examples, and argue in favor of a structure that supports arbitrary nesting of reports. Once the structure clarified we ask the question of whether there is a rhetorical connection holding between the input context, and the report itself, in addition to the boundary-crossing relation. We exhibit a family of examples for which in there is such a coordinating link, which is not allowed by RFC.

As [9] remarks, the picture becomes more complicated when relations comes to cross the boundaries of an embedded speech act such as in example (7) below:

(7)   a.   Fred will go to Dax for Christmas.
        b.   Jane claims that
        c.   Afterwards, he will go to Pau.

Afterwards introduces a veridical relation of *Narration*. With an evidential reading, this example is not problematic: the discourse producer (DP) asserts the content π_c. Consequently, he can use a veridical relation which links it to the context without clash of veridicality. However, with an intensional reading the speaker does not commit to $Narration(\pi_a, \pi_c)$ since he does not assert the content of π_c. But he still can commit to Jane committing to such a relation. To solve this problem, [9] sets up a new paradigm for discourse analysis that examines reported relations against several sources. For instance, example (7) will be analyzed as follows:

The discourse producer is certain of the main eventuallity e_a in a but he does not know anything about the one in c. Jane is attributed to be certain about the main eventuality in c, and, after the source of the *Narration* is identified as being Jane, the picture is completed with the statements of Jane being certain of e_a too, as well as e_a and e_c being in a temporal sequence. Semantically speaking, such examples require some further discussion. First, we cannot always identify a source for a relation. Consider a two level deep embedding as in example (8). Asserting $Narration_{\text{Fred's wife}}(\pi_a, \pi_d)$ in this case would make us unable to distinguish between example (8) and the same without (8-b). With example (8) the writer does not commit to Fred's wife committing that he will go to Pau.

(8)   a.   Fred will go to Dax for Christmas
        b.   Jane told me that
        c.   according to his wife,
        d.   afterwards, he will go to Pau.

Besides, semantically interpreting $Narration_J(\pi_a, \pi_c) \wedge attribution(\pi_b, \pi_c)$ requires some precisions that the framework does not provide. To this end, we may switch to a dialogical framework in which each individual would receive its own SDRS. However, this would make wrong semantic predictions for (8): the discourse should be satisfied in models where neither Jane nor Fred's wife actually said anything, and the speaker is lying. But a if the SDRS for Jane, or Fred's Wife, features a veridical *Narration*, then the semantics of dialog SDRSs (see [10]) commits its owner to the semantics consequences of that relation, which is to be avoided. Of course, the problem of attributing a nested report of depth 2 to someone also remains. To which SDRS should (8-d) belong? Fred's wife? But then what is the content of Jane's SDRS?

Another way to provide an interpretation would be to use [11]'s proposal for the semantics of coherence relations:

$$R(a,b) : [\![C(Speaker(b), K_a) \wedge C(Speaker(b), K_b) \wedge C(Speaker(b), \phi_{R(a,b)})]\!]$$

where C is a commitment relation and R a veridical relation. But again, we have to understand the *Narration*'s producer as being *Jane* in order to make sense of a formula like $Narration_{Jane}(\pi_1, \pi_2)$. Therefore, the interpretation of example (7) would be entailing $C(Jane, \phi_{Narration(\pi_a,\pi_b)})$ while achieving $C(DP, C(Jane, \phi_{Narration(\pi_a,\pi_b)}))$ is what is needed to account for example (8). Our conclusion is that the problem originates from the structure. Attributing the relation to a particular source is misleading and does not allow to account for nested examples like (8). The *Attribution* relation should scope over the whole *Narration relation*. This can be expressed in SDRT with a complex discourse unit representing the embedded content[5]:

$$A = \{\pi_{top}, \pi_a, \pi_b, \gamma, \pi_c\}$$
$$\mathcal{F}(\pi_{top}) = attribution(\pi_b, \gamma) \; \mathcal{F}(\gamma) = Narration(\pi_a, \pi_c) \tag{1}$$

Equation (1) however is missing something as it does not picture any non-embedded left-veridical coherence relation between π_b and a label in the discourse context of the report. It is possible, if not required for the discourse to be coherent, that the speech act of making the report itself is linked to the discourse context with some relation (R). This must hold at least for intensional readings. For instance, there might be a coherence link R between π_a and the speech act of reporting Jane's claim. Since Attribution is subordinating in an intensional reading, R cannot be coordinating without the RFC being violated in example (7). So it seems that R should always be a subordinating like *Background* (a temporal relations which semantically implies that its arguments temporally overlap). Consider the following examples (the square brackets delimit edus):

(9) a. [The train arrived 3 hours late.] [then the company announced that] [as a consequence, the passengers would be refunded]. [As a matter of fact, they never were.]

   b. [John had a deadline at midnight yesterday.] [So we all thought that afterwards he would go to bed.] [But he did not.]

   c. [Yesterday, John fell three times in a row.] [Mary then told him that] [it was probably because he drank to much.] [He did not believe her.]

All these examples involve an intensional attitude report and in each of them, two discourse markers are present. One is a marker of *Narration* or a *Result*, and triggers a relation between the first segment and the matrix clause of the report, the other triggers a relation between the first segment and the embedded clause. Both *Result* and *Narration* are thought to be coordinating relations. So even if we use the subordinating *Background* between π_a and π_b in example (7), we cannot account for these links without violating RFC.

Beside the RFC violation, examples in (9) also show that the intensional and evidential readings behave assymetrically with respect to whether the embedded content may link to something in the factual context or not, as evidential readings do not clash with the RFC in any of the previous examples. We think this is unsatisfactory, because as shown in the previous section, the intensional/evidential

---

[5] Representing SDRS as directed acyclic graphs is confusing in this case, because it does not make it possible to distinguish which complex discourse unit actually hosts the boundary-crossing relation.

readings can be forced by subsequent utterances in the discourse. Attitudes reports must however have a (possibly ambiguous) meaning of their own. Despite discourse structure being known to be non-monotonic, it seems a little counterintuitive that information about subsequent moves, attaching only to the report and not its input context may be essential to the computation of the attachments of the report to its input context.

Finally, the problematic examples we discussed so far all involve explicit discourse markers. We claim that a similar problem can be triggered by implicit rhetorical relations. considered the examples below:

(10)   1.   [The factory blew up.]$_a$ [therefore, John thinks that]$_b$ [there was an accident with dangerous chemicals.]$_c$

      2.   [The factory blew up.]$_a$ [John thinks that]$_b$ [there was an accident with dangerous chemicals.]$_c$ [But sam thinks that]$_d$ [someone lighted a fire.]$_e$

Examples in (10) carry implicit links between a and the reported content b: There is at least one plausible reading for (10) involving a coordinating relation *Result* between a and b and an implicit *Explanation* between a and c: the explosion made John think of a plausible explanation, which is that something happened with dangerous chemicals at the factory which caused the explosion. Example (10) requires implicit explanation relations to make a better sense of the contrast relation that links b and c. The beliefs of John and Sam are fully compatible, unless what John and Sam respectively said is $Explanation(\pi_a, \pi_c)$ and $Explanation(\pi_a, \pi_e)$, in which case they are not.

So far we have shown that RFC conflicts with the discourse structure of our example (at least, in some readings). The next section proposes two different ways of restoring the right predictions.

5   Restoring Accessibility

We have shown that SDRT damages more standard, but essentially correct, accounts of anaphoric links going between modal and factual contexts. An account of attitude reports in DRT for instance, would not have this behaviour. Examples like example (9) would introduce reference to events in the main DRS from the modal context, which is permitted. We would like such a behaviour, but with SDRT treatment of accessibility still applying inside the reported speech. To this end, we could drop the attribution relation, falling back to a DRT like treatment. The structure of one of our problematic report in SDRT would thus be sketched by $R_{coord}(\pi_a, \pi_{att})$ with $\mathcal{F}(\pi_{att}) = K_{\pi_b} \wedge A(x, \phi) \wedge \phi \sim \pi_c$. This structure allows referents in π_c to attach or refer to elements in π_a.[6] This builds on intensional report being "closed" discursive structures. We showed in section 4 that a relation cannot really penetrate the report from the factual context without (a "copy" of) its left argument and itself being embedded under the attitude. Moreover,

---

[6] Such an approach actually needs to slightly modify the syntax of the SDRS language

attachment to the matrix clause and attachment to a complex segment made of both the matrix clause and the report are semantically and dynamically equivalent (they have equivalent semantic meaning and introduce the same constraints on accessibility). This allows us to represent the complete speech act of reporting something with a single discourse unit π_{att}. This approach however requires to adapt SDRT's mecanism for inferring the relations. A reported attitudes present SDRT with a choice between the evidentials and intensionals readings, but this readings are now asymmetric. In the intensional case, the structure involves a "monolitic" constituent π_{att}. In the evidential case, the structure is more respectful of the text's segmentation and the π_{att} constituent is splitted into π_{mat} for the matrix clause segment and π_{emb}. Hence this approach requires to axiomatise, in SDRT's logic of information packaging, the operation of "gluing" together π_{matt} and π_{emb} into π_{att} when performing the intensional choice. This is theoritically speaking feasible, since SDRST's logic of information packaging is designed with the expressivity to describe and access SDRSs logical forms, including labels content.

We propose an alternative, more conservative approach, that makes use of continuation-syle semantics [12]. Continuation style semantics represents a discourse as a $\lambda-$abstraction of type $[\![\Gamma]\!] = \gamma \rightarrow ((\gamma \rightarrow l \rightarrow t) \rightarrow l \rightarrow t)$ where γ is the type of input contexts and l is the type of labels. A discourse thus asks for

i) an input context i of type γ containing the effects of processing the previous discourse.
ii) A continuation o of type $\gamma \rightarrow l \rightarrow t$ representing the discourse to come.
iii) A label π, the label of the SDRS representing the whole discourse.

To represent chunks of an SDRS, a language is used where every $n-$ary σ symbol in SDRT's object language becomes an $n+1$-ary predicate C_σ, the extra argument stands for the label that hosts the predicate: $\mathcal{F}(\pi) = R(\pi_1, \pi_2)$ will be represented with the formula $\exists \pi_1 \exists \pi_2 \exists \pi C_R(\pi_1, \pi_2, \pi)$. The lower-level content of edus can be analougously encoded; for instance, if $\exists .\cdot$, P and x are respectively binary, unary and nullary symbols of the lower-level language, there will be three predicate $C_\exists()$, C_P and C_x of arity respectively 3, 2 and 1 in our language. For instance, $\mathcal{F}(\Pi)(\exists_x P(x))$ is encoded has $\exists \pi, \exists v_1, v_2, C_\exists(v_1, v_2, \pi) \wedge C_x(v_1) \wedge C_P(v_1, v_2)$. In the following, we will refer only to predicate of the representation language. Therefore, for sake of simplicity and readability, we will denote $C_R(\pi_1, \pi_2, \pi_3)$ simply as $R(\pi_1, \pi_2, \pi_3)$ and $C_P(v) \wedge C_x(v)$ simply as $P(x)$.

We assume that a context contains a structural representation of the SDRS for the previous discourse such that the following functions may be defined:

1. $sel_l : \gamma \rightarrow l$ that selects a label for attachment.
2. $\nu : \gamma \rightarrow l \rightarrow \gamma$ that performs the SDRT update operation on the context [4], defined in terms of SDRT's language for inferring relations. Given a label π, it basically picks up a relation and two other labels π_1, π_2 in the context and add the relation $R(\pi_1, \pi, \pi_2)$ to the context.

The semantics representation of an EDU will generaly look like

$$\lambda i o \pi \, \texttt{Some\_Predication}(\pi) \wedge o(\nu(i, \pi))$$

that is, an EDU states that the content of its assigned label π involves a given predication, then relies on SDRT mecanism to update the dicourse context with the newly added information (including the attachment π to some previous label with a coherence relation if needed), then evaluates its continuation on the updated context.

Finally, the composition of a discourse and a sentence is described by the following binder rule:

$$[\![D.S]\!] = \lambda io\pi \, \exists \, \pi_D [\![D]\!] i (\lambda i' \, \exists \pi_S \, [\![S]\!] \, i' \, o \, \pi_S)$$

The main idea is to refine the proposal of [7] of a lexical entry for attitude reports using continuation-style semantics to overcome the right-frontier problems. Since evidential and intensional readings only differ by the way the matrix clause and the embedded content are related, one simple solution is to postpone attachment of the matrix clause until the embedded content has been dealt with and all attachment to previous context have been done. It must however be performed before the following discourse is processed in order to still benefit from the intensional/evidential distinction. This might be done by modifying the continuation of the report in such a way that it proceeds to the attachment of the matrix clause before applying the real continuation.

Let us assume an attitude α in a discourse "$x \, \alpha$ that ϕ" and that syntax delivers us a parse leading to $\alpha(x, \phi)$. We add the following lexical entry for an attitude verbe α, with A a modal operator corresponding to attitude α.

$$[\![\alpha]\!] = \lambda x \lambda s \lambda io\pi_{matt} \exists \, \phi A(x, \phi, \pi_{matt}) \wedge \exists \pi_s \, \phi \sim \pi_s \wedge s \, i \, [\lambda i' o(\nu(\nu(i', \pi_{matt}), \pi_s))] \, \pi_s$$

Consider again the following example from examples (9): [*The train arrived late*]$_a$. [*Then the company annouced that*]$_b$ [*as a consequence, the passengers should be refunded*]$_c$. We assume for a the lexical entry:

$$\lambda io\pi \, \exists x \, train(x, \pi) \wedge Late(x, \pi) \wedge o \, \nu(i, \pi).$$

In this entry the update operation $\nu(i)$ will deliver a context i' containing the structure $\pi_a \mid F(\pi_a) = [x \mid train(x) \wedge late(x)]$, and maybe a relation linking π_a to the previous context. Assuming that the lexical entry for *as a consequence* is

$$[\![as \ a \ consequence]\!] = \lambda s \, \lambda io\pi s \, i \, (\lambda i' Result(sel_L(i'), \pi, sel_L(i')) \wedge o \, i')$$

The entry embedded for the embedded content c is:

$$[\![as \ a \ consequence]\!]([\![Passengers \ be \ refunded]\!]) =$$
$$\lambda io\pi \exists y, z \wedge The\_Passengers(y, \pi) \wedge Be\_Refunded(y, \pi)$$
$$\wedge Result(sel_L(i), \pi, sel_L(i)) \wedge o \, i$$

The lexical entry for *to announce* (our α here) will be given the_company as its first argument and the interpretation of c as its second. Which should yield after beta reduction:

$$\lambda io\pi_{matt} \exists\, \phi A(\text{The\_company}, \phi, \pi_{matt}) \land \exists\pi_s\, \phi \sim \pi_s$$
$$\land\, \exists y, z \land The\_Passengers(y, \pi_s) \land Be\_Refunded(y, \pi_s)$$
$$\land\, Result(sel_L(i), \pi_s, sel_L(i)) \land o(\nu(\nu(i, \pi_{matt}), \pi_s))$$

When composing with $[\![a]\!]$, this entry will receive the context i' containing the structure $\pi_a \mid F(\pi_a) = [x \mid train(x) \land late(x)]$, unmodified, as input context and thus be able to select π_a as first argument for the result relation without RFC violation. Importantly, successive call to the ν function will perform the intensional/evidential choice and choose a relation to link the report to the preceding discourse before processing the continuation.

6   Conclusion

We have shown the necessity of segmenting the matrix clause and its embedded speech reporting clause in a discourse structure account of attitude reports. The discourse structure of segmented reports is not straightforward. We have given a more precise picture of what it should be and why. More specifically, we have discussed the structure of problematic reports involving relations that crosses the boundary of the report and argued in favor of a structure that does not relies on a mechanism attributing each discourse relation to a given individual. Instead we proposed a recursive structure which gives us the possibility to accurately represent the meaning of nested reports, with or without boundary-crossing relations. We have highlighted a family of examples involving attitude reports that clash with the RFC constraint on accessibility. We have proposed a fix for this problem within a continuation-style semantics for SDRT. The solution relies on postponing the SDRT context-update operations after both clauses of the report have been dealt with. We assumed however syntax to deliver us a parse of the form $\alpha(x, \phi)$, for a given attitude α. This is a strong hypothesis. The matrix clause of a report may behave like a parenthetical, in examples such has *The robber had a gun, police says, and resisted arrest*. It would therefore be interesting to see if continuation-style semantics can provide us with a treatment of attitudes more representative of this syntactic flexibility.

References

1. Kaplan, D.: Quantifying in. Synthese 19(1-2), 178–214 (1968)
2. Asher, N.: A typology for attitude verbs and their anaphoric properties. Linguistics and Philosophy 10(2), 125–197 (1987)
3. Maier, E.: Presupposing acquaintance: a unified semantics for de dicto, de re and de se belief reports. Linguistics and Philosophy 32(5) (2010)
4. Asher, N., Lascarides, A.: Logics of Conversation (Studies in Natural Language Processing). Cambridge University Press (June 2005)

5. Kamp, H., Reyle, U.: From Discourse to Logic: Introduction to Model-theoretic Semantics of Natural Language. In: Formal Logic and Discourse Representation Theory. Studies in Linguistics and Philosophy, vol. 42. Kluwer, Dordrecht (1993)
6. Roberts, C.: Modal subordination and pronominal anaphora in discourse. Linguistics and Philosophy 12(6), 683–721 (1989)
7. Hunter, J., Asher, N., Reese, B., Denis, P.: Evidentiality and intensionality: Two uses of reportative constructions in discourse. In: Proceedings of the Workshop on Constraints in Discourse, pp. 99–106. National University of Ireland, Maynooth (2006)
8. Asher, N., Hardt, D., Busquets, J., Sabatier, I.P.: Discourse parallelism, ellipsis, and ambiguity. Journal of Semantics 18, 200–201 (2001)
9. Danlos, L., Rambow, O.: Discourse relations and propositional attitudes. Constraint in Discourse, CID (2011)
10. Lascarides, A., Asher, N.: Agreement, disputes and commitments in dialogue. Journal of Semantics 26(2), 109–158 (2009)
11. Vieu, L.: On the semantics of discourse relations. Constraint in Discourse, CID (2011)
12. Asher, N., Pogodalla, S.: SDRT and Continuation Semantics. In: Bekki, D. (ed.) JSAI-isAI 2010. LNCS (LNAI), vol. 6797, pp. 3–15. Springer, Heidelberg (2011)

ST5: A 5-Valued Logic for Truth-Value Judgments Involving Vagueness and Presuppositions*

Jérémy Zehr

Institut Jean Nicod, UMR 8128,
École Normale Supérieure, France
http://www.institutnicod.org/

Abstract. Both presuppositional and vague expressions may yield non-classical truth-value judgments. Given that expressions of these kinds may combine together, I propose a single logical system intended to deal with them, which would account for our truth-value judgments. The system I propose is based on Cobreros&al's [4] 3-valued system for vagueness, ST, which comes with a notion of assertoric ambiguity that I claim naturally deals with our non-classical judgments for vagueness. I show that the specificities of presuppositions with respect to truth-value judgments can be accounted for within this system if we add two logical values to it. I discuss a specific 5-valued system that I call ST5.

Keywords: vagueness, presuppositions, 5-valued logic, truth judgments.

1 Introduction

In this paper, I will focus on *truth-value judgments* concerning vagueness and presuppositions. I start from the position that we observe *conflicting* judgments for vague sentences as well as for presuppositional sentences in specific situations. For instance, consider the presuppositional sentence (1):[1]

(1)   The amplifiers have stopped buzzing.

---

* I am very grateful to my PhD supervisors, Paul Egré and Orin Percus who gave me their support and their help all through the writing of this paper. I want to thank the participants of the *LANGUAGE* seminar of the Institut Jean Nicod, of the *Paris-Munich Workshop* and of the *Language, Truth and Logic Networkshop* in Princeton, as well as the anonymous reviewers of this article. A part of this work was supported by a 'Euryi' grant from the European Science Foundation ("Presupposition: A Formal Pragmatic Approach" to P. Schlenker). The ESF is not responsible for any claims made here. Acknowledgments to the ANR-10-LABX-0087 IEC and ANR-10-IDEX-0001-02 PSL grants.

[1] Aspectual verbs such as *stop* are well-known to trigger a presupposition. See for instance the article "Presupposition" in the Stanford Encyclopedia of Philosophy [3].

M. Colinet et al. (Eds.): ESSLLI 2012/2013, LNCS 8607, pp. 247–265, 2014.
© Springer-Verlag Berlin Heidelberg 2014

If I'm told (1) and I know that, in fact, the amplifiers have never buzzed, I can say that (1) is *both false and not false*: it is *false* because the amplifiers were *not* buzzing before, and it is *not false* because if (1) were false, it would mean that the amplifiers *were* buzzing before. Similarly, consider the vague sentence (2), involving the vague adjective *loud*:

(2)   The amplifiers are loud.

If I'm told (2) and I find the volume of the amplifiers to be neither clearly loud nor clearly not loud, I can say that (2) is *both true and false*: it is true to some extent, because the amplifiers are not clearly not loud, but it is false to some extent too, because they're not clearly loud either.[2]

My aim here will be to offer a semantics that assigns *logical truth values* to propositions involving vague and presuppositional expressions on the basis of which one could correctly predict the *truth-value judgments* of speakers in regular *and* conflicting-judgment contexts. In Sect. 2, I begin by reviewing truth-value judgments that we find for positive and negative counterparts of sentences involving vague expressions and sentences involving presuppositional expressions. Section 3 presents the 3-valued ST system [5], which has been developed for vagueness and which offers a natural way of accounting for the conflicting truth-value judgments to which vagueness gives rise. I then consider a 5-valued extension of this system, which I call ST5, in order to incorporate presuppositional expressions. In Sect. 4, I consider the interactions between vagueness and presuppositions, by looking at sentences that involve both vague and presuppositional expressions (hybrid sentences). I propose a semantics for presuppositional sentences in ST5 that makes predictions for hybrid sentences and for sentences with iteratively embedded presuppositional expressions. Finally I briefly consider alternative multi-valued systems in Sect. 5 and show why one should prefer ST5 to deal with vagueness and presuppositions.

2   Truth-Value Judgments

By a *truth-value judgment* I here mean any position that a speaker can have toward the *truth* or the *falsity* of a sentence. My use of this notion then refers to the set of combinations of *true* and *false* closed under *not, and, (n)or, both* and *(n)either*.[3]

Each element of this set is a truth-value judgment. It is clear that, as truth-value judgments, some of the elements in the set are so-to-speak "regular": speakers often judge sentences *true, false, not true* or *not false*. But other elements are far less "regular" (*neither true nor false*) and some even sound contradictory:

---

[2] Serchuk & al. [23] conducted several experiments revealing this apparent contradictory characteristic of truth-value judgments for vagueness.

[3] Importantly, the set of *truth-value judgments* is to be distinguished from the set of *logical values* that a system assigns to propositions. There is no necessary one-to-one correspondence between their elements; and the system I will eventually propose exhibits no such correspondence.

both true and false, both true and not true, both false and not false for instance.[4] Yet, I claim that speakers can use these elements to qualify some sentences. That is to say, I claim that speakers can exhibit apparently *conflicting* truth-value judgments. Even though some dialetheists, such as Priest [19], endorse the view that there are true contradictions, Lewis [17] for instance proposed to see underlying ambiguity in judgments of this kind.[5]

In the next two subsections, I present some evidence that speakers have access to these kinds of judgments concerning vagueness and presuppositions. The account I will eventually give for this relies on a notion of assertoric *ambiguity* developed in the 3-valued logic ST [5]. So far, there have been few experiments exploring the *truth-value judgments* of speakers concerning vagueness or presuppositions, I will therefore rely on indirect evidence that speakers have access to conflicting truth-value judgments in the cases of vagueness and of presuppositions.

2.1 Vagueness

In an experiment conducted by Alxatib & Pelletier [2], participants were presented with a series of five men of different heights. For each of these men, participants were shown a particular description that they could choose to label as *true*, *false* or *can't say*. In particular, they were asked to judge whether conflicting descriptions such as (3-a) and (3-b) were true or false.[6]

(3)   a.   This man is both tall and not tall
       b.   This man is neither tall nor not tall

While participants almost unanimously judged these descriptions false when considering clearly not tall and clearly tall men, about half of them judged the conflicting descriptions true when considering the man whose height was average. Other experiments showed similar results (see [20] and [6] for instance).

All these experiments consider the use of a particular vague predicate and all show that for borderline cases of this vague predicate, people can use conflicting descriptions to qualify it. I assume that a speaker can regard (4-a) as respectively *true* or *false* when she *accepts* to qualify the man as respectively *tall* or as *not tall*; and that a speaker can regard (4-a) as respectively *not true* or *not false* when she *refuses* to qualify the man as respectively *tall* or *not tall*. Therefore, on the basis of the results of these experiments, I take it to be plausible that speakers, when asked to evaluate a vague sentence such as (4-a) regarding a borderline-tall man, can judge it *both true and false* or *neither true nor false*; and such judgments are *conflicting truth-value judgments*.

(4)   a.   This man is tall
       b.   This man is not tall

---

[4] Note the italics that distinguish between judging a sentence both *false* and *not false* and judging a sentence *both false and not false*.

[5] See Kooi & Tamminga[13] for support for Lewis' view contra Priest.

[6] The percentage of "can't say" answers proved to be insignificant.

Furthermore, participants gave similar judgments for positive ((4-a)) and negative ((4-b)) counterparts of vague sentences for borderline cases across these
experiments. For this reason, I assume that we can judge *negative* vague sentences about borderline cases in the same way as their positive counterparts (ie.
we can also say that (4-b) is *both true and false/neither true nor false* when the
man is borderline-tall).[7]

2.2   Presuppositions

To my knowledge, there have been very few experiments on *truth-value* judgments concerning presuppositions.[8] Nonetheless, if we focus on what has been
said about truth-value judgments concerning presuppositional sentences when
the presupposition is not fulfilled, we find some clues suggesting that conflicting
truth-value judgments might be accessible. Notably, Strawson [24] argued that
a sentence such as (5) is *neither true nor false* when there is no king of France,
contra Russell [21] according to whom such a sentence is simply *false* in these
circumstances. Von Fintel [8] endorses the former approach, but also admits that
speakers might judge a presuppositional sentence either *true* or *false* when its
presupposition is not fulfilled depending on the meaning of the sentence.[9]

(5)   The king of France is bald

Things get even more intricate when one considers the following pair of presuppositional sentences, when it is known that the amplifiers have never buzzed:[10]

(6)   a.   The amplifiers have stopped buzzing
       b.   The amplifiers have not stopped buzzing

As noted earlier, my intuitions, shared with several speakers I have consulted,
are the following: I can judge (6-a) *both false and not false*. Of course, if I were
talking to someone, I would no doubt add something like "On the one hand, it
is not *false* that the amplifiers have stopped buzzing because for the amplifiers
to have failed to stop buzzing, the amplifiers would have to *have been buzzing*
before; but on the other hand it *is* false to the extent that it can't be true
that the amplifiers have *stopped* buzzing: the amplifiers have *never* buzzed!".

---

[7] These assumptions reflect my intuitions and those of people I've informally surveyed.
[8] Though Abrusán & Szendrői [1] recently explored the judgments of speakers for
some positive and negative counterparts of presuppositional sentences.
[9] In this respect, my distinguishing between truth-value *judgments* and formal *logical*
values is reminiscent of his approach where (5) is semantically *neither true nor false*
but would be judged *false* by speakers.
[10] In Abrusán & Szendrői's experiment, almost no participant gave a *true* judgment
for "the king of France is not bald", but they did for other negative presuppositional
sentences. They explain this contrast by positing that certain linguistic factors affect
speakers' judgments. Taking those factors into account goes beyond the scope of this
paper. All the linguistic pairs of positive and negative counterparts given here will
be reduced to mere logical counterparts in the ST5 system: ϕ and $\neg\phi$.

However, I would never judge this sentence *true* given that the amplifiers were not buzzing before.[11]

By contrast, I can judge the negative counterpart (6-b) *both true and not true*, for the very same reasons. It is not *true* to the extent that the amplifiers have never buzzed; but it *is* true to the extent that the amplifiers have not *stopped* buzzing: the amplifiers were never buzzing in the first place.[12]

2.3   Summary

The important point, ultimately, is that some speakers (such as myself) seem to have access to *conflicting* truth-value judgments both concerning presuppositional sentences (when the presupposition is not fulfilled) and concerning vague sentences (describing borderline cases). Moreover, we see that their judgments are the same concerning the positive and the negative counterparts of vague sentences (describing borderline cases); whereas they differ concerning the positive and the negative counterparts of presuppositional sentences (when the presupposition is not fulfilled). When one tries to sketch a system that would account for the truth-judgments associated with vague sentences as well as the truth-judgments associated with presuppositional sentences, one should ensure that one's system accounts for both this common point and this difference.

The intuitions concerning hybrid sentences, that is to say sentences such as (7-a) or (7-b) that involve both vague and presuppositional expressions, are more complex and, to my knowledge, have never been dealt with.

(7)   a.   The amplifiers have stopped being loud
       b.   The amplifiers are loud and they have stopped buzzing

---

[11] Note that putting stress on the emphasized words can help to bring out these judgments.

[12] An anonymous reviewer has noted that, in justifying the conflicting judgments, I make use of statements like the following, which by all appearances threaten the transitivity of the consequence relation. If you endorse transitivity, it seems that by accepting (i-a) and (i-b), you should conclude that "if the amplifiers have never buzzed, then the speakers used to buzz", which is a contradiction:

(i)   a.   If the amplifiers have never buzzed, then (6-a) is false.
      b.   If (6-a) is false, then the amplifiers used to buzz.

I take the simultaneous acceptance of these sentences to reveal an important fact, namely the ambiguous use of the expression "false". The system I propose offers a natural way to loosen (as in (i-a)) and/or strengthen (as in (i-b)) the meaning of "false".[R1.2] This ambiguity might in fact explain the variation found among speakers for truth-judgments about presuppositional sentences evaluated in situations of presupposition failure: maybe not all speakers have equal access to the loose and to the strong senses of "false".[R1.1]

Not surprisingly, but still interestingly, this approach is reminiscent of the analysis of the sorites paradox and of the Liar paradox advanced by Cobreros&al. [5], who developed the three-valued system that I extend to a five-valued system: in critical cases, one might have to abandon the transitivity of the consequence relation.

To my knowledge, no theory considers such sentences and therefore no theory makes any prediction regarding the semantic status of (7-a) or (7-b): Section 4 tries to sketch an account of such sentences.[13]

3   ST5

3.1   The Original 3-Valued ST System

ST is a trivalent logical system developed to deal with vague predicates [5], and more specifically to account for conflicting judgments such as those diagnosed by responses to "X is tall and not tall".[14] There are two reasons for which I base my 5-valued system on ST: first, ST already comes with an account for vagueness. Hence only half of the work remains to be done. Second, ST comes with a notion of *assertoric ambiguity* that leads to a nice explanation for our conflicting judgments.

Two Notions of Satisfaction. Let's consider as our language $\mathcal{L}$ a non-quantified fragment of monadic first-order logic such that:

Definition 1 (Syntax)

i. *For any predicate $P \in \mathcal{L}$ and any individual name $a \in \mathcal{L}$, Pa is a well-formed formula (wff).*

ii. *For any wff ϕ, $\neg\phi$ is a wff.*

iii. *For any ϕ and ψ such that ϕ and ψ are wff, $[\phi \wedge \psi]$, $[\phi \vee \psi]$ and $[\phi \rightarrow \psi]$ are wff.*

Nothing else is a wff.

$\mathcal{M}$ consists of a non-empty domain of individuals $\mathcal{D}$ and an interpretation function $\mathcal{I}$ such that:

Definition 2 (Semantics)

i. *For any predicate $P \in \mathcal{L}$ and any individual name $a \in \mathcal{L}$, $\mathcal{I}(Pa) = \frac{1}{2}$ iff a is the name of a borderline case for P, $\mathcal{I}(Pa) \in \{0, 1\}$ otherwise.*

---

[13] It is worth noting that supervaluationism has been used independently for vagueness (Lewis [16], Fine [7], Kamp [15]) and for presuppositions (van Fraassen [10]). None of these supervaluationists seems to have specifically entertained a unified treatment of these two phenomena.[R2.1]

[14] ST is a built-in 3-valued version of TCS [4], which assumed bivalent extensions for vague predicates on which it built their trivalent extensions. As I present it here, ST seems to be committed to the existence of a sharp boundary between eg. clearly tall men and borderline tall men, which might sound unrealistic. This point is related to the question of higher-order vagueness, which is much discussed in the literature on vagueness. A discussion of higher-order vagueness goes far beyond the scope of this paper. I will therefore just endorse the assumption that vagueness defines a well defined trivalent extension in the rest of the paper, with no further justification.[R2.3]

ii. For any wff ϕ, $\mathcal{I}(\neg\phi) = 1 - \mathcal{I}(\phi)$.
iii. for two wff ϕ and ψ, $\mathcal{I}(\phi \wedge \psi) = min(\mathcal{I}(\phi), \mathcal{I}(\psi))$,
$\mathcal{I}(\phi \vee \psi) = max(\mathcal{I}(\phi), \mathcal{I}(\psi))$ and $\mathcal{I}(\phi \to \psi) = \mathcal{I}(\neg\phi \vee \psi)$

The system ST owes its name to the definition of two notions of satisfaction:[15]

Definition 3 (Strict and Tolerant Satisfaction)

Strict satisfaction: $\mathcal{M} \models^s \phi$ iff $\mathcal{I}(\phi) = 1$
Tolerant satisfaction: $\mathcal{M} \models^t \phi$ iff $\mathcal{I}(\phi) \geq \frac{1}{2}$

Now, imagine a is the name of a borderline case for P. We have $\mathcal{I}(Pa) = \frac{1}{2}$ and $\mathcal{I}(\neg Pa) = 1 - \frac{1}{2} = \frac{1}{2}$. Hence, we get $\mathcal{I}(Pa \wedge \neg Pa) = min(\frac{1}{2}, \frac{1}{2}) = \frac{1}{2}$ and $\mathcal{I}(\neg(Pa \vee \neg Pa)) = 1 - max(\frac{1}{2}, \frac{1}{2}) = 1 - \frac{1}{2} = \frac{1}{2}$. This leads us to:

i. $\mathcal{M} \models^t Pa$ but $\mathcal{M} \not\models^s Pa$
ii. $\mathcal{M} \models^t \neg Pa$ but $\mathcal{M} \not\models^s \neg Pa$
iii. $\mathcal{M} \models^t Pa \wedge \neg Pa$ but $\mathcal{M} \not\models^s Pa \wedge \neg Pa$
iv. $\mathcal{M} \models^t \neg(Pa \vee \neg Pa)$ but $\mathcal{M} \not\models^s \neg(Pa \vee \neg Pa)$

With P standing for "is tall" and a standing for borderline-tall "John", what we have is that none of "John is tall", "John is not tall", "John is tall and not tall" and "John is neither tall nor not tall"[16] is *strictly* satisfied, but all of them are *tolerantly* satisfied. Cobreros & al. propose to account for the results of Alxatib & Pelletier [2] by assuming that speakers can assert vague sentences either strictly or tolerantly. To this, I add the following bridge principles:[17]

Principle 1 (Truth-Value Judgments). *One can judge a proposition ϕ...*

1. *"true" if $\mathcal{M} \models^t \phi$*
2. *"false" if $\mathcal{M} \models^t \neg\phi$*
3. *"not true" if $\mathcal{M} \not\models^s \phi$*
4. *"not false" if $\mathcal{M} \not\models^s \neg\phi$*
5. *"both true and false" if 1 and 2.*
6. *"neither true nor false" if 3 and 4.*
7. *"both true and not true" if 1 and 3.*
8. *"both false and not false" if 2 and 4.*

It is straightforward that, for borderline-tall John, "John is tall" as well as "John is not tall" can be judged *both true and false* and *neither true nor false*.

No Room for Presuppositions. Now, looking at the bridge principles, it would be ideal if we could add presuppositional propositions ϕ to our language in such a way that, *when the presupposition of ϕ is unfulfilled*:

1. $\mathcal{M} \models^t \neg\phi$ (so that a speaker can judge ϕ *false*)

---

[15] See [5] for a discussion of inference rules in this system.
[16] Here, I regard *neither... nor...* as the negation of a disjunction
[17] In formulating these bridge principles, I use $\mathcal{M}$ to stand for a model determined by the belief state of the speaker.[R2.4]

2. $\mathcal{M} \not\models^s \neg\phi$ (so that a speaker can judge ϕ *not false*)
3. $\mathcal{M} \not\models^t \phi$ (so that a speaker *cannot* judge ϕ *true*)

But the only way in ST to have 1. and 2. is for ϕ to get the value $\frac{1}{2}$, and then we would have $\mathcal{M} \models^t \phi$ and a speaker could judge ϕ *true* as well. More specifically, ST has the following property (see [5]):

Lemma 1 (Duality in ST). *For any wff ϕ, $\mathcal{M} \models^{s/t} \phi$ iff $\mathcal{M} \not\models^{t/s} \neg\phi$*

The solution I propose consists in breaking this duality by adding two logical values to the system: propositions that get one of these two extra values will obey the three constraints above, but propositions that get one of the three initial values will still present the equivalence noted in Lemma 1.

3.2   The ST5 System

In ST, we had three values: $\{0, \mathcal{V} = \frac{1}{2}, 1\}$, and vague predications on borderline cases got the value $\mathcal{V}$. Now, in ST5, we add two more values, $\mathcal{P}^0$ and $\mathcal{P}^1$, such that: $0 < \mathcal{P}^0 < \mathcal{V} < \mathcal{P}^1 < 1$ and such that $\mathcal{P}^0 = 1 - \mathcal{P}^1$. The syntax and the semantics of ST remain unchanged in this extended system, as well as Def. 3 of tolerant and strict satisfactions. By this simple addition, we obtain the following:

Lemma 2 (Duality lost)

- *For any proposition ϕ such that $\mathcal{I}(\phi) = \mathcal{P}^0$:*
    i. $\mathcal{M} \not\models^t \phi$ *and* $\mathcal{M} \not\models^s \phi$ *since* $\mathcal{P}^0 < \frac{1}{2} < 1$.
    ii. $\mathcal{M} \models^t \neg\phi$ *but* $\mathcal{M} \not\models^s \neg\phi$ *since* $1 - \mathcal{P}^0 = \mathcal{P}^1$ *and* $\mathcal{P}^1 \geq \frac{1}{2}$ *but* $\mathcal{P}^1 < 1$.
- *For any proposition ϕ such that $\mathcal{I}(\phi) = \mathcal{P}^1$:*
    i. $\mathcal{M} \models^t \phi$ *but* $\mathcal{M} \not\models^s \phi$ *since* $\mathcal{P}^1 \geq \frac{1}{2}$ *but* $\mathcal{P}^1 < 1$.
    ii. $\mathcal{M} \not\models^t \neg\phi$ *and* $\mathcal{M} \not\models^s \neg\phi$ *since* $1 - \mathcal{P}^1 = \mathcal{P}^0$ *and* $\mathcal{P}^0 < \frac{1}{2} < 1$.

Given that we now have propositions ϕ for which $\mathcal{M} \not\models^s \neg\phi$ but $\mathcal{M} \not\models^t \phi$ (propositions of value $\mathcal{P}^0$), Lemma 1 no longer holds in ST5. Nonetheless, the following holds in ST as well as in ST5:

Lemma 3 (Entailment). *For any wff ϕ, $\mathcal{M} \models^s \phi$ entails $\mathcal{M} \models^t \phi$.*

Now let's stipulate that any simple positive proposition ϕ whose presupposition is *unfulfilled* gets the value $\mathcal{P}^0$. It follows that its negation gets the value $\mathcal{P}^1$. Let ϕ stand for "The amplifiers have stopped buzzing", the bridge principles predict the following:[18]

i. One can judge ϕ *both false and not false* ($\mathcal{M} \models^t \neg\phi$ but $\mathcal{M} \not\models^s \neg\phi$)
ii. One can judge ϕ *neither true nor false* ($\mathcal{M} \not\models^s \phi$ and $\mathcal{M} \not\models^s \neg\phi$)
iii. One can judge $\neg\phi$ *both true and not true* ($\mathcal{M} \models^t \neg\phi$ but $\mathcal{M} \not\models^s \neg\phi$)
iv. One can judge $\neg\phi$ *neither true nor false* ($\mathcal{M} \not\models^s \neg\phi$ and $\mathcal{M} \not\models^s \neg\neg\phi$)
v. One *cannot* judge ϕ *true* ($\mathcal{M} \not\models^t \phi$)

---

[18] Recall that we have $\neg\neg\phi \equiv \phi$.

One should note at this point that presuppositional propositions are propositions that can get a value in $\{\mathcal{P}^0, \mathcal{P}^1\}$, in the same way that vague propositions are propositions that can get the value $\mathcal{V}$. A presuppositional proposition used when the presupposition *is* fulfilled gets a value in $\{0, 1\}$, just like a vague proposition describing a *non*-borderline case. We thus do not predict any non-classical judgment in such contexts (as desired).[19]

4   Hybrid Sentences

4.1   Presupposition Satisfaction in ST5

So far, we have considered situations where presuppositions were simply fulfilled or unfulfilled. But as it turns out, presuppositions can themselves involve vague and presuppositional expressions. Think of sentences such as (8-a) or (8-b) whose presuppositions are (8-a-i) and (8-b-i).

(8)   a.   The amplifiers have stopped being <u>loud</u>
        (i)   The amplifiers were <u>loud</u>
    b.   John knows that the amplifiers have <u>stopped</u> buzzing
        (i)   The amplifiers have <u>stopped</u> buzzing

In situations where the amplifiers were borderline-loud and have never buzzed, (8-a-i) gets the value $\mathcal{V}$ and (8-b-i) gets the value $\mathcal{P}^0$ in ST5. What effect does a presupposition with value $\mathcal{V}$ or $\mathcal{P}^0$ have on the value of the proposition as a whole?

Bearing in mind that the presuppositional propositions in ST5 are the propositions that get one of the values in $\{\mathcal{P}^0, \mathcal{P}^1\}$ in at least one model, I propose that we see the values of these propositions as being determined in the following way:

Definition 4 (Factoring out Presuppositions). *Let us use the notation* ϕ_p *for a proposition whose* assertive *part can be expressed by the proposition* ϕ *and whose* presuppositional *part can be expressed by the proposition* p. *Then:*

- $\mathcal{I}(\phi_p) = \mathcal{I}(\phi)$ *if* $\mathcal{M} \models^s p$
- $\mathcal{I}(\phi_p) = \mathcal{P}^1$ *if* $\mathcal{M} \not\models^s p$ *and* $\mathcal{M} \models^t p$ *and* $\mathcal{M} \models^s \phi$
- $\mathcal{I}(\phi_p) = \mathcal{P}^0$ *if* $\mathcal{M} \not\models^t p$ *or* $[\mathcal{M} \not\models^s p$ *and* $\mathcal{M} \not\models^s \phi]$

---

[19] A reviewer asked whether being borderline can be treated as a case of presupposition failure. ST5 allows us to adopt a liberal understanding of the notion of *presupposition*: one could suggest that any use of a proposition presupposes it to have a classical value (0 or 1). To that extent, ascribing a vague predicate to a borderline case would constitute a case of presupposition failure (for the proposition would get the value $\mathcal{V}$, which is neither 0 nor 1). Percus and I [18] argued for the usefulness of such a position, taking the TCS [4] system as background and building on the account of the sorites paradox by means of presupposition projection presented in my MA thesis [26].

In situations where the amplifiers were borderline-loud, we have $\mathcal{M} \not\models^s$ (8-a-i) but $\mathcal{M} \models^t$ (8-a-i); and in situations where the amplifiers have never buzzed we have $\mathcal{M} \not\models^s$ (8-b-i) and $\mathcal{M} \not\models^t$ (8-b-i). Additionally, let's imagine that the amplifiers are now low and that John *believes* that the amplifiers were buzzing but have stopped. We can then assume that the *assertive* parts are strictly satisfied.[20] Looking at our stipulations, we obtain: $\mathcal{I}((8\text{-a})) = \mathcal{P}^1$ (because $\mathcal{M} \models^t$ (8-a-i) and the assertive part is strictly satisfied) and $\mathcal{I}((8\text{-b})) = \mathcal{P}^0$ (because $\mathcal{M} \not\models^t$ (8-b-i)). So we predict that under such circumstances, a speaker can judge (8-a), "The amplifiers have stopped being loud", *both true and not true* and (8-b), "John knows that the amplifiers have stopped buzzing", *both false and not false*.

As noted earlier, our intuitions for sentences with iteratively embedded presuppositional expressions (henceforth *recursively presuppositional sentences*) and hybrid sentences are somewhat messy and maybe only experimental data can discriminate between theories that make different predictions regarding truth-judgments for these kinds of sentences. Nonetheless, any theory has to make *some* predictions for these sentences, and it doesn't appear to be the case for existing theories, for one simple reason: a majority of these theories only consider *bivalent* presuppositions. As long as a theory of presuppositions treats the presuppositional content as *bivalent*, it can't account for sentences where the presuppositional content is vague. This is precisely the weakness that the definitions above avoid: they let us escape the traditional duality of either "fulfilled" *or* "unfulfilled" presuppositions. Rather, all the conditions above are stated in terms of satisfaction. The first clause states that when a presupposition is strictly satisfied, the whole proposition gets the value of its assertive part: in this situation one would traditionally say that the presupposition is "fulfilled". The second clause considers the case where the presupposition is only *tolerantly* satisfied. To some extent, one could see this as a condition where the presupposition is "partly fulfilled". The whole proposition will be "partly true" if the assertive part is true itself: that's what $\mathcal{P}^1$ stands for. Finally, the third clause states that even if the presupposition is tolerantly satisfied, there is no reason for the whole proposition to be "partly true" if the assertive part is not strictly satisfied; and even less reason if the presupposition is not satisfied *at all*. But still, such a proposition is not merely false, because the presupposition is *not* "fulfilled": that's what $\mathcal{P}^0$ stands for.

Many theories do consider recursively presuppositional sentences. However, none of them deal with hybrid sentences such as (8-a) to my knowledge. Moreover, ST5 is able to make distinctions that other approaches cannot. For example, Karttunen ([12]) proposed to categorize factives (such as *know*) as what he famously called *holes*:

> "If the main verb of the sentence is a hole, then the sentence has all the presuppositions of the complement sentences embedded in it."

---

[20] I take "*X believes* ϕ" to be the assertive part of "*X knows* ϕ". It might well be the case that things are more complex, and that one should consider justified belief for the assertive part. But whatever we take to be the assertive part, the crucial point here is how each part contributes to the value of the whole proposition.

Regarding (8-b), this view provides no way to distinguish between a situation where the amplifiers are still buzzing and a situation where the amplifiers have never buzzed: in the first situation, the complement of the factive is false so it yields a presupposition failure; in the second situation the *inherited* presupposition is *unfulfilled* so it *also* yields a presupposition failure. By contrast, in ST5, we have the tools to make a distinction because the presuppositional part of the whole proposition would have the value 0 in the first case and the value $\mathcal{P}^0$ in the second case. It is not clear whether speakers actually would give different truth-judgments in these two situations for (8-b), and I chose here to treat them equally, as does a theory *à la* Karttunen. But I find it important that ST5 allows more easily than its competitors for the possibility of nuanced judgments for presuppositional sentences, given that it takes the relative "gradedness" of the presuppositions into account.

Because Def. 4 covers all the satisfaction possibilities, it is easy to see that the system is now completely predictive with respect to the kind of proposition (ie. regular, vague, simply presuppositional or even hybrid itself[21]) that appears as a presupposition of the whole sentence.

4.2   Conjunctions, Disjunctions and Implications in ST5

An Example. Finally, because ST5 deals with totally ordered values and defines its connectives in terms of *min* and *max*, it naturally makes predictions for conjunctions, disjunctions and implications combining vague and presuppositional propositions. Consider (7-b) repeated here that conjoins a vague sentence and a presuppositional sentence:

(9)   The amplifiers are <u>loud</u> and they have <u>stopped</u> buzzing

Given that the amplifiers have *never* buzzed, if their volume is somewhere between clearly loud and clearly not loud, the first conjunct gets the value $\mathcal{V}$ and the second conjunct gets the value $\mathcal{P}^0$. Therefore in these circumstances, the whole proposition gets the value $min(\mathcal{V}, \mathcal{P}^0) = \mathcal{P}^0$: it is judged *both false and not false* (for the amplifiers were *not* buzzing before), and it's not judged *true*.

Here is a table summarizing the predictions of ST5 for hybrid conjunctions and disjunctions when the amplifiers (abbreviated as A) are borderline-loud and have never buzzed:

---

[21] As an example of how ST5 deals with hybrid *presuppositions*, consider (i-a), its presupposition being (i-b):

(i)   a.   John knows that the amplifiers have stopped being loud.
       b.   The amplifiers have stopped being loud.

We saw earlier that in cases were the amplifiers were borderline-loud before decreasing in volume, the hybrid proposition expressed by (i-b) gets the value $\mathcal{P}^0$, which prevents it from being even tolerantly satisfied; therefore (i-a) will also get the value $\mathcal{P}^0$ by Def. 4.

Proposition	Value Judgment	
A are loud	$\mathcal{V}$	Both true and false
A are not loud	$\mathcal{V}$	Both true and false
A have stopped buzzing	$\mathcal{P}^0$	Both false and not false
A have not stopped buzzing	$\mathcal{P}^1$	Both true and not true
A are loud & have stopped buzzing	$\mathcal{P}^0$	Both false and not false
A are not loud & have stopped buzzing	$\mathcal{P}^0$	Both false and not false
A are loud & have not stopped buzzing	$\mathcal{V}$	Both true and false
A are not loud & have not stopped buzzing	$\mathcal{V}$	Both true and false
A are loud or have stopped buzzing	$\mathcal{V}$	Both true and false
A are not loud or have stopped buzzing	$\mathcal{V}$	Both true and false
A are loud or have not stopped buzzing	$\mathcal{P}^1$	Both true and not true
A are not loud or have not stopped buzzing	$\mathcal{P}^1$	Both true and not true

Left-Right Asymmetries. In view of these predictions, a word is in order about the alleged left-right asymmetry of presuppositions. Consider the pair of sentences in (10), for which the ST5 truth judgment predictions are clear. In ST5, conjunctions are totally symmetric and (10-a) and (10-b) will get the same value when the amplifiers never buzzed: $min(\mathcal{P}^0, 0) = min(0, \mathcal{P}^0) = 0$. Therefore we predict that both (10-a) and (10-b) will be judged merely *false* when we know that amplifiers have never buzzed.

(10)   a.   The amplifiers have stopped buzzing and they were buzzing before.
        b.   The amplifiers were buzzing before and they have stopped buzzing.

It's been observed since at least Stalnaker [25] and Heim [14] that sentences such as (10-a) have a status that the corresponding reversed sentence (10-b) doesn't. And the standard way of viewing this difference is in terms of presuppositions: (10-a) gives rise to a presupposition that (10-b) doesn't. The point I wish to make is the following. As far as the facts are concerned, it's unclear what *truth-value judgments* speakers would actually give for (10-a) and (10-b). We should, though, distinguish between the question whether (10-a) and (10-b) can give rise to different *truth-value judgments*, and the rather clear intuition that (10-b) is *utterable* in a broader range of conditions that (10-a).

Schlenker ([22]) pointed out that the asymmetry in conditions of use in cases like (10) could be related to a more general property of conjunctions. Indeed, the contrast we observed between (10-a) (which "sounds weird") and (10-b) is in a certain way similar to the one we observe between (11-a) (which "sounds weird" too) and (11-b):[22]

(11)   a.   John lives in Paris and he resides in France.
        b.   John resides in France and he lives in Paris.

---

[22] To insist on the need of distinguishing between giving a *non-classical truth-value* judgment for a sentence and feeling this sentence is "weird", note that you will judge *both* (11-a) and (11-b) completely false if you know John lives in London, but still regard (11-a) as weird.

Schlenker proposes a general constraint that has the effect of ruling out conjunctions where the first conjunct entails the second one. Note that, given the way we proposed to view presuppositional sentences in the previous section, the right conjunct in (10-a) can be regarded as expressing the *presuppositional part* of the left conjunct. Since for a presuppositional proposition to be true in ST5 its presuppositional part has to be true, whenever the left conjunct in (10-a) is true, the right conjunct is too: (10-a) would thus be ruled out by a principle à la Schlenker.

One should note moreover that if the only constraint on the use of (10-a) were for the presupposition of its left conjunct to be fulfilled, then (10-a) should sound totally fine in cases where (10-b) is known to be true, but this is not the case: if we know that the amplifiers used to buzz, (10-a) "sounds weird" in a way in which (10-b) does not. To this extent, the strength of the contrast between (10-a) and (10-b) should not be raised in favor of the view that (10-a) is presuppositional while (10-b) is not: as a matter of fact, we can't use our judgments on (10-a) to clearly distinguish between cases where the presupposition of its left conjunct is fulfilled from cases where it is not.

If one thinks that, nonetheless, these sentences should receive different *truth-value judgments*, a possibility is to revise the semantics of the conjunction operator so that it gives the value $\mathcal{P}^0$ to a conjunction whenever it has a proposition of value $\mathcal{P}^0$ on its left: with such a semantics, and contrary to the option above, (10-a) would come out as presuppositional in ST5 since it would get the value $\mathcal{P}^0$ in at least one model. As Fox [9] and George [11] point out, one can extend this kind of considerations to all the connectives in the system by resorting to a unifying principle in the spirit of the one proposed by Schlenker. However it is not clear whether disjunctions and implications show the same asymmetry (see (12)), and so whether one should or not revise the semantics of the connectives in the system.

(12)   a.   The amplifiers have stopped buzzing or they were not buzzing before.

        b.   The amplifiers were not buzzing before or they have stopped buzzing.

        c.   The amplifiers have stopped buzzing, if they were buzzing before.

        d.   If the amplifiers were buzzing before, they have stopped buzzing.

5   A Discussion of Potential Alternatives

The system that I have described adds two logical values to $\{0, \frac{1}{2}, 1\}$. Would it have been possible to add only one? Not given the semantics for ¬: the semantics for ¬ would force us to include a value corresponding to 1 minus the new additional value; and, since our initial three-valued set was $\{0, \frac{1}{2}, 1\}$, adding a fourth value would then require adding a fifth as well. One might however wonder if one could manage with a different kind of four-valued system in which the value $\frac{1}{2}$ played no role. There are two potential ways of doing this: by making the four values totally ordered, and by making them partially ordered.

Let us consider the first possibility. We would then have a set of four values $\{0, \mathcal{P}, \mathcal{V}, 1\}$, where $\mathcal{P}$ would be a value assigned to propositions describing situations of presupposition failure and $\mathcal{V}$ a value assigned to propositions describing borderline cases. In addition, we would have $\mathcal{P} = 1 - \mathcal{V}$ in order to fit the semantics for $\neg$. But there is a problem with this solution, and it is precisely related to negation. Imagine you have a proposition ϕ_P describing a case of presupposition failure and a proposition ψ_V describing a borderline case: as such, ϕ_P gets the value $\mathcal{P}$ and ψ_V gets the value $\mathcal{V}$. But now $\neg\phi_P$ gets the value $1 - \mathcal{P} = \mathcal{V}$, which is the value of ψ_V. And conversely, $\neg\psi_V$ gets the value $1 - \mathcal{V} = \mathcal{P}$, which is the value of ϕ_P. This has two unwelcome effects: first it predicts that we should observe the same truth judgments for negative counterparts of presuppositional sentences used in case of presupposition failure and for vague sentences used to describe borderline cases; second it predicts that we should observe different truth judgments for affirmative and negative counterparts of vague sentences. We have seen that these predictions are wrong; that excludes this approach.

But what about an alternative assuming a *partial* order — a set of four values $\{0, \mathcal{P}, \mathcal{V}, 1\}$ where $0 < \mathcal{P} < 1$ and $0 < \mathcal{V} < 1$? We would then need to adapt the semantics of our connectives to a partial ordered lattice: negation could semantically contribute as a symmetric operator (ie. for $\mathcal{I}(\phi) = 1$, $\mathcal{I}(\neg\phi) = 0$, for $\mathcal{I}(\phi) = 0$, $\mathcal{I}(\neg\phi) = 1$, for $\mathcal{I}(\phi) = \mathcal{V}$, $\mathcal{I}(\neg\phi) = \mathcal{V}$ and for $\mathcal{I}(\phi) = \mathcal{P}$, $\mathcal{I}(\neg\phi) = \mathcal{P}$), and conjunction and disjunction could respectively semantically contribute as the greatest lower bound and as the least upper bound.[23] But note that in this system, a proposition describing a case of presupposition failure would receive the same value as its negation. We would like to avoid this result given the asymmetry in our truth judgments for presuppositional sentences.

Raising the possibility of partially ordered values does suggest additional alternatives to the system developed here, so I would like to briefly address these. One possibility would be to consider a *partially ordered five*-valued set $\{0, \mathcal{P}^0, \mathcal{P}^1, \mathcal{V}, 1\}$ such that $0 < \mathcal{V} < 1$ and $0 < \mathcal{P}^0 < \mathcal{P}^1 < 1$: positive propositions describing situations of presupposition failure would have the value $\mathcal{P}^0$ and their negation would have the value $\mathcal{P}^1$. In fact, partially ordered systems of this kind give rise to an important problem irrespective of whether or not they incorporate a fifth value. Consider the conjunction and the disjunction in (13).

(13)   a.   The amplifiers are <u>loud</u> and they have <u>stopped</u> buzzing
          b.   The amplifiers are <u>loud</u> or they have <u>stopped</u> buzzing

With either the four-valued or the five-valued version of a partially ordered lattice, in situations where the amplifiers are borderline-loud and have never buzzed, (13-a) would express the conjunction of two propositions that would receive *non-ordered values* and (13-b) would express their disjunction. With

---

[23] A reviewer argued that there are other ways of defining the connectives that might be as legitimate as the standard Dunn-Belnap definition. This is perfectly fair and I am currently exploring a four-valued system with alternative definitions of the connectives. But since there is no place to develop it here, I will focus on standard approaches to the connectives in the rest of this paper.[R1.3]

conjunction being defined as the greatest lower bound and disjunction being defined as the least upper bound, the proposition expressed by (13-a) would get the value 0 and the proposition expressed by (13-b) would get the value 1. Such a system would therefore predict a pure *false* judgment for (13-a) and a pure *true* judgment for (13-b) in those situations, which clearly goes against our intuitions.

One might finally consider a system with still partially ordered values but such that the greatest lower bound and the least upper bound of the values for vagueness and presuppositions are not 0 and 1. With $\mathcal{E}^0$ and $\mathcal{E}^1$ the new $\mathcal{E}$xtra values, we would have a set of six values $\{0, \mathcal{E}^0, \mathcal{V}, \mathcal{P}, \mathcal{E}^1, 1\}$ such that $0 < \mathcal{E}^0 < \mathcal{V} < \mathcal{E}^1 < 1$ and $0 < \mathcal{E}^0 < \mathcal{P} < \mathcal{E}^1 < 1$. In this system, vagueness and presuppositions seem ontologically well distinguished ($\mathcal{P}$ and $\mathcal{V}$ are not ordered with each other), and in critical situations, the conjunction expressed in (13-a) would get the value $\mathcal{E}^0$ (the greatest lower bound of $\mathcal{P}$ and $\mathcal{V}$) and the disjunction expressed in (13-b) would get the value $\mathcal{E}^1$ (the least upper bound of $\mathcal{P}$ and $\mathcal{V}$). But this raises the question of what $\mathcal{E}^0$ and $\mathcal{E}^1$ actually represent. If their existence is motivated only by the existence of conjunctions and disjunctions of propositions describing borderline cases and propositions describing cases of situation failure, this seems a large price to pay. (In addition, the six-valued system I considered here is based on a partially ordered four-valued system which doesn't distinguish between affirmative and negative presuppositional sentences in cases of presupposition failure: a partially ordered *seven*-valued system might then be more adequate.)

In the system that I have settled on, there are five totally ordered values where each value has a clear ontological status. This seems superior to all of the alternatives I considered here. [R1.4,R2.5]

6   Conclusions

ST provides us with a notion of assertoric ambiguity that, along with some bridge principles, lets us explain our conflicting truth-value judgments in case of vagueness. Adding two symmetrical values around $\frac{1}{2}$ has made it possible to capture the difference between *not true* and *false* judgments and between *not false* and *true* judgments by virtue of bridge principles based on ST notions of satisfaction. Moreover, these values lend themselves naturally to an account for the relationship between truth-value judgments for the positive and negative counterparts of presuppositional sentences. Furthermore, we now have a system that incorporates *both* vagueness and presuppositions while *also* accounting for the differences in the judgments they trigger. At the same time, there is clearly more to be said about how the presuppositions of complex sentences depend on the presuppositions of the simple sentences they embed; here we had to add some stipulations. More data would be welcome in order to test the predictions of ST5: we are currently at work on an experimental design for eliciting truth-value judgments for vagueness and presuppositions.[R2.2]

References

1. Abrusán, M., Szendröi, K.: Experimenting With the King of France: Topics, Verifiability and Definite Descriptions. In: Aloni, M., Kimmelman, V., Roelofsen, F., Sassoon, G.W., Schulz, K., Westera, M. (eds.) Amsterdam Colloquium. LNCS, vol. 7218, pp. 102–111. Springer, Heidelberg (2012)
2. Alxatib, S., Pelletier, J.F.: The Psychology of Vagueness: Borderline Cases and Contradictions. Mind & Language 26(3) (2010)
3. Beaver, D.I., Geurts, B.: Article "Presupposition". Stanford Encyclopedia of Philosophy (2011), http://plato.stanford.edu/entries/presupposition/
4. Cobreros, P., Egre, P., Ripley, D., van Rooij, R.: Tolerant, Classical, Strict. Journal of Philosophical Logic 41(2), 347–385 (2011)
5. Cobreros, P., Egre, P., Ripley, D., van Rooij, R.: Vagueness, Truth and Permissive Consequence. In: Achourioti, T., Galinon, H., Fujimoto, K., Martínez-Fernndez (eds.) Volume on Truth. Springer (forthcoming)
6. Égré, P., de Gardelle, V., Ripley, D.: Vagueness and Order Effects in Color Categorization. Journal of Logic, Language and Information 22(4), 391–420 (2013)
7. Fine, K.: Vagueness, Truth and Logic. Synthese 30(3/4), 265–300 (1975)
8. von Fintel, K.: Would You Believe It? The King of France Is Back! Presuppositions and Truth-Value Intuitions. In: Bezuidenhout, Reimer (eds.) Descriptions and Beyond. Oxford University Press (2004)
9. Fox, D.: Two Short Notes on Schlenker's Theory of Presupposition Projection. Theoretical Linguistics 34(3), 237–252 (2008)
10. van Fraassen, B.C.: Presuppositions, Supervaluations, and Free Logic. In: Lambert, K. (ed.) The Logical Way of Doing Things, pp. 67–91 (1969)
11. George, B.R.: Presupposition Repairs: a Static, Trivalent Approach to Predicting Projection. UCLA. MA thesis (2008)
12. Karttunen, L.: Presuppositions of Compound Sentences. Linguistic Inquiry 4(2), 169–193 (1973)
13. Kooi, B., Tamminga, A.: Three-Valued Logics in Modal Logic. Studia Logica 101(5), 1061–1072 (2013)
14. Heim, I.: On the Projection Problem for Presuppositions. In: Barlow, M., Flickinger, D., Westcoat, M. (eds.) Second Annual West Coast Conference on Formal Linguistics, pp. 114–126. Stanford University (1983)
15. Kamp, H.: Two Theories about Adjectives. In: Keenan, E.L. (ed.) Formal Semantics for Natural Languages, pp. 123–155. Cambridge University Press (1975)
16. Lewis, D.: General Semantics. Repr. in Philosophical Papers 1 (1970)
17. Lewis, D.: Logic for Equivocators. Noûs 16(3), 431–441 (1988)
18. Percus, O., Zehr, J.: TCS for Presuppositions. In: Egré, Ripley (eds.) Proceedings of the ESSLLI, Workshop on Three-valued Logics and their Applications, pp. 77–92 (2012)
19. Priest, G.: In Contradiction: A Study of the Transconsistent. Oxford University Press (2006)
20. Ripley, D.: Contradictions at the Borders. In: Nouwen, R., van Rooij, R., Sauerland, U., Schmitz, H.-C. (eds.) ViC 2009. LNCS (LNAI), vol. 6517, pp. 169–188. Springer, Heidelberg (2011)
21. Russell, B.: On Denoting. Mind 14, 479–493 (1905)

22. Schlenker, P.: Be Articulate: A Pragmatic Theory of Presupposition Projection. Theoretical Linguistics 34(3), 157–212 (2008)
23. Serchuk, P., Hargreaves, I., Zach, R.: Vagueness, Logic and Use: Some Experimental Results. Mind & Language 26(5), 540–573 (2011)
24. Strawson, P.F.: On Referring. Mind 59, 320–344 (1950)
25. Stalnaker, R.: Pragmatic Presuppositions. Context and Contents, pp. 47–62. Oxford University Press (1974)
26. Zehr, J.: Le Vague comme Phénomène Présuppositionnel. Université de Nantes. MA thesis under the supervision of Orin Percus (2011)

Reviewers' Comments

Reviewer #1:

Overall, I think this is a quite nice paper. It's very clearly written. Throughout reading it I had a good sense of the goal of the project and the plan for accomplishing it. The formal material is presented clearly, without getting bogged down in unnecessary detail. It is also, as best as I can see, technically correct.

I've got three suggestions. They aren't such that they absolutely need to be addressed in the final version, but they may be food for thought (I wasn't sure whether this should be marked as a 3 or 4 on the form–it definitely can be published as is should the author prefer).

1. I think that the formal account does a really nice job of doing justice to the motivating concerns of the project. However, I have some worries about these motivating concerns. It seems as though the theory is meant to be a predictive, broadly linguistic theory–that is, it's meant to predict actual linguistic behavior. This struck me as worrisome in two ways.

Point 1: First, as you acknowledge, there's a big divide on these linguistic intuitions. As you say, you and some others have access to them, though not everyone does. Does this mean that those with different intuitions ultimately mean something different by certain terms? **(see R1.1 on p. 251)**

Second, in discussing hybrid sentences, you mention that intuitions are murky, and that perhaps we need more empirical research to settle these cases. However, I'd be worried that the empirical research wouldn't be helpful, because all the people surveyed would presumably have the same kinds of murky intuitions. It seems that a good linguistic theory would actually refrain from predictions in these kinds of cases, whereas you say that a theory must make predictions. So maybe you're not giving a linguistic theory after all, but if not it would be good to say what you are doing.

I should note that obviously this gets into very big issues very quickly, so if there's not something reasonably quick you can say here, I wouldn't worry about it.

2.   **Point 2:** In motivating the initial conflicting judgments, you make two arguments. First, that (1) is false because there was no buzzing before. Second, that (1) is not false because if false that would mean there was buzzing before. That is, you seem to accept "if no buzzing, then (1) is false" and "if (1) is false, then buzzing". Given transitivity for the conditional and contraction for the conditional, these imply that there was buzzing, which you don't accept in the case where you say (1) is false and not false. So, it looks like you have to give up transitivity or contraction. That's not necessarily a problem, but it would be interesting to hear which you prefer. **(see R1.2 on p. 251)**

3.   **Point 3:** When you extend your account to conjunctions in 4.2, you make verdicts based on using a definition of conjunction in terms of minimum value. However, my sense here is that you need to do something to justify this. (The justification I have in mind is the kind of intuitive justification you can give

for Strong Kleene truth tables, given their intended application) One might worry that extending the same rules from ST to ST5 is overgeneralizing in a problematic way. **(see R1.3 on p. 260)**

Point 4: Relatedly, the definition in terms of minimum value presupposes that P0 should be lower than V, for instance. I'm not totally convinced by this. I see that $0 < \mathcal{P}^0 < \mathcal{P}^1 < 1$ and $0 < \mathcal{V} < 1$, but I'm not sure I'm convinced that there's any meaning to be attached to the relative orderings of $\mathcal{P}^0$, $\mathcal{P}^1$, and $\mathcal{V}$.**(see R1.4 on p. 259)**

Reviewer #2:

I understand my charge as assessing whether the paper is fit to publish. I judge that it is: it is interesting, developed to an appropriate level of explicitness and rigor, and beautifully written.

Without requiring addressing the comments below as a condition on publication, I offer some reactions and comments that may be of some use to the author either directly or in the future.

Point 1: It might be worth mentioning that Kit Fine's 1975 supervaluation approach to handling vagueness was inspired in part by van Fraassen's theory of presupposition failure. So there are precedents for thinking that formal techniques for handling vagueness and for handling presupposition might converge. **(see R2.1 on p. 252)**

Point 2: For future work, clearly it would be relevant and interesting to get Mechanical Turk data on the judgments people actually give for the more complicated sentences, and see whether the data support the predictions of the model. **(see R2.2 on p. 261)**

Point 3: Re (2)i: Note that the theory assumes that the border between borderline and tall is crisp: the proposition that someone is tall is assigned either to 1/2 or to 1. This is a reasonable compromise, but it is unrealistic: people can be clearly borderline tall, borderline borderline tall, and so on. **(see R2.3 on p. 252)**

Point 4: fn 15: M should be a set of belief states, not a set of beliefs. **(see R2.4 on p. 253)**

Point 5: p. 7: the structure of the dialectic is a bit garbled here. It's perfectly possible to have four truth values: true, false, borderline, and presup-failure, where neither borderline nor presup-failure entails the other. At that point, the argument that if conjunction/disjunction is treated as meet/join, we get undesirable results comes into play.

But in fact, what about a whiskered diamond configuration of six truth values? $5 > 4 > 3 > 1 > 0, 4 > 2 > 1$, but 2 (borderline) and 3 (presup-failure) do not entail each other. The join is still not full truth, and the meet is still not full false. **(see R2.5 on p. 259)**

Author Index